Lecture Notes in Physics

Edited by J. Ehlers, München, K. Hepp, Zürich,
H. A. Weidenmüller, Heidelberg, and J. Zittartz, Köln
Managing Editor: W. Beiglböck, Heidelberg

41

Progress in Numerical Fluid Dynamics

Lecture Series Held at the von Karman Institute
for Fluid Dynamics
1640 Rhode-St.-Genèse, Belgium
February 11–15, 1974

Revised and Updated Version

Edited by H. J. Wirz

Springer-Verlag
Berlin Heidelberg GmbH 1975

Editor
Prof. Dr. H. J. Wirz
von Karman Institute for Fluid Dynamics
Chaussee de Waterloo 72
1640 Rhode-St.-Genèse, Belgium

ISBN 978-3-540-07408-3   ISBN 978-3-540-37926-3 (eBook)
DOI 10.1007/978-3-540-37926-3

PREFACE

    This Lecture Series "Progress in Numerical Fluid Dynamics"
being organized at the von Karman Institute for Fluid Dynamics and
partly supported by the Consultant Exchange Programme of AGARD (Advi-
sory Group for Aerospace Research and Development) reviews recent
developments in the area of computational fluid dynamics.

    More than 130 participants from almost all European countries
and from even further afield attended and contributed during the discussions to the success of the Lecture Series.

    The present material, being revised and updated, comprises
two- and three-dimensional transonic and supersonic flows, the numerical treatment of the incompressible and compressible Navier-Stokes
equations, two- and three-dimensional boundary layer flows, including
unsteady layers, the foundation of the finite element method with applications to fluid dynamics, the treatment of problems of numerical
stability in the presence of boundary conditions and methods to improve
numerical solution using analogue subroutines.

    Expressing my thanks to my colleagues, who did not hesitate
to update and rewrite their contributions, to Mrs N. Toubeau who carefully assembled the material, I finally wish to address my appreciation
to the Springer-Verlag and in particular to Prof. Dr. W. Beiglböck,
Editor of the "Lecture Notes in Physics", for arranging this publica-
tion.

                                    Hans Jochen Wirz
                                    Lecture Series Director

        Rhode Saint Genèse,
           June 29, 1975

TABLE OF CONTENTS

ON THE COMPUTATION OF TWO- AND THREE-DIMENSIONAL STEADY

TRANSONIC FLOWS BY RELAXATION METHODS

by

F. R. Bailey

Ames Research Center

NASA, Moffett Field, Calif., U.S.A., 94035

TABLE OF CONTENTS

LIST OF ILLUSTRATIONS

LIST OF ILLUSTRATIONS (Continued)

ON THE COMPUTATION OF TWO- AND THREE-DIMENSIONAL STEADY

TRANSONIC FLOWS BY RELAXATION METHODS

by

F. R. Bailey
Ames Research Center
NASA, Moffett Field, Calif., U.S.A., 94035

1. INTRODUCTION

Transonic flows are characterized by the presence of adjacent regions of
subsonic and supersonic flow that are usually accompanied by weak shock waves. For
example, sketch A illustrates schematically the development of the transonic flow
pattern over a lifting airfoil for increasing free-stream Mach number. The flow
pattern progresses from a predominantly subsonic flow with embedded supersonic
regions to a predominantly supersonic flow with embedded subsonic regions.

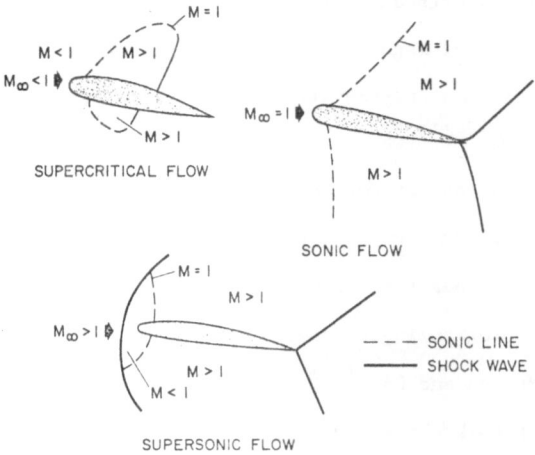

Sketch A

Mathematically, the description of steady transonic flows requires the solution
of "mixed" equations that are elliptic in subsonic regions and hyperbolic in

supersonic regions. The problem is essentially nonlinear, and solutions usually contain discontinuities representing shock waves. The only well-known analytical method for treating transonic flows solves the two-dimensional equations in the hodograph plane, where they are linear. This approach has proven useful for generating shock-free airfoil shapes[1,2] but is not well suited for solving flow fields with embedded shock waves. In the past several years, however, finite difference methods have been developed to treat steady transonic flows with shock waves. These methods follow two distinct approaches — unsteady and relaxation.

In the unsteady methods, the steady flow solution is obtained as a large time asymptotic solution to a real or pseudo time-dependent formulation. The approach is based on the extensive development of finite difference methods for initial value problems[3] and was first introduced by Magnus and Yoshihara[4,5]. For an overview of the unsteady methods, the interested reader is referred to the reviews given in references 6-8.

In the relaxation approach, the steady transonic flow equations are solved by an iteration algorithm in a manner similar to the standard procedure for solving elliptic boundary value problems. The main advantage of the current relaxation methods over the unsteady methods is the smaller amount of computational effort required — about an order of magnitude less.

The first application of a relaxation procedure to transonic flows was by Emmons in the late 1940s[9-11] who solved the density-stream function formulation of the inviscid steady equations of motion that included variations in entropy. Shock waves were explicitly fitted according to the Rankine-Hugoniot jump relations. Unfortunately, the method appears to require a decision process that is too complicated to be well suited for programming on digital computers.

More recently, Murman and Cole[12] introduced a relaxation procedure that automatically accounts for weak shock waves and is well suited for machine computation. The scheme was originally applied to the transonic small disturbance equation with a perturbation potential serving as the single dependent variable. The method has been extended by others[13-15] to include the exact isentropic formulation, in which it is assumed that any shock waves present are sufficiently weak that vorticity can be neglected. The unique feature of the method is that separate difference operators are used in elliptic (subsonic) and hyperbolic (supersonic) regions. In elliptic regions, central difference operators are used to account for the domain of dependence of subsonic flow equations. In hyperbolic regions, upwind difference operators are used to account for the absence of upstream influence in supersonic flow equations. Because of the absence of an entropy inequality, the isentropic flow equation contains discontinuous solutions that are not unique, and both expansion and compression jumps can exist. However, the use of upwind difference operators introduces directionality into the numerical method and creates a positive dissipative truncation error term which acts, in a sense, as a source of entropy production. Thus, expansion jumps are eliminated and uniqueness is restored. The compression jumps evolve naturally in the method and are considered as approximations to the true Rankine-Hugoniot shock jumps.

The present paper discusses recent developments in applying the Murman and Cole scheme to steady, inviscid transonic flow problems in two and three dimensions. Considerable emphasis is placed on two-dimensional methods since they form the basis for three-dimensional algorithms. The basic details of the scheme are described in terms of the original small disturbance formulation of Murman and Cole. In particular, Murman's recent introduction of fully conservative difference operators[16] to obtain the correct shock jumps is discussed. The extension to treat the exact isentropic equation is then covered with special attention given to Jameson's new rotated difference scheme[17] for supersonic flow regions. Following is a brief discussion of axisymmetric methods. Consideration of two-dimensional procedures is concluded by a discussion on comparisons with experiment, emphasizing the effects of viscosity and wind-tunnel walls. Finally, the paper concludes with discussions of the treatment of the three-dimensional small disturbance equation for swept wings and the exact isentropic equation for yawed wings.

2. RELAXATION METHOD APPLIED TO THE SMALL DISTURBANCE EQUATION

2.1 Introduction

We begin the discussion of the transonic relaxation method by considering its application to steady, transonic small disturbance theory for lifting airfoils. The theory is derived under the assumptions that the flow is inviscid, that airfoil slopes are everywhere small so that flow quantities are small perturbations about their free-stream values, and that the free-stream Mach number is near unity. Consequently, the airfoil must be thin and angles of attack small. In practical situations, these assumptions are not always strictly met, particularly at blunt noses, at large angles of attack, and if extensive flow separation is present. Nonetheless, many cases of engineering interest may be adequately described by the theory. Small disturbance theory is a considerable simplication of the exact inviscid theory in both the boundary conditions and governing equations. The resulting governing equation still retains the essential nonlinear, mixed elliptic-hyperbolic character, and its solution contains discontinuous jumps that approximate shock waves.

In the following, we discuss in some detail the transonic small distrubance formulation of Cole[18,19] and the application of the Murman and Cole scheme to lifting airfoils.

2.2 Transonic Small Disturbance Theory

2.2.1 Introduction

The transonic small disturbance equations can be formally derived by an asymptotic expansion procedure[18,19] applied to the inviscid equations of fluid flow in which the airfoil thickness ratio δ (see sketch B), tends to 0 while the

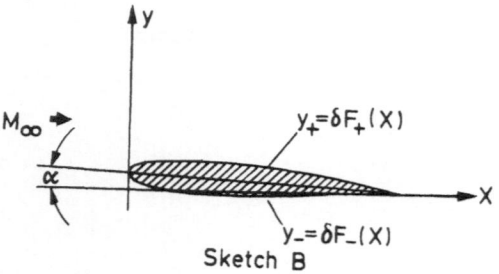

Sketch B

free-stream Mach number, M_∞, approaches 1. Thus, the flow is represented as small disturbances on a uniform stream. The large lateral extent of transonic disturbances is taken into account by use of the scaled coordinate $\tilde{y} = \delta^{1/3}y$. The limit process is $\delta \to 0$, $M_\infty \to 1$ while $K, x, \tilde{y}$ remain fixed where

$$K = (1 - M_\infty^2)/\delta^{2/3}$$

is the transonic similarity parameter. It should be noted that the definitions of K and $\tilde{y}$ are not unique and may be multiplied by functions like $g(M_\infty^2) = 0(1)$ where $g(1) = 1$. In the above, we have followed the scaling of Cole[19], although the more usual scaling is given in reference 20. For a more detailed summary of the theory, the reader is referred to Cole[19].

2.2.2 Transonic Equations

The asymptotic expansions of the form

$$q_x/u_\infty = 1 + \delta^{2/3}u + \delta^{4/3}u_2 + \delta^2 u_3 + \ldots$$

for the x-component of velocity,

$$q_y/u_\infty = \delta v + \delta^{5/3}v_2 + \ldots$$

for the y-component of velocity,

$$p/p_\infty = 1 + \delta^{2/3}p + \delta^{4/3}p_2 + \delta^2 p_3 + \ldots$$

for the pressure, and

$$\rho/\rho_\infty = 1 + \delta^{2/3}\sigma + \delta^{4/3}\sigma_2 + \delta^2\sigma_3 + \ldots$$

for the density, are substituted into the inviscid equations for conservation of mass, momentum, energy, and entropy (but allowing for a jump in entropy across shock waves). Taking into account the shock relations, various integrals are found, of which two give the first-order transonic equations

$$\{Ku - [(\gamma + 1)/2]u^2\}_x + v_{\tilde{y}} = 0 \qquad (2.1a)$$

$$v_x - u_{\tilde{y}} = 0 \qquad (2.1b)$$

Eq. (2.1a) is a transonic version of the continuity equation, and Eq. (2.1b) shows that the flow is irrotational to first order (it turns out to be irrotational to second order also). Thus, a perturbation velocity potential can be introduced by

$$\left.\begin{array}{l} \phi_x \equiv u \\ \phi_{\tilde{y}} \equiv v \end{array}\right\}$$
(2.2)

which yields the governing equation in divergence form as

$$\left\{ K\phi_x - [(\gamma + 1)/2]\phi_x^2 \right\}_x + \phi_{\tilde{y}\tilde{y}} = 0$$
(2.3a)

or alternatively, in nondivergence form, as

$$[K - (\gamma + 1)\phi_x]\phi_{xx} + \phi_{\tilde{y}\tilde{y}} = 0$$
(2.3b)

The essential nonlinearity of the governing equation allows the local formation of either an elliptic equation, representing subsonic flow $[\phi_x < (K/\gamma + 1)]$, or a hyperbolic equation, representing supersonic flow $[\phi_x > (K/\gamma + 1)]$.

2.2.3 Shock Jump and Drag

The shock jump relations, which are of great importance in transonic calculations, are contained in the governing equation in the sense that the weak solution[21] to Eq. (2.1a) and (2.1b) yields a consistent approximation to the usual Rankine-Hugoniot relations — that is, the surface integral forms of Eq. (2.1) integrated across a jump in (u,v) yield

$$\langle Ku - [(\gamma + 1)/2]u^2 \rangle (d\tilde{y})_s - \langle v \rangle (dx)_s = 0$$
(2.4a)

$$\langle v \rangle (d\tilde{y})_s + \langle u \rangle (dx)_s = 0$$
(2.4b)

where

   $\langle \; \rangle$ is the jump

and

   $(\;)_s$ is a shock surface element.

Eqs. (2.4a) and (2.4b) can be recast into the shock polar representation

$$[K - (\gamma + 1)\bar{u}]\langle u \rangle^2 + \langle v \rangle^2 = 0$$
(2.5)

where

   $\bar{u} = (1/2)(u_1 + u_2)$

is the average across the shock. The shock angle relative to the positive y-axis is given by

$$\theta_s = dx/d\tilde{y} = -(\langle v \rangle / \langle u \rangle) \tag{2.6}$$

and the continuity of tangential velocity across the shock yields

$$\langle \phi \rangle = 0 \tag{2.7}$$

The shock jumps given by Eqs. (2.4a) and (2.4b) will give good approximations to the Rankine-Hugoniot jumps for the ranges $0.8 \leq M_\infty \leq 1.3$ and $M_1 \leq 1.3$.

The consistent approximation to the Rankine-Hugoniot shock jumps contained in the transonic small disturbance theory has an important bearing on the calculation of drag, which in a two-dimensional inviscid flow about a closed body can only be due to shock waves. The drag on the airfoil is given by the x-momentum integral

$$D = \oint_C \left\{ (\rho q_x) q_y \; dx - [p + (\rho q_x) q_x] dy \right\} \tag{2.8}$$

where C is a contour enclosing the airfoil. Inserting the small disturbance expansions and taking into account jumps across shock waves yields, to the lowest order (for closed bodies),

$$\frac{D}{\gamma p_\infty} = \delta^{5/3} \left\{ \oint_C \left[uv \; dx + \left(\frac{\gamma + 1}{3} u^3 + \frac{v^2}{2} - K \frac{u^2}{2} \right) d\tilde{y} \right] - \frac{\gamma + 1}{12} \int_S \langle u \rangle^3 \; d\tilde{y} \right\} \tag{2.9}$$

The last integral in Eq. (2.9) is over that part of the shock surface S enclosed in C (see sketch C).

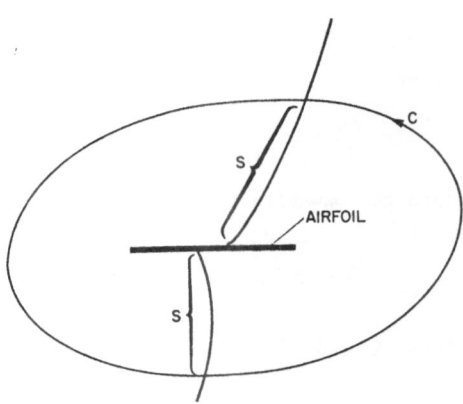

Sketch C

If the contour C is collapsed to the wing surface, $\tilde{y} = 0$, Eq. (2.9) becomes

$$\frac{D}{\gamma p_\infty} = \delta^{5/3} \int_0^1 [(uv)_- - (uv)_+]dx \tag{2.10}$$

which is the usual expression for pressure drag where v is the local body slope.

If the contour C is extended to infinity, the only contribution comes from the shock surface and is the wave drag

$$\frac{D}{\gamma p_\infty} = -\delta^{5/3} \frac{\gamma + 1}{12} \int_S \langle u \rangle^3 \, d\tilde{y} \tag{2.11}$$

Thus, the drag on the airfoil is due to the presence of shock waves and although the drag is a third-order quantity, it may be computed from the solution of the first-order transonic equations. Furthermore, the drag may be computed by (i) integrating surface pressures (assuming leading-edge singularities are integrable), (ii) integrating shock jumps, or (iii) integrating along an intermediate contour, Eq. (2.9).

It is important to note that momentum conservation is satisfied to the first two orders, and in particular the first-order solution has the correct first-order momentum conservation. The drag, however, has been shown to arise from nonconservation of third-order x-momentum across the shock wave. It can also be shown to be equal to that which is due to the entropy jump across weak shock waves. Substituting the transonic expansions into the expression for the jump of entropy across weak shocks yields to lowest order

$$\frac{\Delta s}{R} = -\frac{\gamma + 1}{12} \gamma \delta^2 \langle u \rangle^3 \tag{2.12}$$

Substituting Eq. (2.12) into the Oswatitsch drag relation[22]

$$D = \frac{T_\infty}{q_\infty} \int (\rho q_x)_n \Delta s \, dy \tag{2.13}$$

gives Eq. (2.11). Therefore, drag to lowest order is also due to the jump of entropy across the shocks.

2.2.4 Boundary Conditions

The formulation of the problem is completed by specifying the boundary conditions. The condition that the flow be tangent to the airfoil is linearized by the assumption $\delta \to 0$ and applied on the mean surface $\tilde{y} = 0$ in the form

$$\phi_{\tilde{y}}(x, +0) = F_+'(x) - \frac{\alpha}{\delta} \tag{2.14a}$$

$$\phi_{\tilde{y}}(x, -0) = F_-'(x) - \frac{\alpha}{\delta} \tag{2.14b}$$

For lifting airfoils, the Kutta condition is satisfied by requiring that ϕ_x (pressure) be continuous across the line $\tilde{y} = 0$, $x \geq 1$ and $\phi_{\tilde{y}}$ (flow angle) be continuous across $\tilde{y} = 0$, $x > 1$. The disturbance potential remains single-valued by introducing a cut in the $x, \tilde{y}$ plane ($\tilde{y} = 0$, say) at which ϕ jumps by an amount equal to the circulation Γ defined by

$$\Gamma = -\oint d\phi \tag{2.15}$$

with the integration taken around the airfoil.

The far field boundary conditions require that the perturbation velocities vanish at infinity with the disturbance potential given by

$$\phi = \frac{\Gamma}{2\pi} \theta \tag{2.16}$$

where θ is the angle between the position vector and the positive x-axis.

2.2.5 Some Singularities

Before discussing the small disturbance numerical procedure, it is instructive to look at the singular behavior expected to occur in small disturbance solutions at the airfoil edges. From solutions for wedges, it can be expected that the pressure has a log singularity near sharp edges of a nonlifting symmetric airfoil. This singular behavior is sufficiently weak that it provides no difficulties to the numerical method. In addition, Nonweiler[23] (also see Cole[19]) has shown that, for moderately blunt leading edges, Eq. (2.3) possesses a solution with an integrable singularity. For a nose shape given by $y = x^n$, the pressure approaches infinity in an integrable manner for $n > 0.4$ (slightly blunter than a parabola, $n = 0.5$) to n = 1 (wedge). No drag contribution (other than wave drag) is obtained by small disturbance solutions within this range of bluntness. It appears from numerical calculations that this criterion also holds for the numerical method. The addition of lift (either by angle of attack or camber) compounds the singular behavior at the nose. For example, the linearized boundary condition, because it does not allow the stagnation point to move off the geometric leading edge, causes a loss of expansion around the nose.

There exists still another important singularity, which occurs downstream of a supersonic to subsonic shock wave on a convex curved surface. In an inviscid flow, the shock must be normal to the surface so that the flow remains attached. The curvature of the surface induces an increasing pressure away from the surface in both the supersonic and subsonic regions near the shock. However, the jump conditions imposed by the normal shock cannot accommodate the downstream normal pressure gradient and, consequently, the shock must become oblique immediately off the surface. Their analysis predicts that the shock wave compression will be followed by a rapid expansion. Unfortunately, the scale is unknown, thereby preventing precise modeling of the post-shock expansion. It should be noted that this singularity is a consequence of the inviscid flow assumption and occurs in more exact treatments as well as in small disturbance theory.

2.3 Finite Difference Method

2.3.1 Introduction

The mixed differencing relaxation scheme for solving transonic boundary value problems was introduced by Murman and Cole[12] in their treatment of the small disturbance formulation for nonlifting airfoils. Soon afterwards, their scheme was extended to the lifting case by Krupp[25,26]. Most calculations are obtained by using unequally spaced grids, but in the following, we confine our discussion to evenly spaced grids and refer the reader to the cited references for more details.

The important type-dependent feature of the method is that, in subsonic regions, central difference operators are used to account for the domain of dependence of

elliptic equations; in supersonic regions, backward or upwind difference operators
are used to account for the absence of upstream influence in hyperbolic equations.

2.3.2 Difference Equations

Consider the finite-difference grid shown in sketch D, with Δx and $\Delta \tilde{y}$
assumed constant and velocities defined as

$$
\left.
\begin{array}{l}
u_{i+1/2,j} \equiv (\phi_x)_{i+1/2,j} \equiv \dfrac{\phi_{i+1,j} - \phi_{i,j}}{\Delta x}, \\[2em]
v_{i,j+1/2} = (\phi_{\tilde{y}})_{i,j+1/2} = \dfrac{\phi_{i,j+1} - \phi_{i,j}}{\Delta \tilde{y}}, \quad \text{etc.}
\end{array}
\right\}
\quad (2.17)
$$

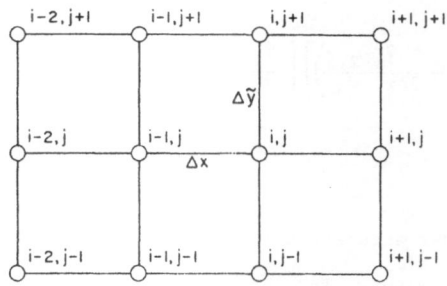

Sketch D

A central difference formula at point i,j approximating the governing equation,
Eq. (2.3), may be written in conservation form as

$$
\frac{\left\{ K\phi_x - [(\gamma + 1)/2]\phi_x^2 \right\}_{i+1/2,j} - \left\{ K\phi_x - [(\gamma + 1)/2]\phi_x^2 \right\}_{i-1/2,j}}{\Delta x}
$$

$$
+ \frac{(\phi_{\tilde{y}})_{i,j+1/2} - (\phi_{\tilde{y}})_{i,j-1/2}}{\Delta \tilde{y}} = 0 \quad (2.18)
$$

Substituting Eq. (2.17) into Eq. (2.18) and factoring yields

$$
\left[K - (\gamma + 1)\left(\frac{\phi_{i+1,j} - \phi_{i-1,j}}{2\Delta x} \right) \right] \left[\frac{\phi_{i+1,j} - 2\phi_{i,j} + \phi_{i-1,j}}{(\Delta x)^2} \right]
$$

$$
+ \left[\frac{\phi_{i,j+1} - 2\phi_{i,j} + \phi_{i,j-1}}{(\Delta \tilde{y})^2} \right] = 0 \quad (2.19)
$$

Eq. (2.19) is a second-order accurate difference operator for a Laplace-type equation in elliptic (subsonic) regions. The operator is stable by a linear stability analysis if the coefficient

$$(V_e)_{i,j} = K - (\gamma + 1)\left(\frac{\phi_{i+1,j} - \phi_{i-1,j}}{2\Delta x}\right) \tag{2.20}$$

is positive.

Similarly, an implicit backward or upwind operator at i,j may be written as

$$\frac{\{K\phi_x - [(\gamma + 1)/2]\phi_x^2\}_{i-1/2,j} - \{K\phi_x - [(\gamma + 1)/2]\phi_x^2\}_{i-3/2,j}}{\Delta x}$$

$$+ \frac{(\phi_{\tilde{y}})_{i,j+1/2} - (\phi_{\tilde{y}})_{i,j-1/2}}{\Delta \tilde{y}} = 0 \tag{2.21}$$

or alternatively

$$\left[K - (\gamma + 1)\left(\frac{\phi_{i,j} - \phi_{i-2,j}}{2\Delta x}\right)\right]\left[\frac{\phi_{i,j} - 2\phi_{i-1,j} + \phi_{i-2,j}}{(\Delta x)^2}\right]$$

$$+ \left[\frac{\phi_{i,j+1} - 2\phi_{i,j} + \phi_{i,j-1}}{(\Delta \tilde{y})^2}\right] = 0 \tag{2.22}$$

Eq. (2.22) is a first-order accurate difference operator for a wave-type equation in hyperbolic (supersonic) regions. The operator is stable by a linear stability analysis if the coefficient

$$(V_h)_{i,j} = K - (\gamma + 1)\left(\frac{\phi_{i,j} - \phi_{i-2,j}}{2\Delta x}\right) \tag{2.23}$$

is negative.

During the iteration procedure, $(V_e)_{i,j}$ is computed at each grid point. If $(V_e)_{i,j} > 0$, the flow at that point is subsonic and the elliptic operator Eq. (2.19) is used. If $(V_e)_{i,j} < 0$ and $(V_h)_{i,j} < 0$, the flow is supersonic and the hyperbolic operator Eq. (2.22) is used. Closer examination is required, however, when the flow crosses the boundary from subsonic to supersonic regions at the sonic line and from supersonic to subsonic regions at a shock wave or sonic line.

As the flow accelerates through sonic velocity from subsonic to supersonic velocities, a point is reached where $(V_e)_{i,j} < 0$ and $(V_h)_{i,j} > 0$ [since $(V_h)_{i,j} = (V_e)_{i-1,j}$] and neither Eq. (2.19) nor Eq. (2.22) is stable. The difficulty is cir-

circumvented by introducing a parabolic point operator[27] for such points by setting $(V)_{i,j} = 0$ to yield

$$\frac{\phi_{i,j+1} - 2\phi_{i,j} + \phi_{i,j-1}}{(\Delta \tilde{y})^2} = 0 \tag{2.24}$$

As the flow decelerates through sonic velocity from supersonic to subsonic flow, a point is reached where $(V_e)_{i,j} > 0$ and $(V_h)_{i,j} < 0$ and both Eq. (2.19) and Eq. (2.22) are locally stable. The original Murman and Cole scheme used the elliptic operator, Eq. (2.19), at this point.

2.3.3 Shock Point Operator

Results obtained using the above difference operators agree well with accepted exact solutions for continuous subcritical and supercritical shock-free flows. Furthermore, results for flows with embedded shock waves have agreed well with experimental data. However, the calculated shock pressure jump on the surface is consistently less than the theoretical value for a normal shock. This fact had been attributed[12] to a smoothing out of the previously discussed reexpansion singularity[24] at the foot of the shock by numerical truncation errors. However, several investigations using progressively finer grid spacing did not show an appreciable increase in the pressure jump with decreasing grid spacing as one would expect. Furthermore, Yoshihara[28] pointed out that no analysis had been presented to show that the calculated jump uniquely satisfies the jump contained in the governing equations. Murman[16] investigated this point further by comparing detached bow-wave solutions with time-dependent solutions[29]. This led to the introduction of the "shock-point operator"[16]

$$\frac{\left\{ K\phi_x - [(\gamma + 1)/2]\phi_x^2 \right\}_{i+1/2,j} - \left\{ K\phi_x - [(\gamma + 1)/2]\phi_x^2 \right\}_{i-1/2,j}}{\Delta x}$$

$$+ \frac{\left\{ K\phi_x - [(\gamma + 1)/2]\phi_x^2 \right\}_{i-1/2,j} - \left\{ K\phi_x - [(\gamma + 1)/2]\phi_x^2 \right\}_{i-3/2,j}}{\Delta x}$$

$$+ \frac{(\phi_{\tilde{y}})_{i,j+1/2} - (\phi_{\tilde{y}})_{i,j-1/2}}{\Delta \tilde{y}} = 0 \tag{2.25}$$

which factors into

$$(V_e)_{i,j} \frac{(\phi_{i+1,j} - 2\phi_{i,j} + \phi_{i-1,j})}{(\Delta x)^2} + (V_h)_{i,j} \frac{(\phi_{i,j} - 2\phi_{i-1,j} + \phi_{i-2,j})}{(\Delta x)^2}$$

$$+ \frac{\phi_{i,j+1} - 2\phi_{i,j} + \phi_{i,j-1}}{(\Delta \tilde{y})^2} = 0 \qquad (2.26)$$

This operator, whose x-differences are the sum of the x-differences of the elliptic and hyperbolic operators, is introduced at each grid point where $(V_e)_{i,j} > 0$ and $(V_h)_{i,j} < 0$.

Murman[16] has shown that this guarantees that the difference equations will give the correct weak solution to Eq. (2.3). For example, consider the normal shock solution for which $(dx)_s$ in Eq. (2.4) vanishes. Assume the shock lies somewhere between grid points $I - 1$ and I, with a uniform supersonic velocity u_1, upstream of the shock and a uniform subsonic velocity u_2 downstream. Sum the hyperbolic operator from $i = -\infty, \ldots, I - 1$, add the shock point operator at $i = I$ and sum the elliptic operator from $i = I + 1, \ldots, \infty$. Because of the cancellation of fluxes between neighboring grid points, the correct shock jump results:

$$\left(Ku - \frac{\gamma + 1}{2} u^2\right)_2 - \left(Ku - \frac{\gamma + 1}{2} u^2\right)_1 = 0 \qquad (2.27)$$

Murman[16] has further generalized this example and shown that, in the limit of vanishing grid spacing, the correct shock jumps are obtained for oblique shocks when arbitrary grid spacing is used. For shock waves that jump from supersonic to subsonic velocities, the upstream velocity is $u_{I-3/2}$ and the downstream velocity is $u_{I+1/2}$ so that the jump is spread over three mesh points.

Now, if the shock point operator is replaced by the elliptic operator in the normal shock example given above, the result is

$$\left(Ku - \frac{\gamma + 1}{2} u^2\right)_2 - \left(Ku - \frac{\gamma + 1}{2} u^2\right)_1 = \left(Ku - \frac{\gamma + 1}{2} u^2\right)_{I-1/2,j}$$

$$- \left(Ku - \frac{\gamma + 1}{2} u^2\right)_{I-3/2,j} \qquad (2.28)$$

The right-hand side can be viewed as a spurious source term that is caused by the noncancellation of fluxes at the shock point, $i = I$. Eq. (2.28) does not guarantee the correct shock jump, since, in general, the source is nonzero. In fact, the right-hand side vanishes only if the shock lies exactly at i = I, since then $u_{I-1/2,j} = u_{I-3/2,j} = u_1$.

A word should be said about calculating shocks when the flow downstream is supersonic. It has been demonstrated[27] that the first-order accurate hyperbolic operator Eq. (2.22) is dominated by dissipation errors, which lead to shocks smeared over 6 to 10 mesh points. Second-order accurate operators have a leading truncation error which is dispersive and often give poorer shock structure (e.g., overshoots and undershoots) and can suffer from instabilities. Hybrid combinations of first- and second-order operators[14,25] generally require so much dissipation for stability that the anticipated increase in sharpness is not realized.

2.3.4 Conservation Form and Consistency

The fact that the addition of the shock point operator gives the correct shock jump is a consequence of writing the difference operators in conservation form. That is, the integral properties of the governing equation are preserved by the difference operators. This is important at shock waves where spurious source terms resulting from the noncancellation of fluxes across mesh boundaries (if nonconservative differencing is used) may not vanish with vanishing mesh spacing. Thus, the calculated shock jump will be in error in proportion to the spurious source. It should be emphasized that, for smooth continuous regions, the choice of conservative or nonconservative differencing is irrelevant, but in regions of discontinuities, conservative form is imperative.

It is axiomatic that finite-difference operators be consistent with the differential equations being approximated, i.e., the difference equations reduce to the differential equations in the limit of vanishing grid spacing. At first glance, it appears that the parabolic operator, Eq. (2.24), and the shock point operator, Eq. (2.26), do not fulfill this requirement. It can be shown[16] that they both are, in fact, consistent with the differential equation, Eq. (2.3), in continuous regions where Eq. (2.3) is valid.

If one expands the coefficient $V = K - (\gamma + 1)\phi_x$ about the sonic value, $V = 0$, and substitutes in Eq. (2.3) the result to lowest order is

$$-\delta x(\gamma + 1)(\phi_{xx})^2 + \phi_{\tilde{y}\tilde{y}} + \ldots = 0 \qquad (2.29)$$

where $\delta x \equiv x - x_{sonic} \leq \Delta x$. As the mesh spacing vanishes, $\delta x \to 0$, showing that the parabolic operator, Eq. (2.24), is, indeed, consistent.

The Taylor series expansion of the shock point operator, Eq. (2.26), shows that its lowest-order approximation to Eq. (2.3) equals

$$2[K - (\gamma + 1)\phi_x]\phi_{xx} + \phi_{\tilde{y}\tilde{y}} + \Delta x \left\{ [K - (\gamma + 1)\phi_x]\phi_{xx} \right\}_x + \ldots = 0 \qquad (2.30)$$

If the flow decelerates through the shock points as a smooth recompression, the solution is continuous and, from the expansion Eq. (2.29), Eq. (2.30) is a consistent approximation with truncation error $O(\delta x)$. If the flow decelerates through the shock point as a shock with finite strength, the governing differential equation, Eq. (2.3), does not hold and the usual consistency condition is not violated. It should be noted, however, that in the limit Δx, $\Delta \tilde{y} \to 0$, the difference equations yield the correct integral properties of the governing equations.

2.3.5 Boundary Conditions

The flow tangency boundary condition on $\phi_{\tilde{y}}$, Eq. (2.14), is applied on the boundary $\tilde{y} = 0$ by writing either the approximation

$$(\phi_{\tilde{y}\tilde{y}})_{i,1} = \frac{2}{\Delta \tilde{y}} \left(\frac{\phi_{i,2} - \phi_{i,1}}{\Delta \tilde{y}} - \phi_{\tilde{y}} \Big|_{\tilde{y}=0} \right) \qquad (2.31)$$

with $j - 1$ placed on the boundary $\tilde{y} = 0$ or by

$$(\phi_{\tilde{y}\tilde{y}})_{i,1} = \frac{1}{\Delta \tilde{y}} \left(\frac{\phi_{i,2} - \phi_{i,1}}{\Delta \tilde{y}} - \phi_{\tilde{y}} \Big|_{\tilde{y}=0} \right) \qquad (2.32)$$

with $j = 1$ placed at $\tilde{y} = \tilde{y}/2$. In his treatment of lifting airfoils, Krupp[26] used a version of Eq. (2.32) for an unequally spaced $\tilde{y}$ mesh. The boundary condition $(\phi_{\tilde{y}})_{\tilde{y}=0}$ is incorporated into the expression for $\phi_{\tilde{y}\tilde{y}}$ at $\tilde{y} = h_1$ (see sketch E).

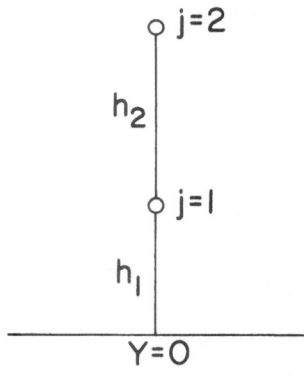

Sketch E

The difference operator, retaining the lowest-order truncation error, is given by

$$(\phi_{\tilde{y}\tilde{y}})_{i,1} = \frac{2}{h_2(h_2 + 2h_1)} (\phi_{i,2} - \phi_{i,1}) - \frac{2}{h_2 + h_1} \phi_{\tilde{y}}\Big|_{\tilde{y}=0}$$

$$+ \frac{h_2^2 - 3h_1^2}{3(h_2 + 2h_1)} \phi_{\tilde{y}\tilde{y}\tilde{y}}\Big|_{\tilde{y}=0} + O(h_2^2) \quad (2.33)$$

which shows that the truncation error is minimized for $h_2 = \sqrt{3}\ h_1$.

The outer boundary condition can be treated by either extending the computational domain to infinity through some transformation or by using a finite-domain and treating the infinity boundary condition in an approximate manner. The latter approach has been used in small disturbance computations reported. Approximate far field solutions have been derived[25,30] by writing Eq. (2.3) as an integral expression for ϕ. The resulting approximations are separated into three terms representing the effect of thickness, lift, and the nonlinearity arising from the term ϕ_x^2. This approximate treatment retains sufficient accuracy if the outer boundary is far enough from the origin, say $|x| \simeq 1.5$, $|\tilde{y}| \simeq 3$.

For lifting airfoils, the circulation Γ which satisfies the Kutta condition must be obtained as part of the solution process for ϕ. A cut is introduced at $\tilde{y} = 0$, $x \geq 1$ across which the potential jump must equal the circulation Γ. Modified difference operators for $\phi_{\tilde{y}\tilde{y}}$ are written that take into account the jump in ϕ

across the cut while maintaining the continuity of ϕ_x and $\phi_{\tilde{y}}$. These difference operators are expressed as

$$(\phi_{\tilde{y}\tilde{y}})_{i,0^-} = \frac{1}{(\Delta\tilde{y})^2} [(\phi_{i,1} - \Gamma_i) - 2\phi_{i,0^-} + \phi_{i,-1}] \qquad (2.34a)$$

$$(\phi_{\tilde{y}\tilde{y}})_{i,1} = \frac{1}{(\Delta\tilde{y})^2} [\phi_{i,2} - 2\phi_{i,1} + (\phi_{i,0^-} + \Gamma_i)] \qquad (2.34b)$$

The iteration procedure for finding the proper circulation proceeds by choosing an initial circulation Γ_0 from which the far field solution is obtained. The circulation at the trailing edge is found from the jump in ϕ, i.e.,

$$\Gamma_{te} = \phi_{i_{te},0^+} - \phi_{i_{te},0^-} \qquad (2.35)$$

On each iterative sweep of the grid, a new value of Γ_{te} is found by linearly extrapolating the ϕ's from above and below. (It is recalled that, in the Krupp procedure, the grid points are at $\pm h_1$ above the airfoil boundary $\tilde{y} = 0$, $0 \leq x \leq 1$.) The new Γ_{te} along with Γ_0 is used to determine the Γ_i, $1 < x < x_{max}$, by interpolation. As the iteration proceeds, new estimates for Γ_0 are made, depending on previous values of Γ_0 and Γ_{te}, until a converged solution for ϕ is obtained and $\Gamma_0 = \Gamma_{te}$. A simplified procedure for converging to the proper circulation has been developed,[31,32] which consists of simply setting $\Gamma = \Gamma_{te}$ along the entire cut during each iteration. This simple method gives both the same results and convergence rate as the method of Krupp.

The treatment of lift introduces an additional iteration into the basic relaxation procedure and can cause a significant increase in computation time. However, certain strategies exist to impove the efficiency of lifting calculations. For example, one can often begin calculations from a previous solution that is not far removed from the desired solution. Another strategy, suggested by Dr. Hall of the Royal Aircraft Establishment, is to update the potential in the entire field (not just the outer boundary) for the first few iterations according to the simple vortex solution. For the following iterations, the vortex solution is replaced by the lifting potential distribution found from the previous iteration.

2.3.6 Relaxation Procedure

Approximating the governing equation, Eq. (2.3), by the appropriate difference operators at each interior grid point and applying the boundary conditions results in a system of simultaneous equations for the value of potential at each interior grid point. The system is, in general, nonlinear in the grid variable $\phi_{i,j}$ because of the upwind difference operator used in supersonic regions. The solution is found by a successive line relaxation along a line of grid points at constant x. The equation system for each vertical line is written as

$$\underset{\sim}{A} \vec{\phi}_i = \vec{f}_i \qquad (2.36)$$

where $\vec{\phi}_i$ is the J dimensional column vector

$$\vec{\phi}_i = \begin{pmatrix} \phi_{i,1} \\ \cdot \\ \cdot \\ \cdot \\ \phi_{i,J} \end{pmatrix}$$

and the $J{\times}J$ dimensional tridiagonal matrix $\underset{\sim}{A}$ and J dimensional vector $\vec{f}_i$ are functions of $\vec{\phi}_{i+1}$, $\vec{\phi}_i$, $\vec{\phi}_{i-1}^+$, $\vec{\phi}_{i-2}^+$ where the superscript $+$ refers to new values. After some initial guess on ϕ, the iteration method consists of successively sweeping the grid from the upstream to downstream boundary, during which, new values of ϕ are obtained from Eq. (2.36) by direct elimination. Convergence is accelerated by the algorithm

$$\vec{\phi}_i^+ = \omega \vec{\tilde{\phi}}_i + (1 - \omega) \vec{\phi} \qquad (2.37)$$

where $\vec{\tilde{\phi}}_i$ is the solution to Eq. (2.36). Switching the difference operators, depending on the values of V_e and V_h with respect to zero, ensures that the matrix $\underset{\sim}{A}$ remains diagonally dominant and nonsingular. Note that in supersonic regions $\underset{\sim}{A}$ is a function of $\phi_{i,j}$ and the elimination algorithm may be successively iterated at columns that contain supersonic points until $\vec{\phi}_i$ satisfies a prescribed error bound before sweeping to the next column. Alternatively, one can consider $(V_h)_{i,j}$ as fixed coefficients and do the elimination only once. The later procedure gives a slightly better overall convergence rate.

The stability and rate of convergence of the relaxation algorithm depend on the choice of relaxation parameters ω in Eq. (2.37). The values of ω depend on the type of flow and acceptable values have been found to be $\omega \simeq 1.9$ in elliptic regions and $\omega \simeq 0.90$ in hyperbolic regions. Because the problem is nonlinear, these values of ω have been determined by numerical experimentation. A clue to the value in elliptic regions comes from standard references on relaxation methods, where it is shown that the line algorithm applied to linear elliptic equations is stable for $0 < \omega < 2$, with the optimum value approaching 2 with increasing grid refinement. For the linear wave equation, the well-known von Neuman stability test based on Fourier analysis (with successive iterations viewed as steps in pseudotime) shows that the supersonic relaxation algorithm is stable only if it is fully implicit, i.e., with $\omega = 1$. In actual transonic calculations, however, experience indicates that convergence difficulties are less likely to occur if ω is set to a value less than 1. It is noted that nonconservative relaxation schemes with both V_e and V_h computed by using central difference operators with data strictly from the previous sweep (thereby linearizing the equation from one sweep to the next) are reported to have no particular stability difficulties with $\omega = 1$. Such a modification is used in both the Garabedian-Korn and Jameson methods for the full potential equation (see section 3).

2.4 Results

Numerous calculated results obtained from the small disturbance relaxation method have been reported by Murman and Krupp, particularly in reference 25. Computation times are reported[26] to be from 3 to 15 minutes on an IBM 360/65 (or 1 to 5 minutes on a CDC 6600) although more recent improvements for converging the circulation have made the lower figure more typical. The bulk of the calculations were performed without incorporating the shock-point operator. Such results are termed "not fully conservative relaxation" (NCR) solutions. In Murman's[16] recent paper he compares some NCR solutions with those obtained using the shock-point operator, termed "fully conservative relaxation" (FCR) solutions. These results are highly significant and some of them are repeated here.

An example of the comparison between NCR and FCR solutions for an embedded shock is shown in Fig. 1. The NCR calculations show that its shock pressure jump does not approach the theoretical normal jump, even as the grid spacing is decreased. The FCR solution, however, shows that as the mesh is refined, the correct jump is obtained, followed by a well-defined reexpansion. The FCR computed shock is stronger and farther aft of the one obtained from the NCR method. This appears typical in comparisons of the two methods. In fact, as the strength of the shock increases, the disparity between the shock locations given by the two methods also increases. For example, the results in Fig. 1 for $M_\infty = 0.872$ show a difference in shock location of about 5 percent chord. The results shown in Fig. 2, on the other hand, indicate that at $M_\infty = 0.909$ the shock is stronger and the difference in location has increased to about 12 percent. Conversely, as the shock strength decreases, the difference in the predicted shock locations also decreases and, in regions of smooth shock-free recompressions, the two solutions are in essential agreement.

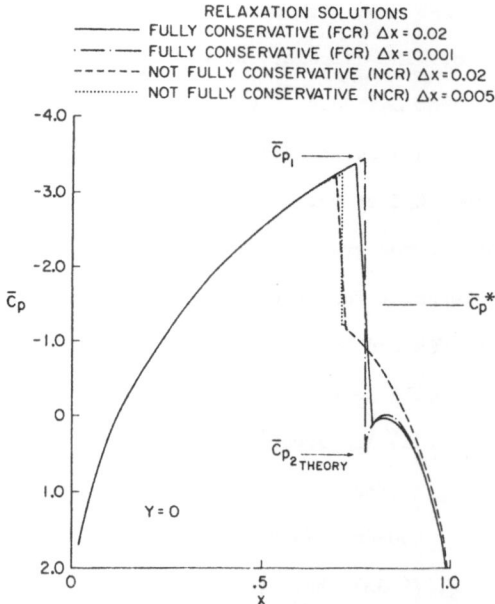

Fig. 1.   Comparison of FCR and NCR solutions for 6 percent parabolic arc airfoil at $M_\infty = 0.872$ [16] (K = 1.8).

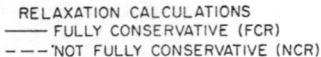

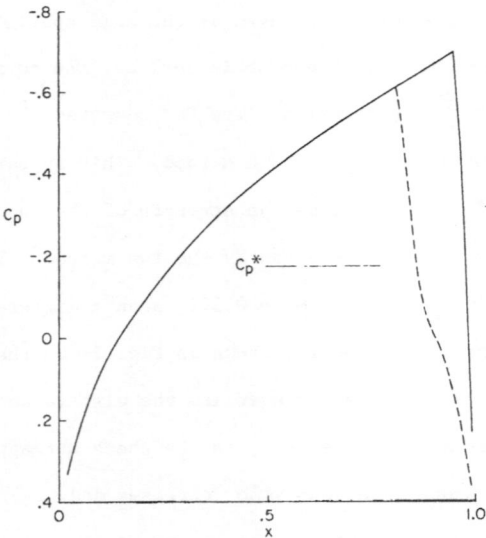

Fig. 2.   Comparison of FCR and NCR solutions for 6 percent parabolic

arc airfoil at $M_\infty = 0.909$[16] $(K = 1.25)$.

Comparative results for flow with a supersonic free stream and a detached bow

shock are shown in Fig. 3.   The calculations were performed using the two relaxation

methods and a time-dependent method,[29] which is also fully conservative.   The FCR

and time-dependent results are in essential agreement, whereas the NCR result shows

too great a shock detachment distance.   Fig. 4 gives the shock jumps for the three

calculations in the form of hodograph plots and illustrates that the NCR method shock

jumps are in considerable error for strong oblique shocks.

Further examples of NCR and FCR calculations for lifting airfoils are given in

following sections.   It should be noted that the existence of singular behaviors at

the airfoil edges to Eq. (2.3) led Krupp[25] to do extensive experimentation with

techniques for calculating surface pressure.   His best results were obtained by

finding ϕ on the airfoil boundary $(\tilde{y} = 0)$ by linear extrapolation.   (Recall from

the boundary condition Eq. (2.33) that ϕ on the boundary is not computed directly.)

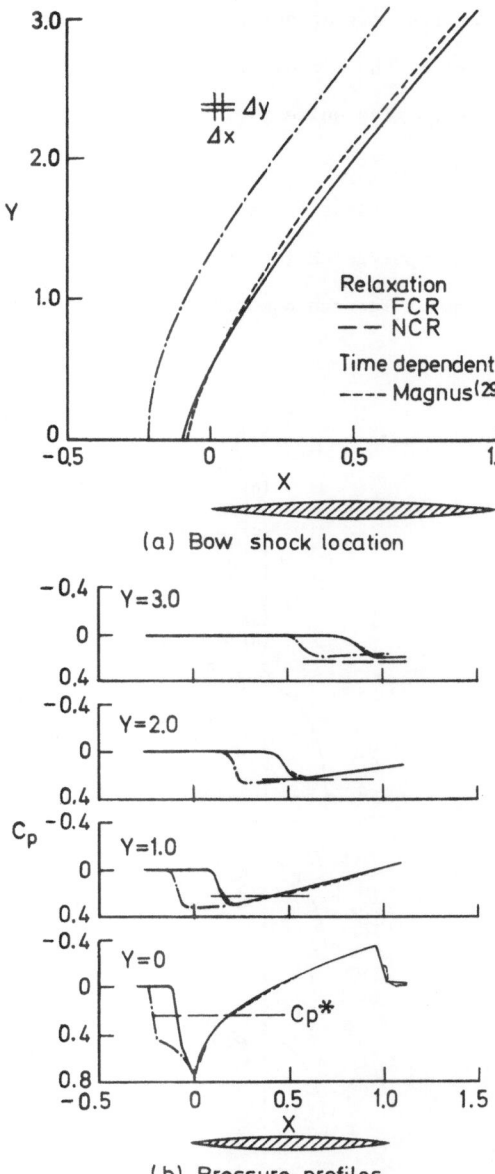

(a) Bow shock location

(b) Pressure profiles

Fig. 3. Comparison of computational methods for 6 percent parabolic arc airfoil with detached bow wave at M_∞ = 1.15.[16]

While local irregularities may appear in the nose region for some solutions, the solution over the rest of the airfoil is unaffected. Thus, small disturbance solutions can be expected to give quite good results for moderately blunt airfoils at small angles of attack.

Small disturbance solutions can also be made to give better agreement with more exact solutions by appropriate choice of the transonic similarity forms which it is recalled are not unique. Through numerical experimentation, Krupp[25] found that the

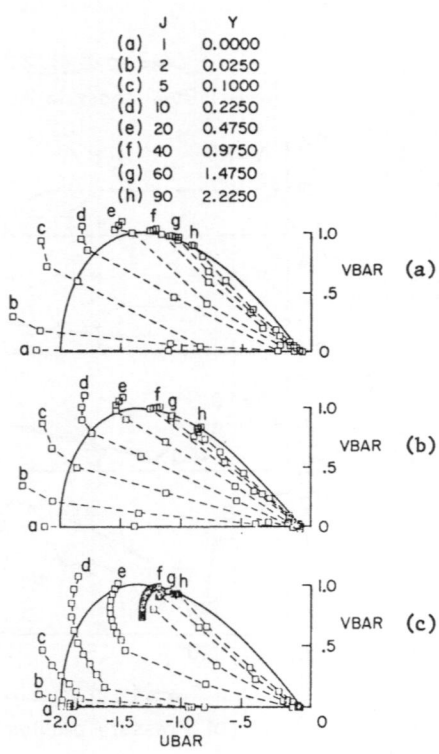

Fig. 4.   Hodograph plots for detached bow wave.

(a)   Fully conservative relaxation.

(b)   Time-dependent.[29]

(c)   Not fully conservative relaxation.[16]

following scaling, when used by the above numerical method, gave improved agreement with full inviscid theory for symmetric shock-free airfoils. These forms are

$$K = \frac{1 - M_\infty^2}{\delta^{2/3} M_\infty}$$

$$\tilde{y} = \delta^{1/3} M_\infty^{2/3} y$$

$$C_p = \frac{-2\delta^{2/3}}{M_\infty^{3/4}} \phi_x$$

3. EXACT ISENTROPIC PROCEDURE IN TWO DIMENSIONS

3.1 Introduction

The equations for irrotational, inviscid, adiabatic flow of a perfect gas in two dimensions may be expressed as a second-order partial differential equation

$$(a^2 - u^2)\phi_{xx} - 2uv\phi_{xy} + (a^2 - v^2)\phi_{yy} = 0 \tag{3.1}$$

for the velocity potential ϕ. The velocity components are

$$u = \phi_x, \qquad v = \phi_y \tag{3.2}$$

and the speed of sound is defined by Bernoulli's law

$$\frac{u^2 + v^2}{2} + \frac{a^2}{\gamma - 1} = \frac{1}{2} + \frac{1}{M_\infty^2(\gamma - 1)} \tag{3.3}$$

There are no assumptions of small disturbances and Eq. (3.1) is exact for subsonic inviscid flow. Furthermore, Eq. (3.1) is assumed valid for transonic flows under the assumption that any shock waves are sufficiently weak to introduce negligible rotation.

The flow equation can be written in the divergence form

$$(\rho\phi_x)_x + (\rho\phi_y)_y = 0 \tag{3.4a}$$

$$\rho = \left[1 - \frac{\gamma - 1}{2}(u^2 + v^2)\right]^{1/\gamma-1} \tag{3.4b}$$

from which a weak solution can be derived that admits jumps analogous to the Rankine-Hugoniot relations. Steger and Baldwin[33] have shown that the component of momentum normal to the isentropic shock is not conserved through the shock. Since the total drag must vanish in a potential flow, the force that is due to the momentum jump at the shock is balanced by the drag on the body. Consequently, the solution to Eq. (3.1) may be used to give an approximation to the drag on an airfoil with shock waves.

Such estimates of inviscid wave drag are valid for relaxation solutions provided the isentropic shock wave is properly captured. However, results reported so far for the exact isentropic equation have failed to give the proper isentropic shock jump because the difference equation is an approximation to the nondivergence form of the differential equation, Eq. (3.1). As pointed out in section 2, conservative difference operators are required to calculate the weak solution correctly. Recall that in the small disturbance procedure, the conservative difference equation factored into coefficient V, multiplying second derivative operators in x. The change of sign of the coefficient V, depending on whether the flow is subsonic or supersonic, is necessary for stability in the Murman-Cole type-dependent scheme. Clearly, application of the appropriate difference operators to Eq. (3.1) will lead to the same stable properties. However, the finite-difference analog to the divergence form, Eq. (3.4) does not possess the straightforward factoring property of the difference analog to the small disturbance divergence form because of the complicated relationship between velocity and density. Further investigation is required to develop a method that gives the correct weak solution, either by conservative differencing or shock fitting. It is important to note, however, that, even though the present methods do not give the correct inviscid shock jumps, their solutions are often in remarkably good agreement with experimental results, as will be shown later.

The principal advantage of using the exact isentropic equation is that one can treat the airfoil surface boundary condition exactly, thereby obtaining more accurate solutions over blunt-nosed, thick airfoils at high angles of attack. Of course, the finite-difference grid must be introduced so that difference operators may be constructed accurately at the boundary and, in general, the airfoil itself must be a coordinate line, or nearly so.

Three relaxation procedures for obtaining solutions to Eq. (3.1) have been developed and are briefly described in the following sections.

3.2 Steger and Lomax Procedure

Steger and Lomax[13] chose to replace the actual airfoil by the blunt-nosed plate control surface shown in sketch F. The nose of the plate is taken to be the

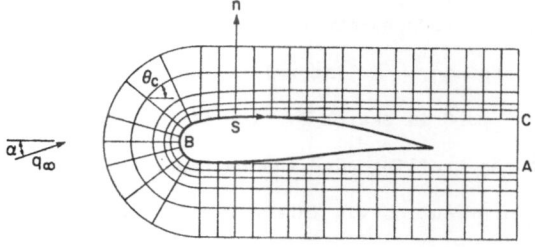

Sketch F

closest possible representation of the airfoil nose (in many cases exact). Over the aft portion of the airfoil, the flow tangency boundary condition is projected onto the control surface by second-order extrapolation. The coordinate system used in the calculation corresponds to the normals to the control surface and their orthogonal members. The far field boundary condition is taken to be a compressible vortex and is applied at some finite distance.

Eq. (3.1) is rewritten in terms of control surface coordinates from which the potential due to a uniform free-stream was substracted to yield

$$\left(\frac{a^2 - u^2}{H}\right)\left[\frac{1}{H}\phi_{ss} + \phi_s\left(\frac{1}{H}\right)_s\right] + (a^2 - v^2)\phi_{nn} - \frac{2uv}{H}\phi_{sn}$$

$$- \frac{1}{HR_c}\left[(a^2 - u^2)\phi_n + \frac{2uv}{H}\phi_s\right] = 0 \quad (3.5a)$$

where

$$a^2 = a_\infty^2 - \frac{\gamma - 1}{2}(u^2 + v^2 - q_\infty^2) \quad (3.5b)$$

$$u = q_\infty \cos(\theta_c - \alpha) + \frac{\phi_s}{H} \qquad (3.5c)$$

$$v = -q_\infty \sin(\theta_c - \alpha) + \phi_n \qquad (3.5d)$$

$$H = \frac{R_c(s) - n}{R_c(s)} \qquad (3.5e)$$

where θ_c is the angle the s-coordinate makes with a horizontal reference line and R_c is the radius of curvature of the control surface. Type-dependent difference operators similar to that of Murman and Cole are written for Eq. (3.5), except that in hyperbolic regions fully second-order accurate operators are used.

The solution is obtained by a line relaxation scheme that sweeps about the airfoil, proceeding from A to C in sketch F. In lifting airfoils cases, the Kutta condition is enforced at the trailing edge and the appropriate value for the circulation is found by operator intervention via interactive computer graphics. Reported solutions were obtained in less than 20 minutes on an IBM 360/67 (6 minutes on a CDC 6600).

3.3 Garabedian and Korn Procedure

The Garabedian and Korn[14,34] approach is to map the exterior of the airfoil conformally onto the interior of the unit circle, with the point at infinity corresponding to the origin. This mapping procedure was introduced by Sells[35] and produces a polar coordinate system that greatly simplifies the fulfillment of the surface boundary condition. Furthermore, it produces a desirable distribution of grid points along the airfoil surface such that the density of grid points is greatest at the leading and trailing edges. The mapping is done numerically and requires the specification of the airfoil surface curvature. For details on the basic mapping procedure, the reader is referred to the paper by Sells[35] and the book by Thwaites.[36]

Assume that the mapping is known and given by

$$x + iy = F(re^{i\theta}) = \sum_{n=-1}^{\infty} a_n r^n e^{in\theta} \qquad (3.6)$$

with modulus

$$f = r^2 |F'(re^{i\theta})| \tag{3.7}$$

In the new (r,θ) coordinate system (θ measured clockwise from the trailing edge back), Eq. (3.1) becomes

$$(a^2 - r^2 f^{-2}\phi_\theta{}^2)\phi_{\theta\theta} - 2r^4 f^{-2}\phi_\theta\phi_r\phi_{\theta r} + r^2(a^2 - r^4 f^{-2}\phi_r{}^2)\phi_{rr}$$

$$+ r(a^2 - r^2 f^{-2}\phi_\theta{}^2 - 2r^4 f^{-2}\phi_r{}^2)\phi_r$$

$$+ f^{-3}(r^2\phi_\theta{}^2 + r^4\phi_r{}^2)(f_\theta\phi_\theta + r^2 f_r\phi_r) = 0 \tag{3.8}$$

This equation is used in the concentric region near the airfoil, $0 < r_o \le r \le 1$ and includes all supersonic points. For the inner concentric region, $0 \le r \le r_o < 1$, a disturbance potential is defined as

$$\Phi = \phi - f_o \frac{\cos(\theta + \alpha)}{r} \tag{3.9}$$

where f_o is a constant of the mapping. Use of the new dependent variable avoids the singularities at infinity, with Φ satisfying the equation

$$(a^2 - r^2 f^{-2}\phi_\theta)\Phi_{\theta\theta} - 2r^4 f^{-2}\phi_\theta\phi_r\Phi_{\theta r} + r^2(a^2 - r^4 f^{-2}\phi_r{}^2)\Phi_{rr} - 2r^3 f^{-2}\phi_\theta\phi_r\Phi_\theta$$

$$+ r(a^2 + r^2 f^{-2}\phi_\theta{}^2 - 2r^4 f^{-2}\phi_r{}^2)\Phi_r$$

$$+ f^{-3}(r^2\phi_\theta{}^2 + r^4\phi_r{}^2)(f_\theta\Phi_\theta + r^2 f_r\Phi_r)$$

$$= f_o f^{-3}(r^2\phi_\theta{}^2 + r^4\phi_r{}^2)\frac{f_\theta\sin(\theta + \alpha) + rf_r\cos(\theta + \alpha)}{r} \tag{3.10}$$

and the vortex boundary values

$$\Phi = \frac{\Gamma}{2\pi}\tan^{-1}[\sqrt{1 - M_\infty{}^2}\tan(\theta + \alpha)] \tag{3.11}$$

at $r = 0$. The Kutta condition requires that $\phi_\theta = 0$ at the trailing edge and that the potential jump across rays $\theta = 0$ and 2π be constant and equal to Γ.

The finite-difference scheme is similar to that of Murman and Cole. However, central difference operators are used to approximate all first derivatives in both subsonic and supersonic regions, and in supersonic regions a weighted average of first- and second-order accurate upwind operators is employed as follows for $\theta > \pi$ (similar expressions are derived for $\theta < \pi$):

$$\phi_{\theta\theta} = \frac{1}{(\Delta\theta)^2} [(\phi_{i,j} - 2\phi_{i-1,j} + \phi_{i-2,j}) + \epsilon(\phi_{i,j} - 3\phi_{i-1,j} + 3\phi_{i-2,j} - \phi_{i-3,j})]$$

(3.12)

and

$$\phi_{\theta r} = \frac{1}{4\Delta\theta\Delta r} [2(\phi_{i,j+1} - \phi_{i-1,j+1} + \phi_{i-1,j-1} - \phi_{i,j-1})$$

$$+ \epsilon(\phi_{i,j+1} - 2\phi_{i-1,j+1} + \phi_{i-2,j+1} - \phi_{i-2,j-1} + 2\phi_{i-1,j-1} - \phi_{i,j-1})] \quad (3.13)$$

By choosing $\epsilon = 0$ the above operators are accurate to the first order in $\Delta\theta$; however, choosing

$$0 < \epsilon = 1 - \lambda\Delta\theta < 1$$

leads to operators accurate to the second order which will be stable over a wide range of diffusive damping parameter, λ, because of the favorable damping given by the leading truncation error term.

The surface tangency condition at $r = 1$ is simply $\phi_r = 0$ and is handled by reflection.

In the relaxation procedure, the computational domain is swept in the direction of flow, i.e., from $\theta = \pi$ to 2π over the upper airfoil surface and then from $\theta = \pi$ to 0 over the lower airfoil surface. During each cycle, the quantities ϕ_θ, ϕ_r, and a are frozen at their values from the previous cycle. At each cycle, a new value of Γ is obtained from the potential jump at the trailing edge and the increment in Γ is used to update ϕ everywhere by use of Eq. (3.11). Solutions are reported to be obtained in 5 to 10 minutes on a CDC 6600.

3.4 Jameson Procedures

Jameson's method as described in reference 15 is essentially the same as that of Garabedian-Korn. However, Jameson avoids the use of two computational domains by writing a perturbation potential about a far field solution, represented by the sum of a doublet and pseudovortex. The resulting reduced potential

$$\Phi = \phi - \frac{\cos(\theta + \alpha)}{r} + \frac{\Gamma\theta}{2\pi} \tag{3.14}$$

is everywhere finite and continuous. The governing equation becomes

$$(a^2 - u^2)\Phi_{\theta\theta} - 2uv\left(r\Phi_{\theta r} + \Phi_\theta - \frac{\Gamma}{2\pi}\right) + (a^2 - v^2)(r^2\Phi_{rr} + r\Phi_r)$$

$$+ (u^2 - v^2)r\Phi_r + (u^2 + v^2)(ur^{-1}f_\theta + vf_r) = 0 \tag{3.15}$$

where

$$\left. \begin{aligned} u &= f^{-1}\left[r\left(\Phi_\theta - \frac{\Gamma}{2\pi}\right) - \sin(\theta + \alpha)\right] \\ v &= f^{-1}[r^2\Phi_r - \cos(\theta + \alpha)] \end{aligned} \right\} \tag{3.16}$$

At the airfoil surface, for $r = 1$, $\phi_r = 0$ so that the flow tangency condition becomes

$$\Phi_r = \cos(\theta + \alpha) \tag{3.17}$$

At the airfoil trailing edge, for $r = 1$ and $\theta = 0$, the mapping is not conformal so that f must vanish. Thus, the Kutta condition that the tangential velocity u be finite at the trailing edge requires that

$$\Phi_\theta = \frac{\Gamma}{2\pi} - \sin\alpha \tag{3.18}$$

at the trailing edge. At the far field boundary, for $r = 0$, the flow approaches the free-stream condition and Φ is given by

$$\Phi = \frac{\Gamma}{2\pi}\left\{\theta - \tan^{-1}\left[(1 - M_\infty^2)^{1/2}\tan(\theta + \alpha)\right]\right\} \tag{3.19}$$

The relaxation procedure is the same as that of Garabedian and Korn except that operators accurate to the first order for $\Phi_{\theta\theta}$ and $\Phi_{\theta r}$ are used in hyperbolic regions. Jameson does not describe the procedure for obtaining the new value of Γ at each cycle; however, it can easily be found from application of Eq. (3.18) at $\theta = 0$ and $\theta = 2\pi$ and $r = 1$. Jameson's procedure is very efficient, with solutions obtained.in about 2 minutes on a CDC 6600.

It should be pointed out that Jameson differs from Murman in that all first
derivatives are evaluated by using central operators with data from the previous
iteration. In supersonic regions, the relaxation parameter, ω, is set equal to 1.

Jameson[17] has recently extended his method to include supersonic free streams.
The circle plane is not well suited for such cases because of the need to distinguish
between the upstream infinity boundary, where Cauchy data are specified, and the
downstream infinity boundary, where no condition is imposed. Jameson suggests that
a more convenient coordinate system is obtained by mapping the circle plane to the
upper half plane by the additional transformation

$$w = \sigma^{1/2} + \frac{1}{\sigma^{1/2}} \tag{3.20}$$

where σ is the circle plane map function. Additional stretching of the coordinates
is used to map the half plane into a rectangle. Thus, as shown in sketch G, the flow
enters through the upper boundary, splits, and leaves through the two sides, with the
airfoil being a segment of the lower boundary.

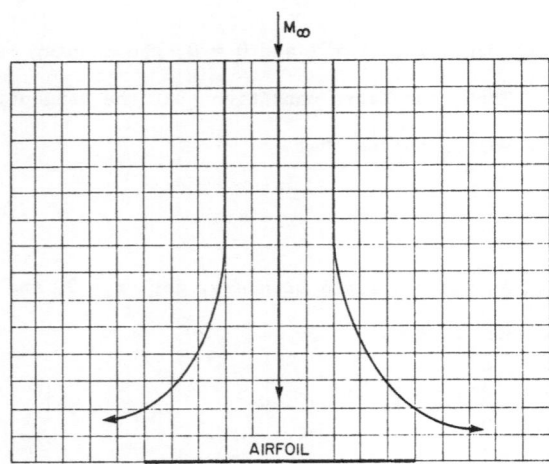

Sketch G

The remaining points on the lower boundary to the right and left of this segment
correspond to points on either side of the slice across which the potential jump
satisfying the Kutta condition is imposed. In the far field, the combined mappings

approach a square-root transformation, and a reduced potential that is finite at infinity is obtained by setting

$$\Phi = \phi - (x^2 - y^2)\cos \alpha - 2xy \sin \alpha \qquad (3.21)$$

where x and y are coordinates in the transformed plane and α is the flow angle at infinity. The far field boundary conditions for subsonic flow at infinity are

$$\Phi = 0 \quad \text{on the top boundary}$$

and $\qquad (3.22)$

$$\Phi = \pm \Gamma/2 \quad \text{on the side boundaries}$$

For supersonic flow at infinity, the conditions are

$$\Phi = 0, \ \Phi_y = 0 \quad \text{on the top boundary} \qquad (3.23)$$

with no condition imposed for supersonic outflow at the sides.

Jameson has considered a similar transformation, which applies the square-root transformation

$$z = w^2 \qquad (3.24)$$

about a point just inside the leading edge (at the center, say, of a circular nose) and generates parabolic coordinates in the physical plane. The airfoil becomes a bump on the lower boundary of the transformed plane. To facilitate a more accurate application of the tangency boundary condition, the coordinate parallel to the lower boundary is displaced to follow the contour of the transformed airfoil and leads to a slightly nonorthogonal coordinate system. The nonorthogonality introduces additional terms in the governing equation, as well as the expression for the surface boundary condition. In addition, care must be taken at the trailing edge to avoid corners in the coordinates and there is no automatic concentration of grid points near the trailing edge. Thus, this transformation does not appear as favorable as the previous one for two-dimensional computations, but, as is discussed later, it is used in Jameson's three-dimensional calculations.

3.5 Jameson's Rotated Difference Scheme for Supersonic Regions

The key to the Murman and Cole scheme is that x-derivatives in the transonic
small disturbance equation are backward differenced in hyperbolic regions. In the
small disturbance formulation, the x-direction is the local streamwise direction and
thus disturbances in hyperbolic regions are prevented from propagating upstream. The
principle that the domain of dependence of the difference equation include the domain
of dependence of the differential equation it approximates is clearly met. The
standard application of the Murman and Cole scheme to the difference equations
approximating the full isentropic equation discussed in this section, however, is
based on backward differencing in a coordinate direction, say x, which is not
aligned with the local stream. Consequently, there will exist points for which
$v^2 < u^2 < a^2 < u^2 + v^2$, that is, the x-component of velocity is subsonic in a supersonic region. Applying the central difference operator at such points allows upstream propagation of disturbances and may lead to instabilities. On the other hand,
the y-coordinate line lies upstream of one of the characteristics (see sketch H), so
that application of the backward difference operator leads to a scheme that uses the
incorrect domain of dependence and may again lead to instability. Fortunately, this
difficulty is of little consequence in many subsonic free-stream applications, since
supersonic flow is confined to a region near the airfoil where the coordinate system
is closely aligned to the stream.

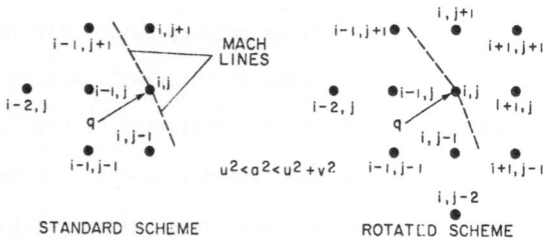

Sketch H

In cases where the flow at infinity is supersonic, the misalignment of the
coordinates with the stream is of greater concern and led Jameson to developing the

"rotated" difference scheme for supersonic flow regions. It was recognized that the
Murman-Cole scheme should be applied to the governing equation written in terms of
coordinates s in the local stream direction and n normal to the stream as follows

$$\left(1 - \frac{q^2}{a^2}\right)\phi_{ss} + \phi_{nn} = 0 \tag{3.25}$$

The derivatives ϕ_{ss} and ϕ_{nn} may be expressed in terms of the coordinate directions
(x,y) locally by a rotation to give

$$\left.\begin{aligned}
\phi_{ss} &= \frac{1}{q^2}\,(u^2\phi_{xx} + 2uv\phi_{xy} + v^2\phi_{yy}) \\
\phi_{nn} &= \frac{1}{q^2}\,(v^2\phi_{xx} - 2uv\phi_{xy} + u^2\phi_{yy})
\end{aligned}\right\} \tag{3.26}$$

where u/q and v/q are the direction cosines. The Murman-Cole scheme is now properly applied by writing backward (or retarded) difference operators in x and y for
the derivatives representing ϕ_{ss} and central difference operators for these representing ϕ_{nn}.

Computation molecules for the standard and rotated schemes are shown in sketch
H. A comparison of the schemes brings out two important features of the rotated
scheme. First, because of the upwind y-difference operators used to approximate ϕ_{ss},
the quantity $\phi_{i,j-2}$ appears. If the tridiagonal solution algorithm is to be used
along the column, i, $\phi_{i,j-2}$ cannot be updated and its value from the previous iteration (old value) will appear in the right-hand side of Eq. (2.36). Second, due to the
central x-difference operators used to approximate ϕ_{nn}, old values appear along the
column i + 1, preventing solution by a simple column-by-column marching procedure.

Extensive analysis of the stability and convergence of the rotated scheme is
presented in reference 17 (also see references 37 and 38). The approach is similar
to that of Garabedian's[39] theory for successive over-relaxation of Laplace's equation, in that iterations are viewed as steps in artificial time (τ). The combination
of new and old values in the difference operators is chosen so that the equivalent
time-dependent equation (neglecting truncation terms) represents a properly posed
problem whose solution approaches the solution of the time-invariant equation. In
particular, the coefficient of ϕ_τ is chosen to vanish and the coefficients of the

remaining time derivatives chosen such that s is the time-like or marching direction in the hyperbolic time-dependent equation.

Jameson further shows that the von Neumann test of local stability for the difference equation

$$\sum_{p,q} (a_{pq} \phi_{i+p,j+q} - b_{pq} \phi^+_{i+p,j+q}) = 0 \tag{3.27}$$

requires

$$\sum_{p,q} a_{pq} = \sum_{p,q} b_{p,q} = 0 \tag{3.28}$$

The superscript + is used to denote new or updated values. Here, Eq. (3.27) is the linear equivalent of rotated scheme applied to Eq. (3.25) and the relaxation parameter is 1.

Upon combining the various conditions resulting from his analysis, Jameson writes the following difference operators.[37,38] The central difference operators contributing to ϕ_{nn} are

$$\left.\begin{array}{l} \phi_{xx} = \dfrac{\phi^+_{i-1,j} - \phi^+_{i,j} - \phi_{i,j} + \phi_{i+1,j}}{(\Delta x)^2} \\[3mm] \phi_{xy} = \dfrac{-\phi^+_{i-1,j+1} + \phi^+_{i-1,j-1} + \phi_{i+1,j+1} - \phi_{i+1,j-1}}{4\Delta x \Delta y} \\[3mm] \phi_{yy} = \dfrac{\phi^+_{i,j+1} - 2\phi^+_{i,j} + \phi^+_{i,j-1}}{(\Delta y)^2} \end{array}\right\} \tag{3.29}$$

The upwind difference operators contributing to ϕ_{ss} when u > 0 and v > 0 (velocity coming from the upper left-hand quadrant) are

$$\left.\begin{array}{l} \phi_{xx} = \dfrac{2\phi^+_{i,j} - \phi_{i,j} - 2\phi^+_{i-1,j} + \phi_{i-2,j}}{(\Delta x)^2} \\[3mm] \phi_{xy} = \dfrac{\phi^+_{i,j} - \phi^+_{i-1,j} - \phi^+_{i,j-1} + \phi^+_{i-1,j-1}}{\Delta x \, \Delta y} \\[3mm] \phi_{yy} = \dfrac{2\phi^+_{i,j} - \phi_{i,j} - 2\phi^+_{i,j-1} + \phi_{i,j-2}}{(\Delta y)^2} \end{array}\right\} \tag{3.30}$$

The central difference operators in Eq. (3.29) are constructed for the simultaneous solution of points along a y-coordinate line. The scheme may easily be modified for simultaneous solution of points along an x-coordinate by the obvious switching of ϕ_{xx} and ϕ_{yy} operator forms in Eq. (3.29). No modification to the upwind operators (3.30) is necessary.

Near the sonic line, a stabilizing term may be required to keep s the marching direction and is given by $\varepsilon(\Delta\tau/\Delta x)[(u/q)\phi_{x\tau} + (v/q)\phi_{y\tau}]$ where

$$\left. \begin{array}{l} \phi_{x\tau} = \dfrac{\overset{+}{\phi}_{i,j} - \phi_{i,j} + \overset{+}{\phi}_{i-1,j} + \phi_{i-1,j}}{\Delta\tau\ \Delta x} \\[3ex] \phi_{y\tau} = \dfrac{\overset{+}{\phi}_{i,j} - \phi_{i,j} + \overset{+}{\phi}_{i,j-1} + \phi_{i,j-1}}{\Delta\tau\ \Delta y} \end{array} \right\} \tag{3.31}$$

for $u > 0$ and $v > 0$. The quantity ε is a small parameter which in practice may often be taken as 0.

3.6 Results

Numerous results obtained using the exact isentropic equations have appeared in the literature.[13,14,15,34,40] Comparisons between the numerical calculations and experimental results are particularly interesting but are complicated by the effects of viscosity and wind-tunnel walls. Such comparisons are given in section 5. Here, we show some theoretical comparisons of interest.

A comparison between the exact hodograph solution[41] and various relaxation solutions is shown in Fig. 5 for an NLR quasi-elliptical airfoil. The Jameson solution is not plotted since it is indistinguishable from the exact solution. Another comparison with the exact hodograph solution[14] for a Garabedian-Korn airfoil is shown in Fig. 6. Both the Garabedian-Korn relaxation solution and the Krupp-Murman solution* show a weak shock wave. In addition, the Krupp-Murman solution shows an under-expansion near the nose. Note that in the small disturbance case, the Mach number was increased by 0.01 to avoid formation of an additional shock wave farther forward.

---

*Calculation is the courtesy of Dr. Murman, NASA Ames Research Center, Moffett Field, Ca.

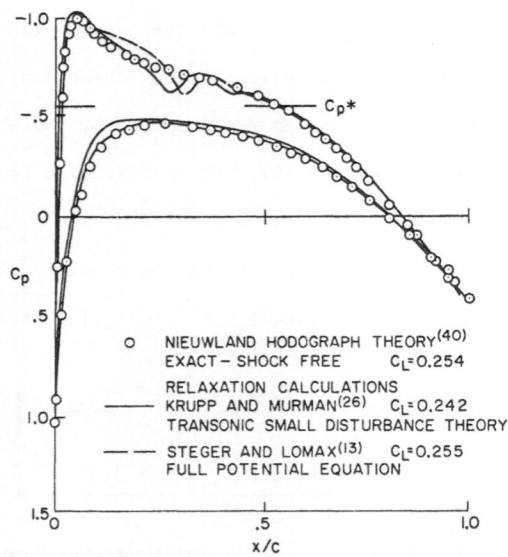

Fig. 5. Comparison of hodograph and relaxation solutions for NLR quasi-elliptical airfoil; δ = 12.12 percent, α = 1.32°, M_∞ = 0.7557.

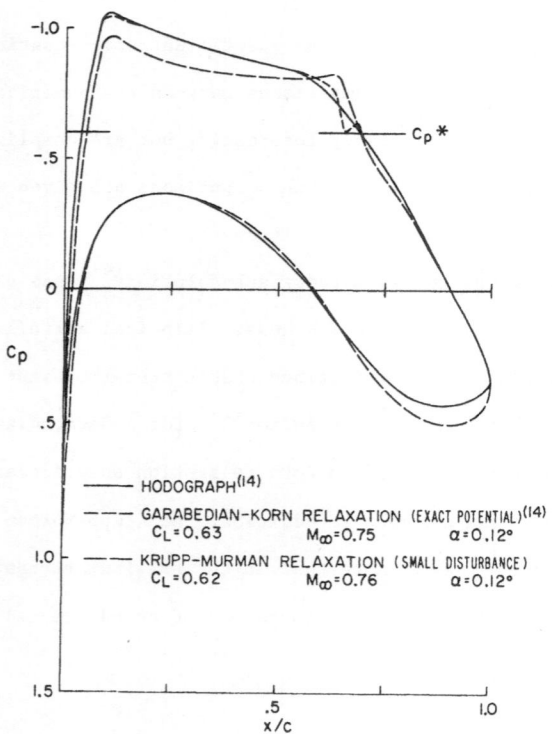

Fig. 6. Comparison of hodograph and relaxation solutions for Garabedian-Korn airfoil.

Superficial airfoils of the Garabedian-Korn type are extremely sensitive to small changes in Mach number, incidence, and inaccuracies in airfoil definition and thus are sensitive indicators of the accuracy of the calculations. It is of particular interest to note that at Mach numbers and angles of attack slightly below the design condition, the solution appears with multiple shock waves.

As mentioned previously, the relaxation methods for the exact equation do not use conservative difference operators and therefore do not fulfill the proper jump conditions. An indication of the error in fulfilling the shock jump is given in Fig. 7, which compares the solution obtained from Jameson's program with that obtained from the conservative time-dependent procedure for the isentropic equations described by Yoshihara.[17] As is expected from similar comparisons in the previous section, the conservative method gives a more nearly correct shock jump, with a shock location further aft.

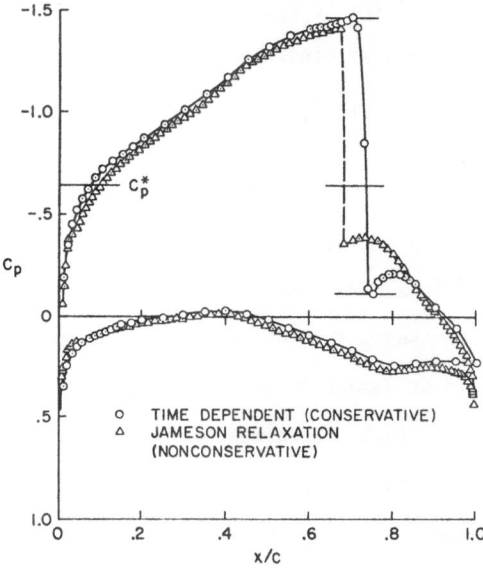

Fig. 7. Comparison of conservative time-dependent and nonconservative relaxation solutions of exact isentropic equation for the NACA 64A410 airfoil at M_∞ = 0.735 and α = 1°.[7]

4. AXISYMMETRIC FLOW

We briefly turn our attention to axisymmetric flows. Such flows are two-dimensional and present no essential numerical difficulties. Because of spatial relief, axisymmetric flows have a smaller transonic Mach number range and have less pronounced transonic effects than corresponding two-dimensional planar flows. Considerable interest exists in developing numerical techniques for transonic flow about optimum bodies, which have longitudinal area distributions that exhibit high drag-rise Mach numbers. Upon application of the transonic equivalence principle (Mach-1 area-rule), such optimized configurations will presumably give a starting point from which to model the area distributions for transonic aircraft.

The transonic small disturbance theory may also be applied to axisymmetric flows over slender bodies and is the basis of axisymmetric relaxation methods reported by Krupp and Murman[26] and Bailey.[42] In the flow field, the only essential modification is the alteration of the governing equation so that the lateral term ϕ_{yy} is replaced by the radial term $(1/r)(r\phi_r)_r$. In addition, because of the logarithmic along the axis, the body boundary condition becomes

$$\lim_{r \to 0} (r\phi_r) = R(dR/dx)$$

where R is the body shape. In the numerical method, the condition is applied on some small cylinder about the axis.

South and Jameson[37] have reported a procedure for calculating axisymmetric flows using the exact isentropic form of the governing equations. The method is a straightforward extension of Jameson's two-dimensional procedure including the rotated difference scheme. Fully conformal coordinates are used for closed spheres and ellipsoids. For blunt-nosed shapes with open tails and sharp corners, body-normal coordinates are used over the nose up to the horizontal tangent, followed by a nonorthogonal "sheared" cylindrical system, which consists of a family of vertical lines and a family of lines parallel to the body surface.

5. COMPARISONS WITH EXPERIMENT

5.1 Introduction

Checks on the accuracy and usefulness of inviscid transonic calculations depend
heavily on comparisons with experimental results obtained by wind tunnel tests.
However, flows with embedded shock waves are extremely sensitive to small changes in
boundary parameters. Thus, significant differences can occur in comparisons of wind
tunnel results and inviscid calculations because of viscous and wind tunnel interference effects. Although the precise isolation of these two effects is not possible,
we shall consider separately their relationship to inviscid relaxation calculations.

5.2 Viscous Effects

We begin by considering, in Figs. 8 and 9, flow about nonlifting, symmetric airfoils with turbulent boundary layers in which the major viscous effect is the interaction between the shock wave and unseparated boundary layer. In Fig. 8, the solution from Jameson's program† shows excellent agreement with data[43] for flow about a
symmetrical Boerstoel airfoil[44] at $M_\infty = 0.834$ and an interference corrected
incidence of -0.06°. The data were obtained in the perforated NAE trisonic tunnel at
a Reynolds number of 20×10^6 based on model chord. The free-stream Mach number correction is negligible. The symbols ● and ■ appearing directly beneath the calculated shock wave designate postshock values computed from the Rankine-Hugoniot and
the isentropic jump conditions, respectively. The difference between these values is
a measure of the error in pressure jump introduced by use of the exact isentropic
flow equation. The difference between the isentropic value and the first computed
subsonic point behind the shock wave is a measure of the failure of the nonconservative difference scheme to satisfy the isentropic jump conditions. Finally, the
difference between the Rankine-Hugoniot value and the first experimental subsonic

---

†Calculations are the courtesy of Dr. Melnik, Grumman Aerospace Corp., Bethpage,
N. Y.

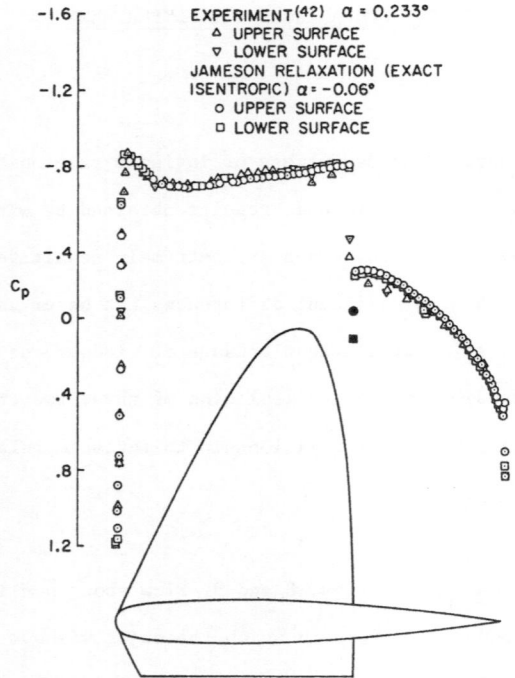

Fig. 8. Comparison Jameson relaxation solution with

data for Boerstoel airfoil; $M_\infty = 0.834$

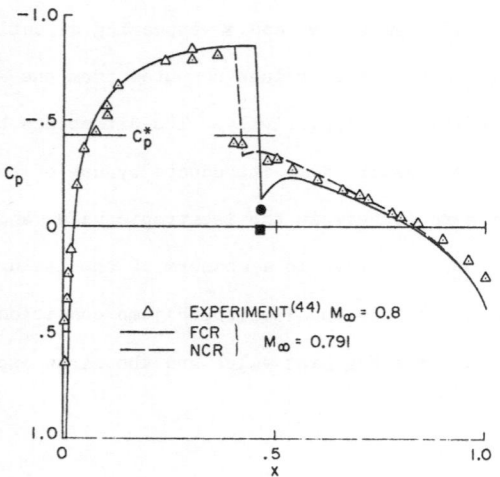

Fig. 9. Comparison of small disturbance solutions with

data for NACA 0012 airfoil at $\alpha = 0°$.

point behind the shock is a measure of the weakening of the experimental shock due to its interaction with the boundary layer.

In Fig. 9, two small distrubance solutions are compared with data[45] for flow about an NACA 0012 airfoil at zero incidence and $M_\infty = 0.791$. The airfoil was tested at an uncorrected Mach number of 0.8 and at a moderate Reynolds number of 3.6×10^6 (boundary layer tripped at $x = 0.14$ to 0.16) in a slotted wall tunnel for which the incidence correction was negligible and the experimentally determined Mach number correction was quite small at -0.009 ± 0.002.[45] Calculations using both the non-conservative (NCR) and fully conservative (FCR) schemes are shown. The comparison between the data and NCR solution is very good, but the FCR solution predicts a stronger shock that is located farther aft. In this case, the symbol ■ represents the postshock value computed from the small disturbance jump condition.

It is suggested that the smaller shock jump predicted by the nonconservative schemes compensates, to a degree, for the weakening of the shock due to its inter-action with the boundary layer. This appears to be quite typical as shown in Fig. 10. Here, a plot of shock pressure ratio versus the Mach number immediately upstream of the shock, M_1, is shown for a sample of nonconservative relaxation calculations and experimental results. Also shown for reference are the normal shock pressure ratios from the isentropic and Rankine-Hugonoit relations. Both calculated and experimental pressure ratios follow the same trend up to the point at which the shock separates the boundary layer ($M_1 \approx 1.3$). This is not a very comfortable state of affairs, however, because the nonconservative shock solution is not unique and does not account for viscosity in a rational manner.

We next consider flows over modern aft-cambered airfoils for which viscous effects are very significant. In Fig. 11, we repeat the results presented by Melnik and Ives[46] for an aft-cambered airfoil designed by Garabedian and Korn[14] at $M_\infty = 0.699$ and a corrected incidence of 3.6°. The wind tunnel data were obtained in the NAE trisonic tunnel at a Reynolds number of 20×10 .[47] The effect of tunnel interference on lift was corrected by a shift in angle of attack as determined by calibration tests. The calculations were obtained using Jameson's program in two versions: the standard one with the trailing edge Kutta-condition imposed and one

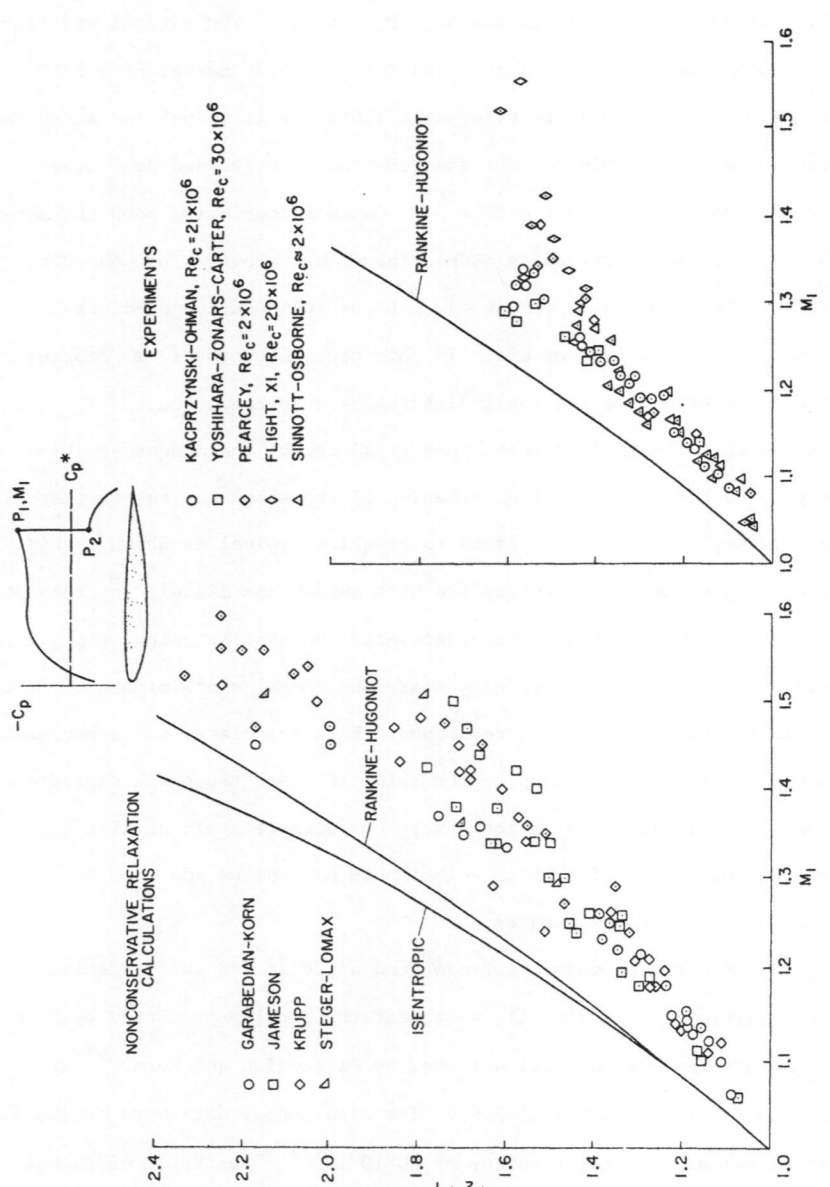

Fig. 10. Shock-wave pressure jump on airfoils in transonic flow.

(a) Computed.   (b) Experimental.

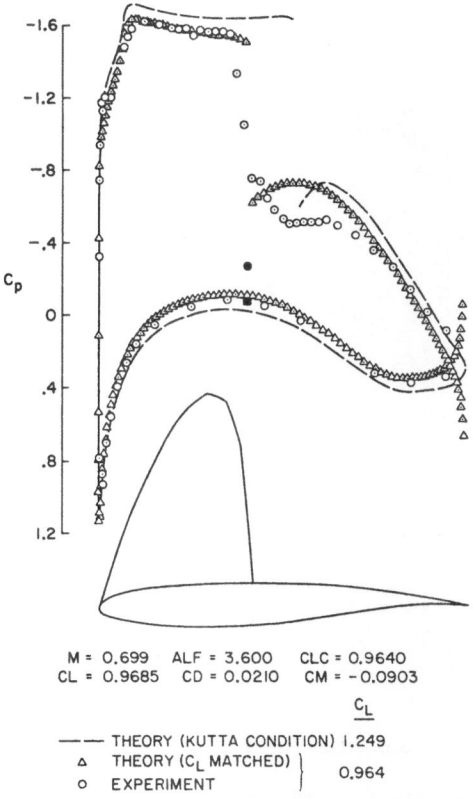

M = 0.699   ALF = 3.600   CLC = 0.9640
CL = 0.9685   CD = 0.0210   CM = -0.0903

$$\underline{C_L}$$

—— THEORY (KUTTA CONDITION) 1.249
△   THEORY (C_L MATCHED) }
○   EXPERIMENT   } 0.964

Fig. 11.   Comparison of Jameson relaxation solution with
data[46] for Garabedian-Korn airfoil.[45]

modified to accept a prescribed circulation constant determined by the experimental
lift. The modified calculation shows better agreement with experiment except at the
trailing edge.

A similar comparison with the FCR small disturbance solutions[16] is shown in
Fig. 12 for the same airfoil at $M_\infty = 0.768$ and a geometric incidence of 1.38°.
The data[40] were also obtained in the NAE tunnel but for a tunnel porosity of 6
percent instead of 20.5 percent.

In both Figs. 11 and 12 we see that the standard method gives a much larger lift
coefficient than the measured value, as well as a large difference in pressure

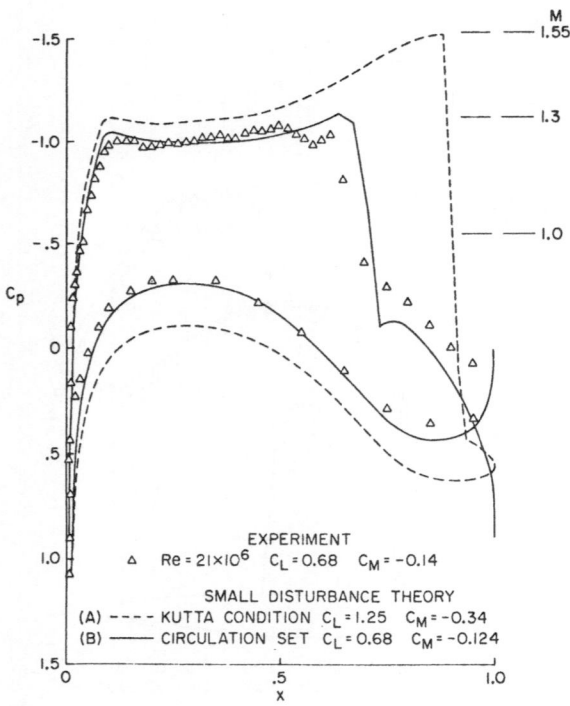

Fig. 12. Comparison of FCR solutions with data[47]

for Garabedian-Korn airfoil;

$\alpha = 1.38°$, $M_\infty = 0.768$.[16]

distribution and shock location. The modified results, on the other hand, show

greatly improved agreement except, of course, at the trailing edge.

It appears that viscous effects are primarily responsible for the disagreement

between the standard calculations and the experimental results. Thus, it is sug-

gested that, when shock wave-boundary layer separation is absent, the principal

viscous effect to account for is the reduction in circulation from the value given by

the inviscid Kutta condition. While this strategy is useful, it does not yield a

predictive procedure. Clearly, what is needed is a technique for solving the turbu-

lent trailing-edge problem, thereby accounting for the reduction in lift and the

displacement effects downstream of the shock wave.

5.3 Wind Tunnel Wall Effects

Although the use of ventilated transonic test sections reduces the magnitude of wall effects, uncertainties in the interference corrections still exist. The classical theory for predicting the corrections is based on linear subsonic theory with compressibility effects accounted for by the Prandtl-Glauert scaling laws.[48] Such an approach is generally inadequate in the transonic regime, particularly since it cannot account for the interference effect on the shock wave location. Thus, recource is often made to experimental calibration studies to predict the correct Mach number and angle-of-attack corrections.

With the availability of transonic finite-difference methods, however, it is now possible to study the nonlinear effects of various wall parameters by replacing the far field boundary with a tunnel wall boundary at the appropriate location. For solid or open jet tunnels, the usual boundary conditions of zero flow angle or constant pressure apply. For ventilated test sections — perforated or porous and slotted walls — simplified homogeneous wall boundary conditions proposed by Baldwin, Turner, and Knechtel[49] can be used.

The applicability of the transonic relaxation technique in simulating wall effects was demonstrated in reference 41, where calculations were made for flow about a parabolic arc of revolution with porous and open jet wall boundaries. Shown in Fig. 13 are results at $M_\infty = 0.99$ for various values of porosity parameter P as compared with data[50] obtained in the Ames 14-foot perforated/slotted wind tunnel at a high Reynolds number of 27×10^6. Note that the shock wave moves forward with increasing porosity. The significance of wall interference near Mach 1 is illustrated in Fig. 14 by comparing experimental measurements with free-air and porous wall calculations of the transonic drag rise.

Murman[51] has extended the relaxation method to study the effect of perforated walls on lifting airfoils. He conducted an interesting numerical experiment in which calculations were made for a lifting airfoil in a simulated wind tunnel and in free air. The resulting wind tunnel lift and moment coefficients were then corrected to free-air conditions according to linear theory.[48] The results are given in Fig. 15

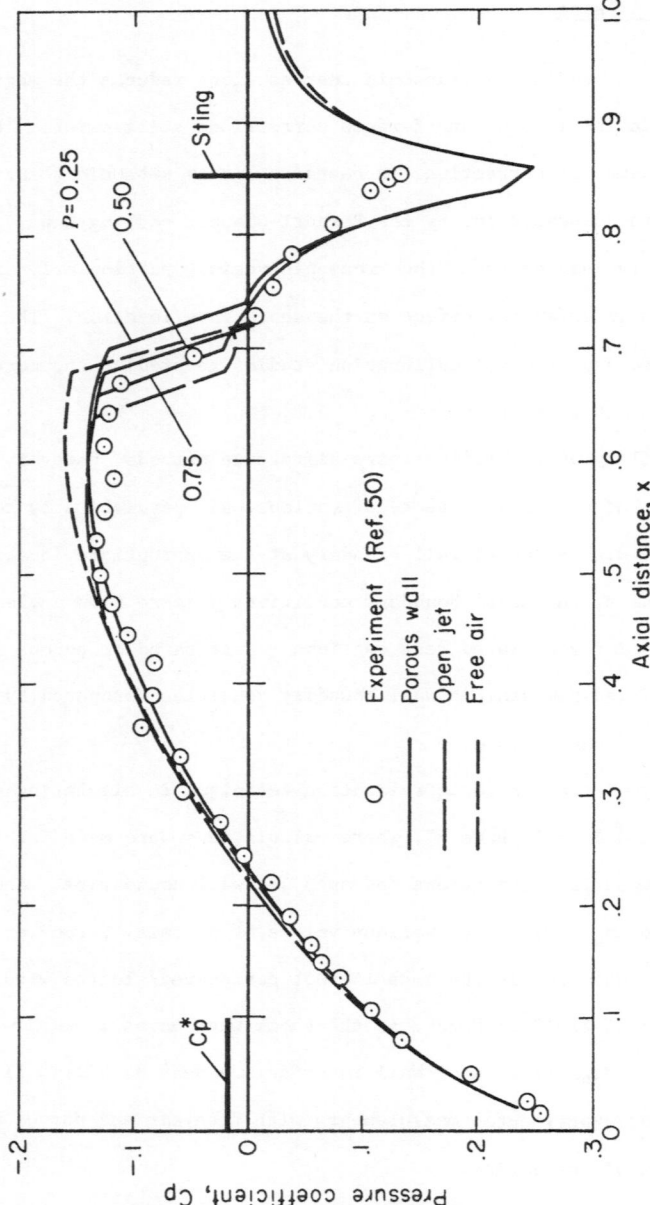

Fig. 13. Effect of wall conditions on body surface C_p for parabolic arc of revolution with sting ($M_\infty = 0.99$ and $f = 10$).

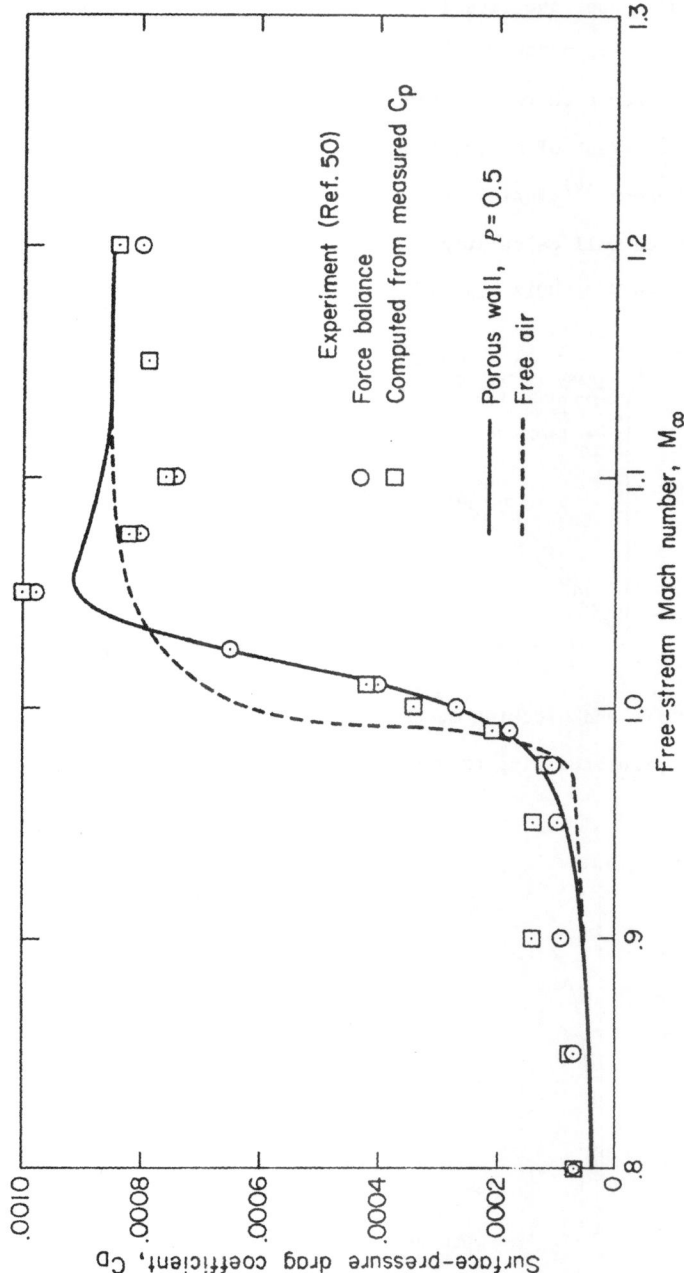

Fig. 14. Variation of surface pressure drag coefficient (based on body length) with Mach number for a parabolic arc of revolution with sting (f = 10).

and show that, in this case, the classical linear method adequately predicts the correction for lift but not moment. This is not entirely surprising because of the effect of wall interference on the location of the surface shock waves.

The interference effect of an ideal slotted wall on a nonlifting airfoil was also calculated by Murman[16] (using the fully conservative method). The comparison of free-air and slotted wall calculations with experimental results[45] obtained in a slotted tunnel for an NACA 0012 airfoil at $M_\infty = 0.80$ is shown in Fig. 16.

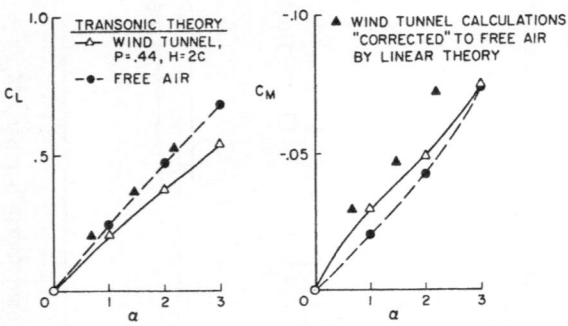

Fig. 15. Lift and pitching moment curves for NACA 0012 airfoil in perforated wind tunnel and free air, M = 0.80.[16]

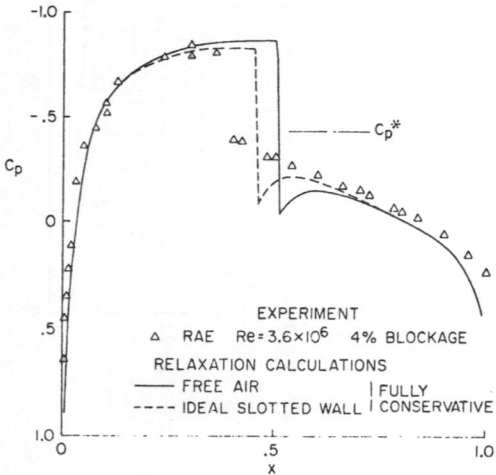

Fig. 16. Comparison of FCR solutions with data,[44] NACA 0012; $\alpha = 0°$, M = 0.8.[16]

6. SMALL DISTURBANCE PROCEDURE IN THREE DIMENSIONS

6.1 Introduction

The success of the Murman and Cole scheme in two dimensions has given rise to interest in applying the scheme to three-dimensional flows. The computational effort required to solve three-dimensional flows is not overly large for modern high-speed computers. For example, useful calculations for transonic flows about wings can be obtained using on the order of 10^5 grid points with converged solutions obtained in about 15 minutes on a CDC 7600 or IBM 370/195.

Possibly the most difficult problem to be overcome in three-dimensional calculations is that of geometry. In particular, exact coordinate transformations, which have proven useful in two dimensions and which force the body to lie along the edge of a finite-difference grid, can be very difficult to apply to a complicated three-dimensional configuration. Thus, most of the three-dimensional calculations reported have been based on small disturbance theory applied to wings. (The exception is the yawed wing calculations of Jameson[38] which are discussed in the next section.) Small disturbance relaxation methods have been reported by Isom and Caradonna[52] and by Ballhaus and Caradonna[53] for transonic flow about a rectangular nonlifting helicopter rotor in hover and by Bailey and Steger[54] and Newman and Klunker[55] for rectangular lifting wings. Transonic flow about lifting swept wings and nonlifting wing-cylinder combinations have been treated by Ballhaus and Bailey.[31,32] The latter procedure is an extension of the Krupp-Murman NCR method discussed in section 2 and is discussed in some detail in this section.

6.2 Basic Formulation

The governing transonic small disturbance equation in three dimensions may be written (see sketch I) in similarity form as

$$[K_1 - (\gamma + 1)\phi_x]\phi_{xx} + K_2\phi_{yy} + \phi_{zz} = 0 \tag{6.1}$$

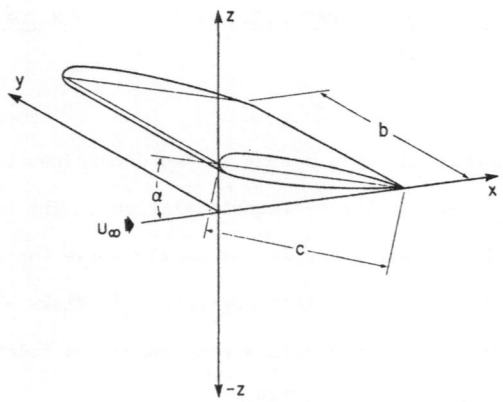

Sketch I

where

$$K_1 = \frac{1 - M_\infty^2}{\delta^{2/3} M_\infty^{4/3}}$$

$$K_2 = \frac{c^2}{b^2 \delta^{2/3} M_\infty^{4/3}}$$

The span coordinate, y, is scaled by the semispan, b, and the vertical coordinate, z, by $(\delta^{1/3} M_\infty^{2/3})^{-1}$. The pressure coefficient is given by

$$C_p = \frac{-2\delta^{2/3}}{M_\infty^{2/3}} \phi_x \qquad (6.2)$$

As in two dimensions, the flow tangency condition at the wing surface is linear-ized and applied on the mean wing plane $0 \le x \le 1$, $-1 \le y \le 1$, $z = 0$. In addition to satisfying the Kutta condition, provision must be made for a trailing vortex sheet downstream of the wing trailing edge. With the small disturbance assumption, the vortex sheet is flat and lies in the plane $z = 0$ with conditions that ϕ_x (pressure) and ϕ_z (flow angle) be continuous across it. The potential ϕ and its second derivative ϕ_{zz}, however, experience a jump across the sheet. Because of the conti-nuity of pressure across the vortex sheet, the jump in potential at any span station, $y = y_0$, is independent of x and is equal to the circulation about the wing section

$$\Gamma(y_0) = - \oint d\phi(x, y_0, z) \qquad (6.3)$$

for any path enclosing the wing section. In the numerical procedure, the Kutta con-dition is satisfied by setting the potential jump across the vortex sheet equal to the value obtained at the trailing edge. A modified difference operator for ϕ_{zz} is written to account for the jump in ϕ and ϕ_{zz}. The operator may be derived by noting that jumps in ϕ occur only at the vortex sheet and only odd functions may jump. Since the jump is independent of x, the solution at the vortex sheet decouples into even, ϕ^e, and odd, ϕ^o, solutions with ϕ^e satisfying Eq. (6.1) and ϕ^o satisfying

$$\phi_{zz}^o = -K_2 \phi_{yy}^o \tag{6.4}$$

At the sheet itself, the odd solution is given by

$$\phi^o(x,y,o^\pm) = \pm \frac{1}{2}\, \Gamma(y) \tag{6.5}$$

Therefore, ϕ_{zz} at the sheet can be written

$$\phi_{zz}\bigg|_{0^\pm} = \phi_{zz}^e\bigg|_{0^\pm} \mp \frac{K_2}{2}\, \Gamma_{yy}(y) \tag{6.6}$$

Approximating Eq. (6.6) at $z = 0^-$ by central difference operators and noting that ϕ^e can be related to ϕ by

$$\phi_{i,j,1}^e = \phi_{i,j,-1}^e = \frac{1}{2}\,(\phi_{i,j,1} + \phi_{i,j,-1})$$

$$\phi_{i,j,0^-}^e = \frac{1}{2}\,(\phi_{i,j,0^-} + \phi_{i,j,0^+}) = \frac{1}{2}\,(2\phi_{i,j,0^-} + \Gamma_j)$$

yields

$$(\phi_{zz})_{i,j,0^-} = \frac{1}{(\Delta z)^2}\,[(\phi_{i,j,1} - \Gamma_j) - 2\phi_{i,j,0^-} + \phi_{i,j,-1}]$$

$$+ \frac{K_2}{2(\Delta y)^2}\,(\Gamma_{j+1} - 2\Gamma_j + \Gamma_{j-1}) \tag{6.7}$$

The outer boundary conditions far from the wing and vortex sheet are given at some finite distance by an approximate analytical expression for the far field.[30] The dominant term in the expression is due to lift and is proportional to the circulation integrated over the span. The conditions at the downstream boundary, i.e., Trefftz plane, are found by iteratively solving Eq. (6.4) with condition Eq. (6.5), along with the rest of the flow field.

6.3 Relaxation Procedure

Consider first the extension of the Murman-Cole scheme to three-dimensional supersonic regions with the additional spanwise derivative approximated by the central operator

$$
\phi_{yy} = \frac{\phi_{i,j+1,k} - 2\phi_{i,j,k} + \phi_{i,j-1,k}}{(\Delta y)^2}
\tag{6.8}
$$

and the remaining derivatives approximated by the appropriate operators given in section 2. A direct extension of the two-dimensional, fully implicit, marching procedure to three-dimensional supersonic regions requres a successive plane relaxation scheme (in contrast to the two-dimensional line scheme) in which new values of ϕ are obtained simultaneously in an x = constant plane before proceeding to the next downstream plane. Recall that in the two-dimensional method, new values of ϕ are obtained along an x = contant line by solving Eq. (2.36) by direct elimination. In three dimensions, the analogous procedure would require the direct solution of a similar equation with the elements of $\underset{\sim}{A}$ being themselves tridiagonal matrices. Efficient methods for the direct solution of this matrix equation have yet to be developed. However, this equation can be solved by relaxation. Thus, the fully implicit supersonic relaxation procedure is to pause at each x-constant plane and do a line relaxation within that plane, until all values of ϕ in the plane satisfy some error criterion, before moving to the next downstream plane. This fully implicit method ($\omega = 1$) is stable for the linear wave equation. It should be noted that if one does not pause, but rather sweeps immediately from one plane to the next, the scheme is no longer fully implicit. In addition, the von Neuman stability test shows that this procedure is unstable for small wave numbers when applied to the linear wave equation. However, such a scheme has been used successfully in three-dimensional calculations even though the linear stability criterion was violated. The reason for this is unclear, although contributing factors may be the influence of the boundaries, the grids are not sufficiently fine, the nonlinearity of the equation, and the use of a relaxation parameter less than 1.

Line relaxation schemes for supersonic regions that are not fully implicit, but that satisfy the stability test, can be derived by the method outlined by Jameson[17, 37,38] (see section 3.5). For example, one can use operators in supersonic regions given by

$$
\left.\begin{array}{l}
\phi_{xx} = \dfrac{2\phi^+_{i,j,k} - \phi_{i,j,k} - 2\phi^+_{i-1,j,k} + \phi_{i-2,j,k}}{(\Delta x)^2} \\[3mm]
\phi_{yy} = \dfrac{\phi^+_{i,j-1,k} - \phi^+_{i,j,k} - \phi_{i,j,k} + \phi_{i,j+1,k}}{(\Delta y)^2} \\[3mm]
\phi_{zz} = \dfrac{\phi^+_{i,j,k-1} - 2\phi^+_{i,j,k} + \phi^+_{i,j,k+1}}{(\Delta z)^2}
\end{array}\right\}
\tag{6.9}
$$

and sweep immediately from one plane to the next.

The supersonic procedures outlined above have been used successfully in actual three-dimensional calculations and they all converge to the same solution with nearly the same efficiency.

In subsonic regions the extension to three dimensions requires only the additional spanwise operator Eq. (6.8) and either plane or line relaxation may be used with $\omega \simeq 1.8$.

6.4 Nonrectangular Planforms

Consider next the application of the relaxation method to nonrectangular planforms (e.g., including sweep and taper). To facilitate the use of a high density of grid points at the leading edge and to retain the same number of chordwise grid points at each span station, it is convenient to use a coordinate transformation to map the planform into a rectangle. The transformation, valid for wings with finite tip chords, is given by

$$
\left.\begin{array}{l}
\xi(x,y) = \dfrac{x - x_{\ell.e.}(y)}{c(y)} \\[3mm]
\eta = y \\[2mm]
z = z
\end{array}\right\}
\tag{6.10}
$$

where $x_{\ell.e.}(y)$ is the value of x at the leading edge and $c(y)$ is the ratio of local chord to root chord. The governing small disturbance equation can then be written in terms of the new independent variables ξ, η, z in the form

$$\left(K_1 - \frac{(\gamma+1)}{c} \phi_\xi\right)\frac{\phi_{\xi\xi}}{c^2} + K_2\xi_y^2\phi_{\xi\xi} + 2K_2\xi_y\phi_{\xi\eta} + K_2\xi_{yy}\phi_\xi + K_2\phi_{\eta\eta} + \phi_{zz} = 0 \qquad (6.11)$$

In transformed coordinates, the pressure coefficient becomes

$$C_p = \frac{-2\delta^{2/3}\phi_\xi}{cM_\infty^{2/3}} \qquad (6.12)$$

The transformation is not continued beyond the tip of tapered wings. In this case, $c(y)$ is held fixed beyond the tip and a smooth juncture is fit between the tapered and untapered regions.

At the wing root, the boundary condition is the required symmetry about the plane $y = 0$ and leads to

$$\phi_y = \phi_\eta + \xi_y\phi_\xi = 0 \qquad (6.13)$$

Straightforward incorporation of the root boundary condition into the finite-difference analog of Eq. (6.11) has led to generally unsatisfactory results, both with respect to stability and accuracy. In reference 31, an interpolated or "skewed" scheme is discussed, which essentially leads to differencing the untransformed equation at the wing root and in general works well. It has been recently suggested by Dr. Hall of RAE that one modify the transformation such that the η coordinate lines are normal to the symmetry plane. This is accomplished by turning the coordinate lines over a distance very small with respect to the wing semispan so that the deviation from the wing planform is not great. This approach is easily implemented and works quite well.

In writing the finite-difference operators for Eq. (6.1), it is recognized that in supersonic flow regions the domain of dependence of the finite-difference equation will contain the domain of dependence of the differential equation if upwind difference operators are used for x-derivatives and central difference operators are used for the y- and z-derivatives. This is taken into account in the difference analog to Eq. (6.8) by using central difference operators for ϕ_{zz} and the derivatives multiplied by K_2 (i.e., the term ϕ_{yy}) and by using upwind operators for the remaining ξ-derivatives.

Because central operators are used in supersonic regions for ξ-derivatives contributing to ϕ_{yy} and these contain points downstream of a $\xi = $ constant plane, it is no longer possible to use a fully implicit supersonic marching scheme plane by plane. Stable calculations have been obtained, however, by the line relaxation procedure using essentially the operators given in section 2. A modified line relaxation scheme based on Jameson's method, which satisfies the linear stability test, has also been used. For this method, the ξ-derivative operator contributing to ϕ_{xx} is given by

$$\phi_{\xi\xi} = \frac{2\phi_{i,j,k}^+ - \phi_{i,j,k} - 2\phi_{i-1,j,k}^+ + \phi_{i-2,j,k}}{(\Delta\xi)^2} \tag{6.14}$$

the operators contributing to ϕ_{yy} are given by

$$\left.\begin{aligned}
\phi_{\xi\xi} &= \frac{\phi_{i-1,j,k}^+ - \phi_{i,j,k}^+ - \phi_{i,j,k} + \phi_{i+1,j,k}}{(\Delta\xi)^2} \\[2mm]
\phi_{\eta\eta} &= \frac{\phi_{i,j-1,k}^+ - \phi_{i,j,k}^+ - \phi_{i,j,k} + \phi_{i,j+1,k}}{(\Delta\eta)^2} \\[2mm]
\phi_{\xi\eta} &= \frac{\phi_{i+1,j+1,k} - \phi_{i-1,j+1,k}^+ - \phi_{i+1,j-1,k} + \phi_{i-1,j-1,k}^+}{4\Delta\xi\,\Delta\eta}
\end{aligned}\right\} \tag{6.15}$$

and the z-derivative operator is given by

$$\phi_{zz} = \frac{\phi_{i,j,k-1}^+ - 2\phi_{i,j,k}^+ + \phi_{i,j,k+1}^+}{(\Delta z)^2} \tag{6.16}$$

It should be pointed out that, to avoid central differences of $\phi_{\xi\xi}$ across shock waves and thus computing sharper shock jumps, the earlier method given in reference 31 delayed upwind differencing until the flow normal to the local sweep was supersonic, and then the entire $\phi_{\xi\xi}$ term was upwind differenced. This procedure, however, can lead to erroneous results.

6.5 Results

Subcritical ($M_\infty = 0.752$) and supercritical ($M_\infty = 0.853$) results using the NCR method are shown in Figs. 17 and 18 for flow about a constant-chord, 23.75° sweptback wing with a Lockheed C-141 airfoil section (11.4 percent thick streamwise) at 2°

62

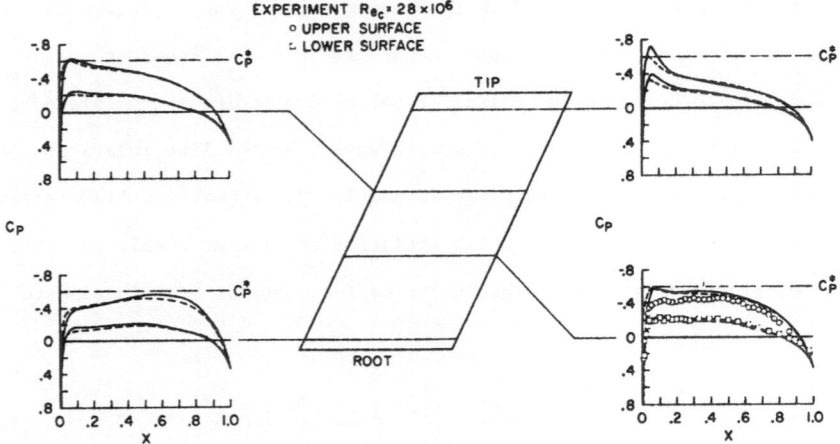

Fig. 17. C_p distribution on C-141 swept panel model,
$M_\infty = 0.752$, $\alpha = 2°$. [31]

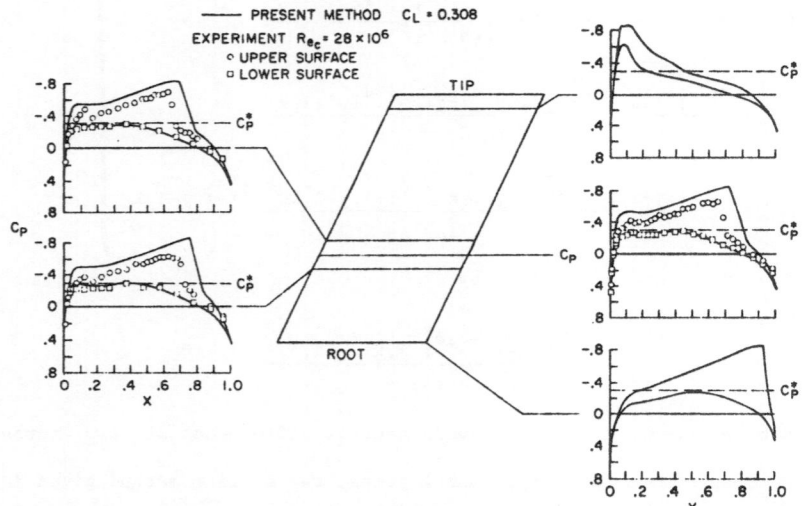

Fig. 18. C_p distribution on C-141 swept panel model,
$M_\infty = 0.853$, $\alpha = 2°$. [31]

angle of attack. The calculation at $M_\infty = 0.752$ is compared in Fig. 17 with both experimental results[56] at Re = 28×10^6 and results obtained by a subsonic panel method.[57] The relaxation results agree well with those obtained by the panel method but both calculations are in poor agreement with experiment on the upper surface. It is suggested that this disagreement is due primarily to the loss in lift caused by viscous effects at the trailing edge. Calculated results for the super-critical case shown in Fig. 18 are in even greater disagreement with experiment. Such a trend is to be expected, however, because of the shock-induced amplification of the trailing edge viscous effect. As in two dimensions, good agreement with experiment can be accomplished only through some consideration of viscous effects.

For the next example, consider the NCR calculation about a simulated C-141 wing given in reference 58. The wing had a leading-edge sweep of 25.6°, a taper ratio of 0.373, an aspect ratio of 8, a constant profile of 11.4 percent thick streamwise (the profile of the actual wing at 40 percent span) and a linear twist from 4° at the root to -1° at the tip. The calculated upper surface isobars for $M_\infty = 0.825$ is shown in Fig. 19. A shock wave swept at about 15° is indicated by the heavy band of coalesced isobars. In Fig. 20, the upper surface pressure distribution at the 40 percent span station is compared to wind tunnel and flight test results.[59] The wing angle of attack was adjusted to match the experimental leading-edge pressure gradient and peak pressure coefficient. The comparison between flight and wind tunnel data illustrates the importance of viscous scale effects on the surface pressure. Note that, in this case, a better prediction for the shock location is obtained from the calculation than from the wind tunnel experiment.

A final example is shown in Fig. 21 for flow at $M_\infty = 0.908$ about a 30° swept wing with a 6 percent biconvex section on a straight cylinder and on a symmetrically indented cylinder based on Mach-1 area-ruling. Note that the area-ruling eliminates the embedded shock waves on the wing. These NCR calculations were carried out on an equally spaced cylindrical coordinate system with the spacing adjusted such that $\Delta r = \Delta x/\tan \Lambda$, where Λ is the sweep angle (see reference 32).

While the above results are encouraging, they are considered to be interim results and much needs to be done to improve the numerical procedure, particularly

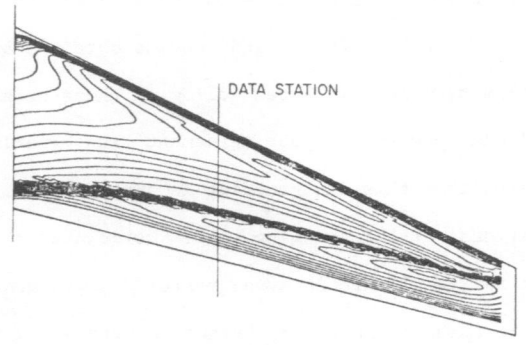

Fig. 19. Calculated upper surface isobars for
simulated C-141 wing at $M_\infty = 0.825$.[58]

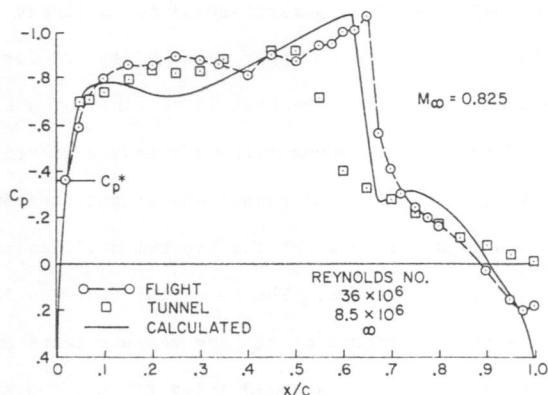

Fig. 20. Wind tunnel, flight and computed upper surface
C_p distribution for C-141 wing at
$y/b \approx 0.4$ and $M_\infty = 0.825$.[58]

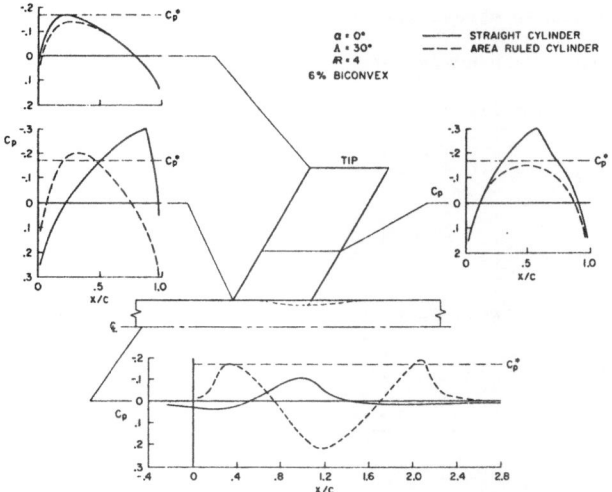

Fig. 21. C_p distribution on cylinder-wing combination at $M_\infty = 0.908$.[32]

with regard to developing a fully conservative method. This leads to the next topic — the capture of shock waves that are more oblique to the stream than the ones shown in these examples.

6.6 Swept Shock Waves

Transonic wings are nearly all swept; thus, it is of interest to investigate how well the transonic small distrubance equations approximate the jump across swept shocks that may be encountered on swept wings of large aspect ratio. The underlying basis for using wings with swept leading and trailing edges is generally derived by considering a sheared wing of infinite span and constant section. The principal assumption is that the velocity component parallel to an edge is constant or that the perturbation velocity parallel to an edge is zero. Under these conditions, a shock, if it occurs, would also have to be parallel to an edge, and its strength would depend only on the component of velocity normal to the edge. Thus, from the two-dimensional weak solution derived in section 2, the small disturbance normal shock solution written in terms of unscaled perturbation velocities normal to the leading edge is

$$u_{n_1} + u_{n_2} = \frac{2}{\gamma + 1}\left(\frac{1}{M_n^2} - 1\right) \tag{6.17}$$

Now, the three-dimensional streamwise perturbation component of velocity (normalized by u_∞) and free-stream Mach number are related to the two-dimensional normal perturbation components (normalized by $u_\infty \cos \theta$) by

$$\left. \begin{array}{l} u = u_n \cos^2 \theta \\[2mm] M_\infty = \dfrac{M_n}{\cos \theta} \end{array} \right\} \tag{6.18}$$

where θ is the angle of sweep with respect to the y-axis. Thus, Eq. (6.17) may be rewritten as

$$u_1 + u_2 = \frac{2\cos^2 \theta}{\gamma + 1} \left(\frac{1}{M_\infty^2 \cos^2 \theta} - 1 \right) \tag{6.19}$$

which represents the jump in streamwise velocity across a vertical shock swept parallel to the edge of an infinite sheared wing.

Consider next a nonsimilarity version of the three-dimensional transonic equation given by

$$\left[(1 - M_\infty^2)u - M_\infty^2 \frac{\gamma + 1}{2} u^2 \right]_x + v_y + \omega_z = 0 \tag{6.20}$$

$$v_x - u_y = 0$$

$$w_y - v_z = 0$$

The weak solution to Eq. (6.20) is given by

$$\left\langle 1 - M_\infty^2 u - M_\infty^2 \frac{\gamma + 1}{2} u^2 \right\rangle \cos \alpha_1 + \langle v \rangle \cos \alpha_2 + \langle w \rangle \cos \alpha_3 = 0 \tag{6.21}$$

$$\langle v \rangle \cos \alpha_1 - \langle u \rangle \cos \alpha_2 = 0$$

$$\langle w \rangle \cos \alpha_2 - \langle v \rangle \cos \alpha_3 = 0$$

where $\cos \alpha_1$, $\cos \alpha_2$, and $\cos \alpha_3$ are the direction cosines of the shock with respect to the x, y, and z axes, respectively. If the shock is assumed vertical, i.e., $\cos \alpha_3 = 0$, and the wing is an infinite sheared wing, the condition that must be met across the shock is expressed by

$$u_1 + u_2 = \frac{2}{\gamma + 1} \left(\frac{1}{M_\infty^2 \cos^2 \theta} - 1 \right) \tag{6.22}$$

The disagreement between Eq. (6.19) and Eq. (6.22) shows the breakdown of traditional transonic small disturbance theory in its ability to calculate swept shocks. The difference is illustrated in Fig. 22 for $M_\infty = 0.85$, showing that the transonic small disturbance equation is a very poor model for flows with shocks that are more than $25°$ oblique to the free stream.

The question arises: is it possible to modify Eq. (6.20) such that a more accurate shock jump can be obtained in regions of swept shocks? Such a modified equation is considered in reference 58. Terms that appear in the perturbation form of the isentropic equations (see, e.g., reference 60) and are neglected in the transonic equation are added to Eq. (6.17) to give

$$\left[(1 - M_\infty^2)u - \left(\frac{3 - \gamma}{2}\right)M_\infty^2 v^2 - M_\infty^2 \frac{\gamma + 1}{2} u^2\right]_x + \left[v - M_\infty^2(\gamma - 1)uv\right]_y + w_z = 0 \qquad (6.23)$$

The additional terms (underlined) are higher-order terms in the usual expansion process, so that they should not have a significant influence in continuous regions

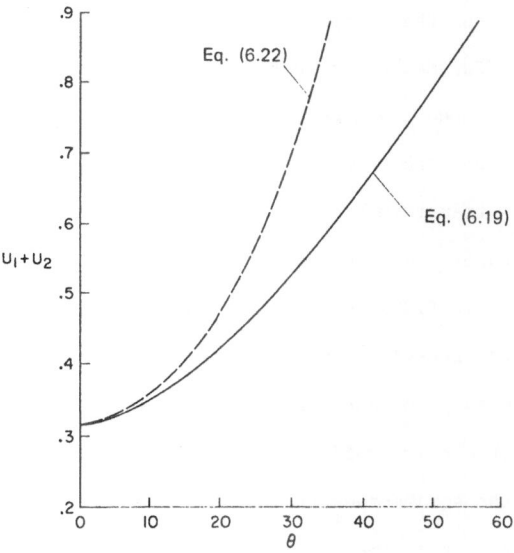

Fig. 22. Values of $(u_1 + u_2)$ from Eq. (6.22) and Eq. (6.19) for $M_\infty = 0.85$

of the flow. The weak solution to Eq. (6.23) with $w = 0$ is given by

$$u_1 + u_2 = \frac{2 \cos^2 \theta}{\gamma + 1}\left[\left(\frac{1}{M_\infty^2 \cos^2 \theta} - 1\right) + 2\left(\frac{v_2}{\tan \theta} + u_2\right)\tan^2 \theta\right]$$

$$v_2 - v_1 = -(u_2 - u_1)\tan \theta \qquad\qquad (6.24)$$

If we consider the special case when $v_2 = -u_2 \tan \theta$, which exists when the component of the perturbation velocity parallel to the shock is zero (i.e., the special case when the condition of simple sweep theory applies), we see that Eq. (6.24) reduces to Eq. (6.13), the sweep theory result. It is concluded, therefore, that Eq. (6.23) provides a better model for approximating highly swept shocks as may occur on three-dimensional swept wings.

7. JAMESON'S EXACT ISENTROPIC PROCEDURE FOR YAWED WINGS

7.1 Introduction

In reference 38, Jameson presents a scheme for solving the exact isentropic equation in three dimensions for transonic flow about a yawed wing. The impetus for such calculations arises from the increasing interest in the yawed wing aircraft concept of R. T. Jones.[61] The method is restricted to straight leading edges but allows for a curved and tapered trailing edge. Many of the concepts used in Jameson's two-dimensional procedure are incorporated in the yawed wing scheme — notably the square-root plus shearing transformation in planes normal to the span and the rotated differencing scheme for supersonic flow regions.

The coordinate system is fixed to the wing leading edge and the yaw angle is introduced by rotating the free-stream velocity vector. The vortex sheet is assumed to lie flat in the wing mean plane and trail in the shadow of the wing. The need for asymptotic far field solutions is avoided by stretching the coordinates to infinity. At downstream infinity, the square-root transformation collapses the region influenced by the vortex sheet to the line of points containing the sheet.

7.2 Transformed Equation

The exact isentropic equation in three-dimensions is expressed as

$$(a^2 - u^2)\phi_{xx} + (a^2 - v^2)\phi_{yy} + (a^2 - w^2)\phi_{zz} - 2uv\phi_{xy} - 2vw\phi_{yz} - 2uw\phi_{xz} = 0 \quad (7.1)$$

where

$$u = \phi_x$$
$$v = \phi_y$$
$$w = \phi_z$$
$$a^2 = a_0^2 - \frac{\gamma - 1}{2} q^2$$
$$q^2 = u^2 + v^2 + w^2$$

Following the coordinate system shown in sketch I, the square-root transformation in the x-z planes is given by

$$\left. \begin{array}{c} x + iz = \frac{1}{2} (X_1 + iZ_1)^2 \\ y = Y_1 \end{array} \right\} \quad (7.2)$$

If the wing surface is represented by

$$Z_1 = S(X_1, Y_1) \quad (7.3)$$

the shearing transformation, which displaces the vertical coordinates to be parallel to the wing surface, is expressed as

$$X = X_1, \quad Y = Y_1, \quad Z = Z_1 - S(X_1, Y_1) \quad (7.4)$$

A reduced potential Φ, from which the singularity at infinity has been removed, is introduced as

$$\Phi = \phi + \left\{ \frac{1}{2} [X^2 - (Z + S)^2]\cos \alpha + X(Z + S)\sin \alpha \right\}\cos \theta + Y \sin \theta \quad (7.5)$$

where θ is the yaw angle and α the angle of attack in the cross plane normal to the leading edge.

Under the above transformations, the governing equation takes the form

$$A\Phi_{XX} + B\Phi_{YY} + C\Phi_{ZZ} + D\Phi_{XY} + E\Phi_{ZY} + F\Phi_{XZ} = H \quad (7.6)$$

Here, the terms A through H are expressed as functions of the local speed of sound, the orthogonal velocity components in (X_1, Y_1, Z_1) space, the surface slopes and the mapping modulus given by

$$h^2 = X^2 + (Z + S)^2 \tag{7.7}$$

The condition that flow normal to the surface vanishes is expressed as

$$\Phi_Z = - \frac{(S \cos \alpha - X \sin \alpha)\cos \theta + U_1 S_X + h^2 V_1 S_Y}{1 + S_X{}^2 + h^2 S_Y{}^2} \tag{7.8}$$

where

$$U_1 = \Phi_X + (X \cos \alpha + S \sin \alpha)\cos \theta$$

$$V_1 = \Phi_Y + \sin \theta$$

7.3 Numerical Method

As in Jameson's two-dimensional method, the proper domain of dependence in supersonic regions is taken into account by using a rotated differencing scheme. Let Eq. (7.1) be expressed in the nondivergence canonical form

$$(a^2 - q^2)\phi_{ss} + a^2(\Delta\phi - \phi_{ss}) = 0 \tag{7.9}$$

where s denotes the stream direction and $\Delta\phi$ denotes the Laplacian

$$\Delta\phi = \phi_{xx} + \phi_{yy} + \phi_{zz} \tag{7.10}$$

The streamwise derivative can be expressed in terms of the x, y, and z derivatives as

$$\phi_{ss} = \frac{1}{q^2} (u^2\phi_{xx} + v^2\phi_{yy} + w^2\phi_{zz} + 2uv\phi_{xy} + 2vw\phi_{yz} + 2uw\phi_{xz}) \tag{7.11}$$

where u/q, v/q, and w/q are the direction cosines. In the numerical procedure, the velocity components are computed using central difference operators evaluated from the previous cycle, and their values determine whether the flow is subsonic or supersonic at a given point. At subsonic points, all second derivatives are approximated by central difference operators. At supersonic points, all second derivatives contributing to ϕ_{ss} in the first term of Eq. (7.9) are approximated by upwind difference operators, while those contributing to $\Delta\phi - \phi_{ss}$ are approximated by central operators. Note that when the governing equation is written in curvilinear coordinates, such as Eq. (7.6), only the principal part need be rotated since the domain of dependence is determined by coefficients of second derivatives. Furthermore, the term H contains only first derivatives, which are always approximated by central operators.

An analysis of the three-dimensional scheme based on an equivalent time-dependent equation as discussed in section 3.4 is presented in reference 38. The analysis leads to central and upwind difference operators of the form given in Eqs. (3.27) and (3.28).

The body boundary condition is satisfied by incorporating an additional row of points behind the boundary at which Φ is assigned values that satisfy Eq. (7.8). Points on the surface are treated as field points. In addition, special treatment is required at the vortex sheet.

Jameson uses a line relaxation scheme in which the subsonic points are over-relaxed and the relaxation parameter is set equal to 1 at supersonic points. The line algorithm can be used in any coordinate direction as long as the iteration sweeps downstream. Jameson divides the x-z plane into three strips. The method then proceeds by marching towards the surface in the central strip, and outwards with the flow in the left-hand and right-hand strips.

7.4 Results

Fig. 23 shows results reported in reference 38 for flow about a wing with 30° yaw at M_∞ = 0.866 and 3° incidence. The planform is shown in Fig. 23a and the upper surface pressure in Fig. 23b.

8. CONCLUDING REMARKS

The rapid advancement of relaxation methods over the past few years has led to a useful and efficient tool for the analysis of transonic flows. Two important advances have been made in the past year. One is the fully conservative differencing scheme that allows the correct calculation of the small disturbance shock jump. The other is the rotated differencing scheme that allows stable calculations regardless

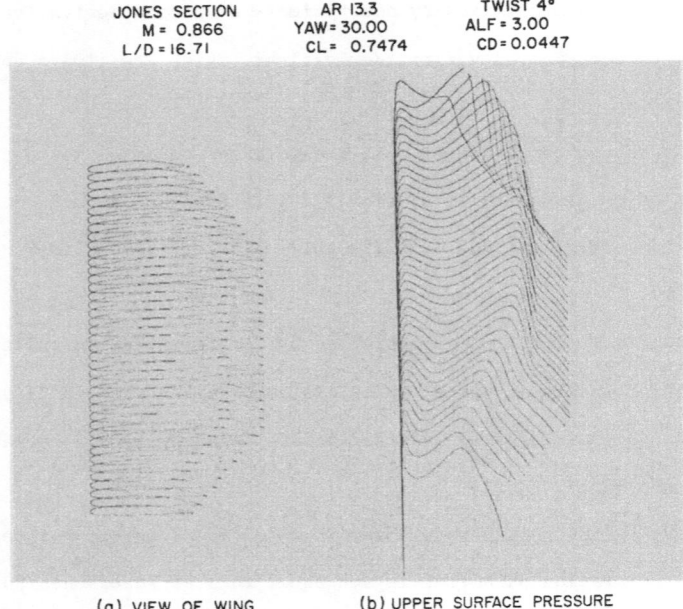

Fig. 23. Yawed wing results for $M_\infty = 0.866$, $\alpha = 3°$, $\theta = 30°$.[38]

    (a) View of wing.   (b) Upper surface pressure.

of the alignment between the flow and the finite-difference grid. However, further improvements are clearly needed to obtain complete solutions of transonic flow problems.

While procedures for the exact isentropic formulation allow more accurate treatment of the boundary conditions, further improvement is required to calculate correct shock jumps, either by conservative difference operators or by explicitly fitting the shock waves as discontinuities and imposing the correct jump conditions. Because the exact isentropic jump and particularly the small disturbance jump become increasingly larger than the Rankine-Hugoniot jump with increasing shock Mach number, it is suggested that techniques be investigated for incorporating the Rankine-Hugoniot jump rather than the jump appropriate for the isentropic equation. Although such an approach is not rigorous, it may lead to a better approximation of the pressure distribution obtained from the exact inviscid equations of motion. It is yet unclear how the correct computation of shock jumps (whether isentropic or not) will affect the apparent general good agreement between experimental jumps and those calculated by the nonconservative methods.

It is clear that viscosity plays a very important role in transonic flows and there remains the formidable task of incorporating turbulent viscous effects, particularly over the aft end of lifting airfoils. Although the development of a complete theory may take some time, more phenomenological methods based on modifications of the boundary layer approach with empirical adjustments may be developed for the short term. The urgent need is for predictive methods, however crude, that yield results for real flows.

Finally, we consider the additional complications of three-dimensional flows, which are centered about the increased difficulty of satisfying the boundary condition on complicated three-dimensional configurations. The small disturbance formulation considerably simplifies the boundary conditions and appears a reasonable near-term approach. In particular, it can be applied to wings on simulated fuselages to aid in the design of wing-fuselage junctures for minimum interference. However, efforts are needed to improve solutions near blunt leading edges and to improve the calculation of swept shocks, either through modifications to the equation or shock fittings.

REFERENCES

1. Nieuwland, G. Y., "Transonic Potential Flow Around a Family of Quasi-Elliptical Aerofoil Sections," National Lucht-en-Ruimteraartlaboratorium (NLR)
   Technical Report T. 172 (Netherlands), 1967.

2. Korn, D. G., "Computation of Shock-Free Transonic Flows for Airfoil Design,"
   Courant Institute of Math. Sci., NYO-1480-125, Oct. 1969.

3. Richtmyer, R. D., and Morton, K. W., Difference Methods for Initial-Value
   Problems, Interscience Publishers, Second ed., 1967.

4. Magnus, R., Gallaher, W., and Yoshihara, H., "Inviscid Supercritical Airfoil
   Theory," Transonic Aerodynamics, AGARD Conference Proceedings No. 35, Sept.
   1968, pp. 3-1, 3-3.

5. Magnus, R., and Yoshihara, H., "Inviscid Transonic Flow Over Airfoils," AIAA
   Paper 70-47, Jan. 1970.

6. Murman, E. M., "Computational Methods for Inviscid Transonic Flows With
   Embedded Shock Waves," Numerical Methods in Fluid Dynamics, AGARD Lecture
   Series No. 48, 1972, pp. 13-1, 13-34.

7. Yoshihara, H., "A Survey of Computational Methods for 2D and 3D Transonic
   Flows With Shocks," Advances in Numerical Fluid Dynamics, AGARD Lecture Series
   No. 64, 1973, pp. 6-1, 6-35

8. Nieuwland, G. Y., and Spee, B. M., "Transonic Airfoils: Recent Developments
   in Theory, Experiment, and Design," Annual Review of Fluid Mechanics, Annual
   Reviews, Inc., Vol. 5, 1973, pp. 119-150.

9. Emmons, H. W., "The Numerical Solution of Compressible Fluid Flow Problems,"
   NACA TN 932, 1944.

10. Emmons, H. W., "The Theoretical Flow of a Frictionless, Adiabatic, Perfect
    Gas Inside of a Two-Dimensional Hypersonic Nozzle," NACA TN 1003, 1946.

11. Emmons, H. W., "Flow of a Compressible Fluid Past a Symmetrical Airfoil in a
    Wind Tunnel and in Free Air," NACA TN 1746, 1948.

12. Murman, E. M., and Cole, J. D., "Calculation of Plane Steady Transonic Flows,"
    AIAA Journal, Vol. 9, No. 1, 1971, pp. 114-121.

13. Steger, J. L., and Lomax, H., "Transonic Flow About Two-Dimensional Airfoils
    by Relaxation Procedures," AIAA Journal, Vol. 10, No. 1, 1972, pp. 49-54.

14. Garabedian, P. R., and Korn, D. G., "Analysis of Transonic Airfoils," Comm.
    Pure Appl. Math., Vol. XXIV, 1971, pp. 841-851.

15. Jameson, A., "Transonic Flow Calculations for Airfoils and Bodies of Revolu-
    tion," Grumman Aerospace Corp. Report 390-71-1, 1971.

16. Murman, E. M., "Analysis of Embedded Shock Waves Calculated by Relaxation
    Methods," Proceedings of AIAA Computational Fluid Dynamics Conference, Palm
    Springs, Ca., July 1973, pp. 27-40.

17. Jameson, A., "Iterative Solution of Transonic Flows Over Airfoils and Wings,
    Including Flows at Mach 1," to appear in Comm. Pure Appl. Math.

18. Cole, J. D., and Messiter, A. F., "Expansion Procedures and Similarity Laws for Transonic Flow," _Zeit. ang. Math. u. Physik._, Vol. 8, No. 1, 1957, pp. 1-25.

19. Cole, J. D., "Twenty Years of Transonic Flow," Boeing Scientific Research Laboratories Document D1-82-0878, July 1969.

20. Ashley, Holt, and Landahl, Martin, _Aerodynamics of Wings and Bodies_, Addison-Wesley Publishing Co., Inc., 1965.

21. Lax, Peter D., "Weak Solutions of Nonlinear Hyperbolic Equations and Their Numerical Computation," _Comm. Pure Appl. Math._, Vol. VII, No. 1, 1954, pp. 159-193.

22. Oswatitsch, K., _Gas Dynamics_, Academic Press, 1956.

23. Nonweiler, T. R. F., "The Sonic Flow About Some Symmetric Half-Bodies," _J. Fluid Mech._, Vol. 4, Pt. 2, 1958, pp. 140-148.

24. Oswatitsch, K., and Zierep, J., "Das Problem des senkrechten Stosses an einer gekrummten wand," _Zeit. ang. Math. u. Mech._, Vol. 40, Supplement, T143-144, 1960.

25. Krupp, J. A., "The Numerical Calculation of Plane Steady Transonic Flows Past Thin Lifting Airfoils," Boeing Scientific Research Laboratories Report D180-12958-1, June 1971.

26. Krupp, J. A., and Murman, E. M., "The Numerical Calculation of Steady Transonic Flows Past Thin Lifting Airfoils and Slender Bodies," AIAA Paper 71-566, June 1971.

27. Murman, E. M., and Krupp, J. A., "Solution of the Transonic Potential Equation Using a Mixed Finite Difference System," _Lecture Notes in Physics_, Vol. 8, Springer-Verlag, 1971, pp. 199-206.

28. Yoshihara, Y., "A Survey of Computational Methods for 2D and 3D Transonic Flows With Shocks," GDCA-ERR-1726, Convair Aerospace Div., General Dynamics, Dec. 1972.

29. Magnus, R. M., "The Direct Comparison of the Relaxation Method and the Pseudo-Unsteady Finite Difference Method for Calculating Steady Planar Transonic Flow," TN-73-SP03, General Dynamics - Convair Aerospace Div., 1973.

30. Klunker, E. B., "Contributions to Methods for Calculating the Flow About Thin Lifting Wings at Transonic Speeds - Analytical Expression for the Far Field," NASA TN D-6530, 1971.

31. Ballhaus, W. F., and Bailey, F. R., "Numerical Calculation of Transonic Flow About Swept Wings," AIAA Paper 72-677, June 1972.

32. Bailey, F. R., and Ballhaus, W. F., "Relaxation Methods for Transonic Flow About Wing-Cylinder Combinations and Lifting Swept Wings," _Lecture Notes in Physics_, Vol. 19, Springer-Verlag, 1972, pp. 2-9.

33. Steger, J. L., and Baldwin, B. S., "Shock Waves and Drag in the Numerical Calculation of Isentropic Transonic Flow," NASA TN D-6997, 1972.

34. Bauer, E., Garabedian, P., and Korn, D., "Supercritical Wing Sections," _Lecture Notes in Economics and Mathematical Systems_, Vol. 66, Springer-Verlag, 1972.

35. Sells, C., "Plane Subcritical Flow Past a Lifting Airfoil," Proc. Roy. Soc. (London), Vol. 308A, 1968, pp. 377-401.

36. Thwaites, B., Incompressible Aerodynamics, Oxford Press, 1960.

37. South, J. C., and Jameson, A., "Relaxation Solutions for Inviscid Axisymmetric Transonic Flow Over Blunt or Pointed Bodies," Proceedings AIAA Computational Fluid Dynamics Conference, Palm Springs, Ca., 1973, pp. 8-17.

38. Jameson, A., "Numerical Calculation of the Three Dimensional Transonic Flow Over a Yawed Wing," Proceedings AIAA Computational Fluid Dynamics Conference, Palm Springs, Ca., 1973, pp. 18-26.

39. Garabedian, P. R., "Estimation of the Relaxations Factor for Small Mesh Size," Math. Tables Aids Comp., Vol. 10, 1956, pp. 183-185.

40. Kacpryznski, J., Ohman, L., Garabedian, P., and Korn, D., "Analysis of Flow Past a Shockless Airfoil in Design and Off-Design Conditions," Paper presented at AIAA 4th Fluid and Plasma Dynamics Conference, June 21-23, 1971, Palo Alto, Ca. (see NRC Aeronautical Report LR-54 (Canada)), June 1972.

41. Lock, R. C., "Test Cases for Numerical Methods in Two-Dimensional Transonic Flows," AGARD Report No. 575, Nov. 1970.

42. Bailey, F. R., "Numerical Calculation of Transonic Flow About Slender Bodies of Revolution," NASA TN D-6582, 1971.

43. Kacprzynski, J. J., "Wind Tunnel Tests of a Boerstoel Shockless Symmetrical Airfoil - 0.11-0.75-1.375," National Aircraft Establishment (NAE) Project Report 5x5/0061 (Canada), July 1972.

44. Boerstoel, J. W., "A Survey of Symmetrical Transonic Potential Flows About Quasi-Elliptical Airfoil Sections," National Lucht-en-Ruimtevaartlaboratorium (NLR) Technical Report T.136 (Netherlands), Jan. 1967.

45. Osborne, J., "A Selection of Measured Transonic Flow Pressure Distributions for the NACA 0012 Aerofoil: Provisional Data From an NPL Tunnel," Aerodynamics Dept., Royal Aircraft Establishment, June 1971.

46. Melnik, R. E., and Ives, D. C., "On Viscous and Wind-Tunnel Wall Effects in Transonic Flows Over Airfoils," AIAA Paper 73-660, July 1973.

47. Kacpryzynski, J., "A Second Series of Wind Tunnel Tests of the Shockless Lifting Airfoil No. 1," Report 5x5/0062 NRC (Canada), June 1972.

48. Garner, H. C. et al., "Subsonic Wind Tunnel Corrections," AGARDograph 109, Oct. 1966.

49. Baldwin, B., Turner, J., and Knechtel, E., "Wall Interference in Wind Tunnels With Slotted and Porous Boundaries at Subsonic Speeds," NACA TN 3176, 1954.

50. Taylor, R. A., and McDevitt, J. B., "Pressure Distributions at Transonic Speeds for Parabolic-Arc Bodies of Revolution Having Fineness Ratios of 10, 12, and 14," NACA TN 4234, 1958.

51. Murman, E. M., "Computation of Wall Effects in Ventilated Transonic Wind Tunnels," AIAA Paper 72-1007, Sept. 1972.

52. Isom, M. P., and Caradonna, F. X., "Subsonic and Transonic Potential Flow Over Helicopter Rotor Blades," AIAA Paper 72-39, Jan. 1972.

53. Ballhaus, W. F., and Caradonna, F. X., "The Effect of Planform Shape on the Transonic Flow Past Rotor Tips," _Aerodynamics of Rotary Wings_, AGARD Conference Proceedings No. 111, Paper 17, Sept. 1972.

54. Bailey, F. R., and Steger, J. L., "Relaxation Techniques for Three Dimensional Transonic Flow About Wings," AIAA Paper 72-189, Jan. 1972.

55. Newman, P. A., and Kunker, E. B., "Computation of Transonic Flow About Finite Lifting Wings," _AIAA Journal_, Vol. 10, No. 7, July 1972, pp. 971-973.

56. Cahill, J. F., and Stanewsky, E., "Wind Tunnel Tests of a Large-Chord, Swept-Panel Model to Investigate Shock-Induced Separation Phenomena," Air Force Flight Dynamics Lab. Report AFFDL-TR-69-78, Oct. 1969.

57. Saaris, G. R., and Rubbert, P. E., "Review and Evaluation of a Three Dimensional Lifting Potential Flow Computational Method for Arbitrary Configurations," AIAA Paper 72-188, Jan. 1972.

58. Lomax, H., Bailey, F. R., and Ballhaus, W. F., "On the Numerical Simulation of Three-Dimensional Transonic Flow With Application to the C-141 Wing," NASA TN D-6933, 1973.

59. Cahill, J. F., Teron, S. L., and Hofstetter, W. R., "Feasibility of Testing a Large-Chord, Swept-Panel Model to Determine Wing Shock Location at Flight Reynolds Number," AGARD Proc. 83, Paper 17, April 1971.

60. Liepmann, H. W., and Roshko, A., _Elements of Gas Dynamics_, John Wiley and Sons, New York, 1967.

61. Jones, R. T., "Reduction of Wave Drag by Antisymmetric Arrangement of Wings and Bodies," _AIAA Journal_, Vol. 9, 1971, pp. 114-121.

A CRITICAL REVIEW OF NUMERICAL SOLUTION OF NAVIER-STOKES EQUATIONS

by

Sin-I Cheng

PRINCETON UNIVERSITY

Department of Aerospace and Mechanical Sciences

(Separately distributed as Lecture Notes on Numerical
Solutions in Fluid Dynamics by v. Karman Institute
of Fluid Dynamics, AGARD, NATO, Belgium, February 1974
and as AMS Report #1158, Princeton University)

This research was conducted under the sponsorship
of the Office of Naval Research under Contract No.
N00014067-A-0151-0028.

February 1974

A CRITICAL REVIEW OF NUMERICAL SOLUTION OF NAVIER-STOKES EQUATIONS

by

Sin-I Cheng
Princeton University
Department of Aerospace and Mechanical Sciences

I. INTRODUCTION

This article is concerned primarily with the various practical problems
encountered in using high speed electronic computers to obtain approximate
solutions of various fluid flow problems, rather than with the general mathematical theory of difference approximations of partial differential equations.
It is important to realize that, in the mathematical formulation of a given
physical problem, the boundary conditions are as important as the partial
differential equations which describe the phenomenon. There are complicated
practical problems involved in discretizing the differential formulation (both
the differential equations and the boundary conditions) into an appropriate
difference formulation for numerical solution. Several aspects of the discretization procedure involve behavior quite different from those encountered
in the more familiar differential analysis.

Fluid dynamicists usually ignore the question of convergence in asymptotic
differential approximations obtained via perturbation arguments. They often
consider the computational solution of the resulting system of ordinary dif-

ferential equations as routine (albeit tedious), even though such systems are often ill-posed, especially for multi-eigenvalue problems. Under difficult circumstances, heuristic local treatments or various ad hoc methods are often introduced. Such hit-or-miss approaches have been carried over to the direct computational solution of the partial differential equations systems of fluid dynamics, resulting in much disappointment. While it may not be crucial to appreciate all of the details of the underlying mathematical theory, it is important to be aware of the implications of some fundamental mathematical results concerning the difference approximations of a partial differential equation. Accordingly, a brief review of these mathematical aspects will be outlined prior to the discussion of the practical art of numerically integrating the partial differential equations system of fluid dynamics.

Within the continuum description, the fluid will be considered to be homogeneous and to possess two independent thermodynamic (or state) variables, i.e., the density ρ and the internal energy e per unit mass. There is an algebraic equation of state $p = p(\rho, e)$ relating the thermodynamic pressure p to the density ρ and internal energy e. Let u_i (i = 1,2,3) be the velocity vector of a fluid element in a three-dimensional space x_i. u_i, ρ and e are the five dependent variables and will be considered as functions of x_i and t. The Eulerian description of the time rate of change of these variables is represented by a set of five partial differential equations, expressing the conservation of mass, momentum, and energy (written here in divergence form) as:

$$\frac{\partial \rho}{\partial t} + \frac{\partial}{\partial x_j} \left[\rho u_j \right] = 0 \quad , \tag{1.1}$$

$$\frac{\partial (\rho u_i)}{\partial t} + \frac{\partial}{\partial x_j} \left[\rho u_i u_j + p \delta_{ij} - \tau_{ij} \right] = 0 \quad , \tag{1.2}$$

$$\frac{\partial (\rho e)}{\partial t} + \frac{\partial}{\partial x_j} \left[\rho u_j (e + u_i u_i/2) + p u_j - q_j - u_i \tau_{ij} \right] = 0 \quad . \tag{1.3}$$

When the surface stress τ_{ij} is related linearly to the strain rate as

$$\tau_{ij} = \mu \left(\frac{\partial u_i}{\partial x_j} + \frac{\partial u_j}{\partial x_i} \right) + (\kappa - \frac{3}{2} \mu) \frac{\partial u_j}{\partial x_j} \quad , \tag{1.4}$$

and when the heat transfer vector q_j is linearly related to the temperature gradient as

$$q_j = - k \frac{\partial \theta}{\partial x_j} = - \frac{\gamma}{Pr} \mu \frac{\partial e}{\partial x_j} \quad , \tag{1.5}$$

the system of equations(1.1-1.3) will be referred to as the Navier-Stokes equations for a compressible fluid. The Prandtl number Pr and the specific heat ratio γ are properties of the fluid and are both of O(1). The shear viscosity coefficient μ is assumed to be a known algebraic function of temperature (or internal energy). The bulk viscosity coefficient κ is often taken as zero, or is otherwise absorbed in μ. In dimensionless form, a Reynolds number may be defined in terms of some characteristic length L_o and velocity U_o as $Re_o = \rho_o U_o L_o / \mu_o$, where subscript o indicates that the quantity is to be evaluated at some reference state. For most fluid dynamics applications, the Reynolds number is very large.

The divergence form of the Navier-Stokes equations (1.1)-(1.3) may be written as

$$\frac{\partial v}{\partial t} + \frac{\partial}{\partial x_j} F(v, \frac{1}{Re} \frac{\partial v}{\partial x_k}) = 0 \quad , \tag{1.6}$$

posed as an initial value problem for the vector unknown v having the five scalar components ρ, ρu_i, and ρe. F is the flux of v, given by the nonlinear

quantities in the square brackets of Equations (1.1)-(1.3). When physically
meaningful, initial and boundary data are prescribed, Equation (1.6) is
expected to give a satisfactory description of the temporal development of
the flow field at later times. This expectation is mathematically justifiable.
The integration of this equation system is needed, for example, in weather
forecasting, and in the determination of the temporal development of blast
waves, hurricanes, and turbulent fluctuations (where the gravitational field
and the coriolis forces are included where necessary). In most aeronautical
applications, steady state (or quasi-steady state) problems, where the temporal
dependence is neglected, are more often of primary interest. Thus Equation
(1.6) becomes

$$\frac{\partial}{\partial x_j} F(v, \frac{1}{Re} \frac{\partial v}{\partial x_k}) = 0 \qquad , \qquad (1.7)$$

which is to be solved as a boundary value problem. The boundary conditions
must, of course, be independent of time. But it is not clear how such boundary
conditions should be specified to provide the required steady state solution
or, indeed, any solution at all. Physical intuition often provides some meaningful guidance, but not all that is needed.

The stress and the heat conduction terms give rise to the second (and
highest) order partial derivatives, with coefficients proportional to Re^{-1}.
The steady state Navier-Stokes Equation (1.7) generally assumes elliptic behavior. When Re becomes large, the flow field may be divided into sub-regions.
In the region sufficiently far away from any solid boundary, the inviscid
approximation, obtained by dropping terms in (1.6) or (1.7) containing Re^{-1},
is a valid approximation, known as Euler's Equation:

$$\frac{\partial v}{\partial t} + \frac{\partial}{\partial x_j} F(v) = 0 \qquad\qquad (1.8a)$$

$$\frac{\partial}{\partial x_j} F(v) = 0 \qquad\qquad\qquad (1.8b)$$

The time-dependent inviscid Equation (1.8a) remains hyperbolic, and is posed as an initial value problem, as is Equation (1.6). The steady state Equation (1.8b), however, can be purely elliptic (subsonic), purely hyperbolic (supersonic), or mixed (i.e., with both elliptic and hyperbolic regions). The boundaries between different regions of a mixed problem will depend on the solution, and thus are not known before hand. This situation arises in the supercritical transonic inviscid flow problem, for example.

In the regions near a solid boundary, or near where there is large shear stress or heat conduction, some or all of the stress terms contained in $F(v, \frac{1}{Re} \frac{\partial v}{\partial x_k})$ have to be kept, despite the large Re. If this viscous region should extend along a coordinate surface (x_1, x_2) such that the lateral extent (along x_3) of this viscous layer is small (compared with its physical extent along the (x_1, x_2) surface), then Prandtl's boundary layer theory applies. Only the highest order partial derivative in this lateral direction $(\partial^2/\partial_3^2)$ will survive in the limit of very large Re. This asymptotic limit at large Re gives the boundary layer equations, which are parabolic. However, not all viscous layers are sheet-like, and, hence, not all are amenable to Prandtl's boundary layer approximation. For viscous layers like the near-wake and the interaction region between a shock wave and a boundary layer, the full Navier-Stokes Equation (1.7) will have to be used, and the problem becomes elliptic, at least in a significant portion of the flow field of interest.

The change of the mathematical character of the flow field in different
regions when the Reynolds number is large is both a blessing and a cause for
concern. It is a blessing in that, historically, it enabled the development
of fluid dynamics in the forms of the inviscid (or perfect) fluid theory and
the boundary layer theory. But it is also the fundamental difficulty in the
analysis of the mixed flow regions which are characteristic of most interaction
flow problems. Now there are significant differences in the numerical integration of the three types of partial differential equations. A method that has
proved successful for one type need not be so for another. It is, therefore,
important to recognize the type of partial differential equation at hand before
formulating a difference approximation for its numerical integration. Clearly,
then, there are difficulties in the numerical integration of mixed problems.
Such difficulties are quite different from those encountered in the asymptotic
analysis of mixed flow problems. In a few examples, these difficulties have
been successfully resolved with appropriate cautionary measures. But there
is no theorem to guarantee similar success in other problems.

If the elliptic steady state Navier-Stokes Equations (1.7) could be
integrated for a given large but finite Re, why should the difficulties arising
from the asymptotic limit of Re → ∞ concern us? An obvious answer might be
that the asymptotic form of the partial differential equations system is much
simpler than the full system. However, a more fundamental reason is that, at
large but finite Reynolds numbers, the asymptotic behavior of the flow in
different regions bears strongly on the appropriateness of the difference
formulation and on the numerical integration of the Navier-Stokes equations
when the resolution (or the number of meshes per linear dimension of the field
of computation) is severely limited.

For a flow problem in three space dimensions, an average of 30 meshes per linear dimension will give rise to 3×10^4 nodel points; and thus 1.5×10^5 words of storage space will be needed for the 5 unknowns at each point. This storage space should preferably be provided in the core of the computer unit for ready access. Such a requirement will stretch the core memory capacity of most of the currently available large computers, such as the CDC 6600 or the IBM 360-91. The solution of the full Navier-Stokes Equations (1.7) for a well-posed boundary value problem will need hours of computation in such machines. Parallel computers in advanced stages of development, like the ILLIAC IV and the STAR, cannot promise much improvement in this regard. To extend the core memory capacity of these parallel machines, a hierarchy of external storage devices will be provided. Frequent reference to such external storage, however, will greatly increase the time required for data management, due to the slowness of the input-output devices connected to the central processing unit. This input-output slowness is crucial in parallel computers, where the promised large gain in arithematic speed can be obtained only for specific modes of "parallel" or "vector" computations in which a huge amount of data must be properly processed and continuously fed into the arithematic unit(s).

The concept of parallel use of an array of mini-computers might appear to relieve the difficulty associated with any such specific mode of high speed arithematic operations. The benefit is likely to be illusory, however, at least for present applications. The use of such a system simply transfers to the users the tremendous problem of optimal coordination of the operations of the array of mini-computers, and the problem of data management among the diverse "internal" and "external" storage facilities. The users are generally not equipped with the expertise of the computer scientists who designed the software which manages the business of the central processing unit of large computers. Thus the users will be left to derive only whatever speed advantage each individual program may provide. It should be remembered that, without order(s) of magnitude increase in the "overall processing speed" of computation, an increase in the number of mesh points in a linear dimension for the integration of a problem can easily escalate the computer processing time from hours to days or months. It appears that, even with a slight projection into the future, no more than a couple hundred mesh points per linear dimension can possibly be considered in the integration of the system of equations of fluid dynamics. With such a limitation on the resolution of computational solutions, the integration of the full Navier-Stokes equation for the flow field over a vehicle (the external flow problem) seems futile. The field of computation is thus much like the test section of a windtunnel, in that, without a "full scale" facility, computers (like windtunnels) should be used at present to study only the components (local flow fields) of a large flow problem. For such purposes, various asymptotic forms of the Navier-Stokes equations should be used for different parts of the flow field. The full Navier-Stokes system should be called upon only for the study of those flow problems that cannot be consistently treated with the simplified flow equations, including notably those mixed interaction flow problems. Under practical limitations of resolution and of computer time, it is particularly important to delineate the varying nature of the flow regimes in the different treatments of the Navier-Stokes equations, and to consider the various possibilities of how the boundary conditions may be implemented. To quote the 1960 statement of Forsythe and Wasow: "The numerical solution of partial differential equations is no easy matter. Almost every problem arising out of the physical sciences requires original

thought and modifications of existing methods." This statement is equally
true today, particularly for the type of flow problems under consideration
here.

II. FUNDAMENTAL CONCEPTS

Consider now the problem of solving a partial differential equation,
subject to a set of initial and boundary data, through numerical integration.
A difference formulation, as an approximation to the differential problem,
is obtained by replacing the differential coefficients by appropriate difference
quotients. There will be some "errors" in the approximate formulation of the
equation and of the initial and boundary data. When these "errors" vanish as
the mesh sizes $\Delta t \rightarrow 0$ & $\Delta x \rightarrow 0$ in some manner, the difference approximation is
said to be "consistent" with the differential problem. The solutions of this
difference formulation provide a sequence of approximate solutions, which, in
the limit of $\Delta t, \Delta x \rightarrow 0$, is supposed to "converge" to the solution of the differential problem in some sense; i.e., the "error" of the solution, as a
measure of the departure of each member of the sequence of approximate solutions from the solution of the differential problem, tends to "zero." This
convergence is, however, not guaranteed for a consistent approximate difference
formulation. Various aspects of this situation will be considered in the
following sections.

2.1 Well-Posed Differential Problem

Whether the problem is to be integrated analytically or numerically, the
differential problem should not only possess a unique solution, but should also

possess "neighboring solutions." This means that when the initial-boundary
data is slightly perturbed, the differential problem should still provide a
solution which, hopefully, departs from the unperturbed solution of the
problem only slightly. This is primarily a physical requirement, expounded
by Hadamard, to insure that a mathematical formulation describes a physical
situation reasonably. Mathematically speaking, the solution of the differential problem is said to vary continuously with the data; and the differential problem is said to be "well-posed." A given partial differential
equation is well posed only when (among other things) the boundary conditions are properly specified. For example, the Laplace equation in two
variables x and y,

$$\frac{\partial^2 u}{\partial x^2} + \frac{\partial^2 u}{\partial y^2} = 0 \qquad , \qquad (2.1)$$

is well-posed when the value of the function $u(x,y)$ is specified on a
boundary enclosing the domain of interest (Dirichlet Problem).
Now the function

$$u(x,y) = n^{-a} \sin nx \cosh ny \qquad (2.2)$$

is an exact solution of the Laplace equation with the initial data

$$u(x,0) = n^{-a} \sin nx$$

$$\frac{\partial u}{\partial y} (x,0) = 0$$

This set of initial data is small everywhere on x with $a > 0$ and n sufficiently large. If the Laplace equation is to be solved when $u(x,0)$ and
$\frac{\partial u}{\partial y} (x,0)$ are specified, then a small perturbation of the initial data can
introduce perturbations of the type (2.2) onto the solution of the problem.

This perturbation (2.2) is not small in the immediate neighborhood of $y = 0$, despite the small error in the initial data when n is large and a is positive. While the perturbation (2.2) does vanish at $y = 0$ for any value of n (including $n \to \infty$), the value of $u(x,y)$ given by (2.2) at some small but finite value of y becomes infinitely large as $n \to \infty$. Thus the Laplace equation is not 'well-posed" (or is "ill-posed") when $u(x,0)$ and $\frac{\partial u}{\partial y}$ $(x,0)$ are specified (Cauchy Problem). If we should proceed to integrate this "ill-posed" problem, the perturbed initial data would be expected to contain components like (2.2), and the numerical solution would not converge to the desired solution even if $\Delta x \to 0$ (i.e., $n \to \infty$).

If the gradient of $u(x,y)$ is specified over a closed boundary (Neumann Problem), or if the gradiant is specified on part of a closed boundary and the value of $u(x,y)$ is specified on the remainder, the problem of solving the Laplace equation is well-posed (provided that some integral conditions are met). Ill-posed problems will result otherwise, i.e., either when Dirichlet or Neumann conditions are specified only on an open boundary, or when Cauchy conditions are used anywhere. This statement is applicable to elliptic partial differential equations in general. Parabolic equations are well-posed under similar conditions, but only on an "open" boundary and when integrated in the "positive" direction. Hyperbolic problems are well-posed only when Cauchy conditions are specified on an appropriate "open" portion of the boundary. It becomes difficult, then, to specify the boundary conditions that will render a mixed differential problem well posed, even before attempting to formulate a difference approximation of the problem for numerical integration. From this point of view, the algebraic complexities of the full Navier-Stokes equations system, either for the time-dependent hyperbolic problem (1.6) or for the steady state elliptical problem (1.7), may well be tolerated to facilitate the formulation of a well-posed problem.

2.2 Well-Posed Difference Problem

To provide a convergent numerical solution, it is not only required that the differential problem be well-posed for a specific or selected class of initial data, but also that the difference problem be well-posed for a more general class of initial data. This is because the perturbations implicit in the numerical solution of the approximate difference formulation need not fall within the class of the initial data for which the differential problem is well-posed. The function

$$u(x,t) = \exp\left[i\alpha(x + t)\right] \qquad (2.3)$$

satisfies the first-order hyperbolic equation

$$\frac{\partial}{\partial t} u - \frac{\partial}{\partial x} u = 0 \qquad (2.4)$$

with the initial value

$$u(x,0) = \exp(i\alpha x).$$

The complex notation with $i = \sqrt{-1}$ is used here for simplicity, to mean that both the real and imaginary parts of the expressions should be valid simultaneously. A wide class of functions $u(x,0)$ can be formed by superposing various trigonometric initial data, corresponding to various choices of values of the constant α. Each component can possess an arbitrarily assigned amplitude. By summing up the component solutions of different α's, the solution of the problem with generalized initial data is obtained. Any number of the component solutions can be perturbed, with a correspondingly small perturbation on the solution. The differential problem is thus well-posed.

Suppose now the forward-time, centered-space difference algorithm is used to provide a difference approximation to (2.4) as:

$$\frac{U_j^{u+1} - U_j^n}{\Delta t} = \frac{U_{j+1}^n - U_{j-1}^n}{2\Delta x} \tag{2.5}$$

$$U_j^o = \exp(i\alpha x)$$

where $U_j^n = U(j\Delta x, n\Delta t)$.

An exact solution of the difference problem (2.5) is

$$U_j^n = U(j\Delta x, n\Delta t) = (1 + i\frac{\Delta t}{\Delta x}\sin\alpha\Delta x)^n \exp(i\alpha x) \quad , \tag{2.6}$$

where $n = t/\Delta t$. In the limit of $\Delta t \to 0$ & $\Delta x \to 0$, the difference solution U_j^n, (2.6), converges uniformly to the solution $u(x,t)$, (2.3) of the differential problem (2.4). The same holds true for all components, and for their sum, with generalized initial data. Now when the difference problem (2.5) is computed for any small Δt and Δx, the computation is always unstable, as is well known. It is apparent that some components of the perturbations introduced by the computation of the difference form (2.5) cannot be represented by trigonometric data, and grow out-of-bounds during the calculation.

The Euler's equation (1.8a) for inviscid gas dynamics is easily cast into a Cauchy-Kowaleski type quasi-linear hyperbolic equation system

$$\frac{\partial u}{\partial t} + A(u)\frac{\partial u}{\partial x_j} = 0, \tag{2.7}$$

where $A(u) = \frac{\partial F}{\partial u}$. If the initial data $u(x, t = 0) = f(x)$ and $A(u)$ are analytic, then the solution $u(x,t)$ for all x and t is analytic. The requirement of analyticity of the initial data might not appear to be very restrictive, in view of the fact that, according to the Weierstrass approximation theorem, any continuous function within a closed interval can be approximated arbitrarily closely by analytic functions (including polynomials and sinusoidal functions). But an arbitrarily close approximation of the initial data does not promise an arbitrarily close approximation of the solution. The examples (2.I) and (2.5) given above illustrate this point, both for the differential and the difference equations.

Equations (1.8a) and (2.7) for inviscid gas dynamics are well-posed for a fairly broad class of initial data. Even if it is presumed that a consistent difference approximation possesses a solution that converges uniformly to the solution of the differential problem, stable computation is not guaranteed. The instability of the computation is attributed to the fact that perturbations on the initial data, introduced by the computational procedure, are beyond the class of perturbations expressible in terms of piecewise analytic data. For a difference problem to be well-posed, its solution must vary continuously with a much wider class of perturbations on the initial data. This is the crux of the concept of computational stability.

Computational stability in general calls for the boundedness of all of the perturbations in the computed solution. Then, when the magnitudes of the perturbations in the initial data are made arbitrarily small (in the limit of vanishingly small mesh sizes), the resulting perturbations in the computed solution will likewise vanish. The computed neighboring solutions based on a consistent difference formulation will then converge to the solution of the differential problem; i.e., stability and consistency imply convergence. This is the essence of the equivalence theorem of Lax. Success in obtaining a convergent approximate solution through computation based on a given difference formulation therefore depends on:

(i) the consistency of the difference formulation with the well-posed differential problem, and

(ii) the stability of the difference formulation.

Here, the difference formulation means, collectively, all of the difference relations connecting the values of functions at different time levels and at all mesh points in the interior of, and on the boundary of, the field of computation.

2.3 Computational Stability

Computational stability is a characteristic of a set of difference equations, rather than of a particular difference algorithm indicating how a differential coefficient in the differential equation is to be replaced by a difference quotient. Thus it is incorrect to refer to an algorithm as stable or unstable. The same algorithm when applied to various differential equations can lead to different difference equations with entirely different stability characteristics. For example, the forward time and centered-space difference algorithm applied to the simple wave equation (2.4) leads to an always unstable difference equation (2.5). But, when the same algorithm is applied to the heat diffusion equation, the resulting difference equation is stable if

$s = \frac{\Delta t}{\Delta x^2} \leq \frac{1}{2}$. More simple examples are given in Tables I and II.

Slightly different algorithms, applied to the same differential equation, may yield difference equations with quite different stability behavior. For the simple wave equation (2.4), the forward-time and backward-space difference algorithm will yield the always unstable difference equation

$$L(u) = (\frac{\partial}{\partial t} + c\frac{\partial}{\partial x})u = 0$$

$L_\Delta(u)$ WITH $r = c\frac{\Delta t}{\Delta x} > 0$	$e_t = L_\Delta(u) - L(u)$	STABLE
1) $(u_j^{n+1} - u_j^n) + r(u_{j+1}^n - u_j^n)$	$O(\Delta t, \Delta x)$	UNSTABLE
2) $(u_j^{n+1} - u_j^n) + r(u_j^n - u_{j-1}^n)$	$O(\Delta t, \Delta x)$	IF $r \le 1$
3) $(u_j^{n+1} - u_j^n) + \frac{r}{2}(u_{j+1}^n - u_{j-1}^n)$	$O(\Delta t, \Delta x^2)$	UNSTABLE
4) $u^{n+1} - \frac{u_{j+1}^n + u_{j-1}^n}{2} + \frac{r}{2}(u_{j+1}^n - u_{j-1}^n)$	$O(\Delta t, \Delta x^2)$	IF $r \le 1$
5) $u_j^{n+1} - u_j^{n-1} + r(u_{j+1}^n - u_{j-1}^n)$	$O(\Delta t^2, \Delta x^2)$	ALL r
MOST IMPLICIT SCHEMES		ALL r

TABLE I

$$L(u) = \left(\frac{\partial}{\partial t} - \nu \frac{\partial^2}{\partial x^2}\right) u = 0$$

$L_\Delta(u)$ WITH $s = \nu \Delta t / \Delta x^2$	$e_t = L_\Delta(u) - L(u)$	STABLE
1) $(u_j^{n+1} - u_j^n) - s(u_{j+1}^n - 2u_j^n + u_{j-1}^n)$	$O(\Delta t, \Delta x^2)$	IF $s \leq \frac{1}{2}$
2) $\left(u_j^{n+1} - \dfrac{u_{j+1}^n + u_{j-1}^n}{2}\right) - s(u_{j+1}^n - 2u_j^n + u_{j-1}^n)$	$O(\Delta t, \Delta x^2)$	UNSTABLE
3) $(u_j^{n+1} - u_j^{n-1}) - 2s(u_{j+1}^n - 2u_j^n + u_{j-1}^n)$	$O(\Delta t^2, \Delta x^2)$	UNSTABLE
4) $(u_j^{n+1} - u_j^{n-1}) - 2s(u_{j+1}^n - u_j^{n+1} - u_j^{n-1} + u_{j-1}^n)$	$O(\Delta t^2, \Delta x^2)$	ALL s
MOST IMPLICIT SCHEMES		ALL s

TABLE II

$$\frac{U_j^{n+1} - U_j^n}{\Delta t} - \frac{U_j^n - U_{j-1}^n}{\Delta x} = 0 \qquad . \qquad (2.8)$$

The forward-time and the forward-space difference algorithm will provide the difference equation

$$\frac{U_j^{n+1} - U_j^n}{\Delta t} - \frac{U_{j+1}^n - U_j^n}{\Delta x} = 0 \qquad , \qquad (2.9)$$

which is stable if $\Delta t/\Delta x \leq 1$. And, as mentioned previously, the forward-time and centered-space difference algorithm, Equation (2.5), is always unstable. The choice of a difference algorithm which yields a stable difference equation is not trivial.

A partial differential equation representing a physical principle may be written in different (but equivalent) forms in terms of different subsidiary variables. Moreover, a partial differential equation of higher order can sometimes be written as an equivalent system of lower order equations. When the same difference algorithm is applied to discretize these equivalent differential forms, the resulting difference equations are not equivalent, and may possess widely different stability behavior. Consider the simplest case of the second-order wave equation

$$\frac{\partial^2 \phi}{\partial t^2} = C^2 \frac{\partial^2 \phi}{\partial x^2} \qquad (2.10)$$

which is differentially equivalent to a system of two first-order wave equations. We may write the system in terms of two different pairs of variables as:

$$\begin{cases} \dfrac{\partial \phi}{\partial t} = v \\[2mm] \dfrac{\partial v}{\partial t} = C^2 \dfrac{\partial^2 \phi}{\partial x^2} \end{cases} \qquad (2.11a)$$

and

$$
\begin{cases}
\dfrac{\partial v}{\partial t} = C \, \dfrac{\partial u}{\partial x} \\[3mm]
\dfrac{\partial u}{\partial t} = C \, \dfrac{\partial v}{\partial x}
\end{cases}
\tag{2.11b}
$$

When the forward-time and centered-space difference algorithm is used, the following difference equation systems result:

$$
\begin{cases}
\dfrac{\Phi_j^{n+1} - \Phi_j^n}{\Delta t} = V_j^n \\[4mm]
\dfrac{V_j^{n+1} - V_j^n}{\Delta t} = C^2 \, \dfrac{\Phi_{j+1}^n - 2\Phi_j^n + \Phi_{j-1}^n}{\Delta x^2}
\end{cases}
\tag{2.12}
$$

and

$$
\begin{cases}
\dfrac{V_j^{n+1} - V_j^n}{\Delta t} = C \, \dfrac{U_{j+1}^n - U_{j-1}^n}{2\Delta x} \\[4mm]
\dfrac{U_j^{n+1} - U_j^n}{\Delta t} = C \, \dfrac{V_{j+1}^n - V_{j-1}^n}{2\Delta x}
\end{cases}
\tag{2.13}
$$

System (2.12) is always unstable for any choice of Δt and Δx (as is easily verified by v. Neumann Analysis), while system (2.13) is stable if $C \, \Delta t/\Delta x \leq 1$. Note also that the similar difference equation (2.5) for the first-order wave equation (2.4) is always unstable. Thus, it is not a matter of trivial consequence to rewrite a partial differential equation into equivalent, but different, forms before discretization with a given difference algorithm.

The equations of fluid dynamics represent the three conservation laws of mass, momentum, and energy. They can be expressed in terms of a great number of dependent and independent variables, in terms of particular combinations of

such variables, and in various coordinate systems. Second-order equations
may be split into first-order systems. (For the moment, the question of nonlinearity is put aside). When applied to these various physically and differentially equivalent systems of partial differential equations, a given
difference algorithm will yield various difference forms with quite different stability and other computational behaviors.

The complete difference formulation of a fluid dynamics problem calls for
the discretization not only of the differential equations, but also of the
boundary conditions. The set of difference relations connecting the values
of various functions at mesh points neighboring the boundary is generally
different from the set of recursive relations for the interior points derived
from the differential equations. This set of boundary difference relations
may be unstable, while the recursive difference relations for the interior
points are stable. Furthermore, apparently trivial modifications of the difference formulation of the boundary conditions often lead to substantial
changes in the stability behavior.

In view of such a complicated situation and of the frequent experience
of severe computational instability, it is highly desirable to be able to
analyze the stability behavior of a given difference formulation; but there
is no simple means available except the so-called "energy analysis." "Energy
analysis" attempts to establish a finite bound on the solution (over the
entire net or mesh space) in some suitably-chosen norm; if this is possible,
then the formulation is, by definition, stable. When such a bound is established, the proof of convergence, existence, and uniqueness follows trivially.
For a nontrivial boundary value problem, such a proof is very difficult and

tedious, even for a simple equation. Such proofs are available for the Navier-Stokes equations for an incompressible fluid, but only for periodic boundary conditions - a case which is really not that much different from a pure initial value problem. With rather complicated boundary conditions, it is impractical, if not impossible, to ascertain the stability property of a difference formulation of a fluid dynamics problem via such an approach. At present, it is a practical art to draw both from experience with similar problems and from inferences of model analysis in formulating the recursive difference relations for the interior points. The formulation of the boundary conditions is approached on an individual (i.e., ad hoc) basis and modified where necessary. The entire algorithm is then tested in actual computation for its stability. Considerable work will be involved before stable computation is achieved; by then, quite a few modifications may have been introduced. It is opportune to check if the final difference formulation is consistent with the differential problem to be solved, with regard to both the differential equations and the boundary conditions.

It may well be that the difference boundary conditions which prove successful in computation are not consistent with the correct physical boundary conditions, or that some spurious terms are introduced into the differential equations that fail to vanish in the limit of $\Delta t, \Delta x \to 0$. The v. Neumann stability analysis for the local linearized model will most likely impose some restrictions on Δt and Δx for the computation to remain stable. This restriction should be observed by all approximate solutions, viewed as successive members in a Cauchy sequence converging toward the solution of the differential problem. The limit process in the t-x space is not to be taken in any arbitrary manner; this restriction should be considered while investigating the consistency of the difference formulation.

Certain difference algorithms are often referred to as "unconditionally stable." This means that when such an algorithm is used to discretize a certain type of differential equation for the solution of pure initial value problems (or periodic boundary value problems), there will not be, according to the v. Neuman stability analysis of the linear equations, restrictive conditions on the choice of Δt (or on the iterative steps) for a given set of Δx. When such an algorithm is used in the numerical solution of non-periodic boundary value problems, even for the same particular type of equations, computational instability will often result, especially for complicated boundary conditions and for non-linear equations. Even if no question of stability should arise, the apparent advantage of permitting the use of quite large time steps Δt need not lessen the overall computing time, while inevitably decreasing the accuracy of the computed solution. Indeed, under such circumstances, it is advisable to verify the consistency conditions for both the equation and the boundary condition.

A case to illustrate this point is the following. The integration of the simple heat diffusion equation

$$\frac{\partial u}{\partial t} = \frac{\partial^2 u}{\partial x^2} \tag{2.14}$$

with the formally second-order accurate, centered-time, centered-space algorithm of DuFort-Frankel:

$$\frac{U_j^{n+1} - U_j^{n-1}}{2\Delta t} = \frac{U_{j+1}^n - 2U_j^n + U_{j-1}^n}{\Delta x^2} + \frac{U_j^{n+1} - 2U_j^n + U_j^{n-1}}{\Delta t^2} \left(\frac{\Delta t}{\Delta x}\right)^2 \tag{2.15}$$

is "unconditionally stable" for any positive values of $s = \frac{\Delta t}{\Delta x^2}$, so that Δt

can be made as large as Δx (or larger) without leading to computational instability. Most other explicit difference algorithms, when applied to the
heat diffusion equation, will impose a stability limit like $s \leq {}^1/2$. This
restriction on Δt is particularly severe at small Δx. Now, in integrating
Equation (2.15), it is tempting to use as large a Δt as is practical, usually comparable to Δx, to save computing time. Indeed, this is often
credited as the "merit" of the Dufort-Frankel scheme. Equation (2.15) is,
however, consistent with the heat diffusion equation only when $\frac{\Delta t}{\Delta x} \to 0$ as $\Delta x \to 0$.
Otherwise, it is consistent with the wave equation (2.15a), having the wave
speed $\frac{\Delta x}{\Delta t}$.

$$\left(\frac{\Delta t}{\Delta x}\right)^2 \frac{\partial^2 u}{\partial t^2} + \frac{\partial u}{\partial t} = \frac{\partial^2 u}{\partial x^2} \tag{2.15a}$$

With $\frac{\Delta t}{\Delta x} = O(1)$, the computed solution is expected to display waves comparable
in magnitude to the actual solution, and therefore, loses much of its value
as an approximation to the solution of the diffusion problem (2.14). Even
with $\Delta t / \Delta x^2 = s \leq 1/2$, for example, the computed solutions will still display
oscillations, albeit of smaller amplitude. An averaged solution (taken over
the waves) is not necessarily a meaningful approximation of the solution to
the diffusion problem with Dirichlet boundary conditions - if a Neumann
boundary condition is imposed, instability will often result. The qualitative statements mentioned here should not be generalized; the simple example
is given above only to drive home the point that every individual problem
should be carefully examined according to the fundamental principles. Our
current understanding of the numerical integration of partial differential
equations does not warrant any simple generalizations applicable to the
complicated situations of fluid dynamics.

III. STABILITY ANALYSIS

In the numerical integration of the Navier-Stokes Equations, an outstanding example of a complicated partial differential equations system, it
is expected that quite serious practical difficulties will be encountered.
Such difficulties can be grouped under the following three (not necessarily
independent) considerations:

(i) Computational Stability - All disturbances should remain bounded
in the computation. Otherwise, the value of some quantity would eventually
become so large as to be beyond the capability of any computer, and no results would be obtained. Hence, stability is often referred to as "computa-
bility."

(ii) Convergence Rate - The solution, at some later time T or at the
asymptotic steady state, should be obtained with a reasonable amount of computational work; i.e., the number of time or iterative steps in the solution
must not be too large, and the computational work for each step not excessive,
so that results can be obtained within a reasonable amount of time (and hence
cost).

(iii) Accuracy - For it to be useful, the solution eventually obtained
must be in some sense approximate the physical results in question. The criterion for an adequate approximation is, however, subject to judgment. The
accuracy criterion imposes limitations on the fineness of the resolution
(both temporally and spatially) which in turn determines the convergence rate.

Computational stability is clearly the most pressing problem, since it
is the first one to be encountered in an attempt to get any solution. Much
work has been devoted to this question. As explained in the previous chapter,

its fundamental nature is essentially understood, but there are quite a few
subtle aspects in its implementation, even for simple examples. The practical
difficulty encountered in achieving a stable computation for the complicated
system of the Navier-Stokes Equations is expected to be formidable. The
various heuristic approaches that promise to guide the formulation of a stable
difference problem will be reviewed in the following chapter. Generally
speaking, with some hard work a stable computation can usually be achieved,
as may be verified in actual computation. It is important, however, to bear
in mind that the convergence rate and the accuracy of the formulation should
not be seriously compromised in an all-out effort to achieve stability of the
computation. The objective of the computation is to obtain a valid approximation to a given physical problem. Therefore, the following review is intended
to bring out primarily the mathematical assumptions and the physical implications of various approaches when they are applied to the solution of different
types of practical problems.

3.1 v. Neumann Stability Analysis

A vector unknown $U(t,x_j)$ of dimension p is to be calculated over
mesh spacings Δx_1, Δx_2, Δx_3 for successive increments of Δt from the initial
values of $U(t=0,x_j)$, based on a system of linear difference equations.

The general form of the linear difference relations may be that some
linear combinations of the values of the function U^{n+1} at a group of neighboring mesh points are given by some other linear combinations of U^n at various
neighboring points. If only the U^{n+1} evaluated at a single mesh point is involved in the difference equations, the unknown values of U^{n+1} at any given

mesh point can be determined without reference to the advanced values of U^{n+1} at other mesh points. Such difference equations are explicit. If the advanced values of U^{n+1} at more than one mesh point are involved, a set of recursive difference relations written for all the mesh points have to be solved simultaneously, so that the advanced values of all the mesh points in the entire field of computation are obtained at the same time. Such difference equations are implicit. Sometimes it is preferable to solve simultaneously for the advanced values at special groups of mesh points in succession, such as by rows, columns, diagonals, blocks, or bands. Such difference equations are by nature partially implicit and partially explicit. The organization of the special group may change from one group to the next, and such different groups are often applied in alternating sequence, or in some special order. These techniques are then referred to as alternating direction methods. The difference algorithms that may be employed to represent a differential problem are indeed very numerous when these specific details are considered.

If all of the coefficients of the difference equations are constant, and if the system of equations is to be solved under periodic boundary conditions (or under the presumption that the boundary is so far away as to exert no influence on the solution, i.e., the pure initial value problem), the solution of the system of equations can be extended periodically beyond the field of computation, with both U^n and U^{n+1} represented by Fourier series. The linearity of the difference equation system permits the treatment of each Fourier component separately. Thus, by replacing U by $V(k_j)$ exp $\{ik_j x_j\}$ in the system of difference equations, and cancelling the common factor in each equation, an equation

$$H_1 V^{n+1}(k_j) = H_0 V^n(k_j) \tag{3.1}$$

results. Here i is the complex number used to represent the sinusoidal functions with wave numbers k_1, k_2, k_3 in the x_1, x_2, x_3 directions respectively
and $V(k_j)$ is the amplitude of the particular wave component under consideration. Each of the Fourier components may be considered either as a part of
the proper solution U, or as a small perturbation (or error) superposed on
the solution U. H_1 and H_0 are matrix operators depending on the constant
coefficients of the difference equations and on Δt and Δx_j. On the assumption that H_1 can be inverted, Equation (3.1) becomes

$$V^{n+1}(k_j) = G(\Delta t, \Delta x_j, k_j) V^n(k_j) \tag{3.2}$$

where

$$G(\Delta t, \Delta x_j, k_j) = (H_1)^{-1} H_0$$

Equation (3.2) gives the evolution of each Fourier component (interpreted
either as a part of the solution or as a perturbing error). Accordingly,
$G(\Delta t, \Delta x_j, k_j)$ is called the amplification matrix of the system of difference
equations. The condition that the solution U be uniformly bounded requires
each and every component to be so bounded. Since $\|V^n\| \le \|G\|^n \cdot \|V^q\|$, it is
necessary and sufficient that $\|G\|^n$ be so bounded for all wave components
and for all $n = T/\Delta t$, where T is the time period for which U is to be cal-
culated, with some choice of small but positive Δt. Now $\|G\|^n \ge R^n(\Delta t, \Delta x_j, k_j)$,
where R is the spectral radius of G, i.e., the largest eigenvalue of G.
Hence for such initial-periodic boundary value problems, the v. Neumann condition requires that all of the eigenvalues of the amplification matrix G be

$\leq 1+O(\Delta t)$. This is a necessary condition for computational stability. (The eigenvalues of the amplification matrix are often most conveniently obtained by the direct substitution of $u^{n+1} = \lambda u^n$ into the difference equation to obtain the determinant which vanishes when λ takes up the eigenvalues.) The v. Neumann condition becomes sufficient for the stability of the stated problem when the matrix G is normal. Both this sufficiency aspect and the additional term $O(\Delta t)$ are without much practical significance for the present consideration, as will shortly become clear. The important points to recognize from the above are the physical implications of the various conditions under which the v. Neumann stability analysis is formulated.

3.2 Local Linearization

The application of the v. Neumann analysis for the stability of the numerical integration of a system of nonlinear partial differential equations such as (1.7) calls for quite a few important additional assumptions and approximations:

(i) The nonlinear difference (or differential) equation is linearized by considering the solution as the sum of a small perturbation (or variation) superposed on the local solution $u(x,n\Delta t)$ of the problem. By substituting the perturbed solution into the difference equation and keeping only the terms involving the first power of the perturbation, the result is the equation of the first variation. The coefficients in this equation of the first variation depend on the solution of the differential problem, and therefore, vary with x and t.

(ii) The coefficients are assumed to be slowly varying, so that these coefficients can be replaced by the constant local values at various mesh points. The system of equations of the first variation then becomes linear

with constant coefficients, at each mesh point. The coefficients (and hence

the difference relations) vary from mesh point to mesh point, however.

(iii) The stability behavior of the computation at each mesh point is

assumed to be independent of its neighbors, so that the v. Neumann stability

analysis may be applied locally at every mesh point to find the local stability

limit on Δt based on the local linear difference equation of the first vari-

ation with constant coefficients.

(iv) The local stability limit is evaluated at every mesh point in the

interior with the local computed value $U^n(x)$ rather than the genuine solution

$u(x, n\Delta t)$. The most restrictive of the local stability limits, computed over

all the interior points, is then taken as the stability limit on Δt for the

integration of the difference problem.

The concept of local linear stability analysis applied to fluid dynamics

problems is probably what led v. Neumann to develop the Fourier method for

constant coefficient linear difference equations. This method is still the

most valuable practical tool. It should be noted that a slight difference

in the linearization procedure can lead to slightly different linearized e-

quations of the first variation. They will then give slightly different local

linearized stability criteria. Consider the following example, taken from

Richtmeyer and Morton's book. P.P.201-206:

$$\frac{\partial u}{\partial t} = \frac{\partial^2}{\partial x^2} (u^5) \tag{3.3a}$$

$$= \frac{\partial}{\partial x} (5u^4 \frac{\partial u}{\partial x}) \tag{3.3b}$$

with the initial condition

$$u_o = u(x,t=0) = \Psi[v(x_o-x)]$$

and the boundary conditions

$$u(0,t) = \Psi[v(vt+x_o)]$$

$$u(L,t) = \Psi[v(vt-L+x_o)]$$

The following is a solution representing a running wave with constant wave velocity v

$$u(x,t) = \Psi[v(vt-x+x_o)] \quad , \tag{3.4}$$

where the function Ψ is given implicitly as the inverse of

$$\frac{5}{4} (u-u_o)^4 + \frac{20}{3} u_o(u-u_o)^3 + 15 u_o^2 (u-u_o)^2 + 20 u_o^3 (u-u_o)$$

$$+ 5 u_o^4 \ln(u-u_o) = v(vt-x+x_o) \quad . \tag{3.5}$$

The solution u is shown in Fig. 1 with a relatively sharp front and approximately a quartic curve far downstreams. It may be interesting to note that Equation (3.3) stands as a heat diffusion equation with variable diffusivity $5u^4$, rather than as an equation describing the steady propagation of a nondecaying wave.

Let Equation (3.3) be discretized with a forward-time difference, and with the spatial derivative evaluated as a weighted average of centered differences at the advanced and initial time steps, with weights θ and $(1-\theta)$ respectively:

$$U_j^{n+1} - U_j^n = \frac{\Delta t}{\Delta x^2} \{\theta[\delta^2(U^5)]_j^{n+1} + (1-\theta)[\delta^2(U^5)]_j^n\} \tag{3.6}$$

where the second order spatial difference operator is defined by

$$[\delta^2(\ )]_j^n = (\ )_{j+1}^n - 2(\ )_j^n + (\ )_{j-1}^n \tag{3.7}$$

The parameter θ can be chosen at convenience. (3.6) is a nonlinear equation. The linearized approximation

$$(U^5)_j^{n+1} - (U^5)_j^n = 5(U^4)_j^n (U_j^{n+1} - U_j^n)$$

gives the equation of first variation of (3.6) as

$$(U_j^{n+1} - U_j^n) - \frac{5\theta\Delta t}{\Delta x^2} \left[(U^4)_{j+1}^n (U_{j+1}^{n+1} - U_{j+1}^n) \right.$$

$$\left. - 2(U^4)_j^n (U_j^{n+1} - U_j^n) + (U^4)_{j-1}^n (U_{j-1}^{n+1} - U_{j-1}^n) \right]$$

$$= \frac{\Delta t}{\Delta x^2} \left[(U^5)_{j+1}^n - 2(U^5)_j^n + (U^5)_{j-1}^n \right] . \tag{3.8}$$

Note that the equation is linear in the unknown $(U_j^{n+1} - U_j^n)$, if the values of the function U at all of the spatial mesh points at the time level n are known. Otherwise, the equation retains its nonlinear form. Alternatively, it is appropriate to linearize in many other ways. A particularly simple one is to take

$$\left[\delta^2(U^5) \right]_j^{n+1} - \left[\delta^2(U^5) \right]_j^n$$

$$= 5(U^4)_j^n \left[(\delta^2 U)_j^{n+1} - (\delta^2 U)_j^n \right]$$

and

$$\left[\delta^2(U^5) \right]_j^n = 5(U^4)_j^n (\delta^2 U)_j^n .$$

Then the equation of the first variation of (3.6) becomes

$$(U_j^{n+1} - U_j^n) - 5\frac{\theta\Delta t}{\Delta x^2} (U^4)_j^n \left[(U_{j+1}^{n+1} - U_{j+1}^n) - 2(U_j^{n+1} - U_j^n) + (U_{j-1}^{n+1} - U_{j-1}^n) \right]$$

$$= 5 \frac{\Delta t}{\Delta x^2} (U^4)_j^n \left[U_{j+1}^n - 2U_j^n + U_{j-1}^n \right] \qquad (3.9)$$

This last equation is indeed the same as that which would result if the effective diffusivity $5u^4$ in Equation (3.3b) is treated as a constant before and during discretization.

With all U_{j+1}^n, U_j^n, and U_{j-1}^n taken to be constant, the v. Neumann stability analysis for Equation (3.8) will require the following for all wave numbers k:

$$\left| \frac{1 - (1-\theta)s}{1 + \theta s} \right| \leq 1 \quad , \qquad (3.10)$$

where s is the complex expression

$$s = 5(U^4)_j^n \frac{\Delta t}{\Delta x^2} \left[2 - (\alpha+\beta) \cos k\Delta x - i(\alpha-\beta) \sin k\Delta x \right]$$

$$\text{with} \quad \alpha = (U_{j+1}^n/U_j^n)^4$$

$$\text{and} \quad \beta = (U_{j-1}^n/U_j^n)^4 \quad .$$

The restriction on the value of $\frac{\Delta t}{\Delta x^2}$ can be computed for all k from (3.10) for the values of α and β at every interior mesh point. This is a very tedious process. The v. Neumann stability analysis for Equation (3.9) leads to the same relation (3.10); but with $\alpha=\beta=1$. This provides an explicit limit on $\frac{\Delta t}{\Delta x^2}$, such that when $\theta < 1/2$,

$$5(U^4)_j^n \cdot \frac{\Delta t}{\Delta x^2} \leq \frac{1}{2(1-2\theta)} \quad , \qquad (3.11)$$

with no limit on $\frac{\Delta t}{\Delta x^2}$ if $\theta \geq 1/2$. This is the well-known result for the simple heat equation.

To test the usefulness of the local linear stability criterion, computations were carried out with Equation (3.8), taking $\theta = 0.4$ and $\frac{\Delta t}{\Delta x^2} = 0.001$. The parameters v and U_0 were chosen as $v\frac{\Delta t}{\Delta x} = 0.075$ and $5U_0^4 \frac{\Delta t}{\Delta x^2} = 0.005$. The last value is much less than 2.50, as required by Equation (3.11) for local computational stability. As the computation proceeds, the values of U increase with t over the entire field of computation. According to the local stability criterion (3.11), we would expect instability to appear in the form of rapidly increasing amplitudes of oscillation when and where the values of (U_j^n/U_0) exceed $(500)^{1/4} \sim 4.7$. This was what happened, as is illustrated in Figure (1). The computed points lie very close to the analytical solution except at the foot of each wave front, where the solution undergoes a rapid change; and in the region $U_j^n/U_0 \stackrel{\sim}{>} 5$, where the computed solution oscillates, signalling the onset of computational instability.

It is remarkable that the simple local criterion deduced for the difference Equation (3.9) provides highly satisfactory guidance for the integration of Equation (3.8), even though α and β generally differ from unity. When the stability boundary (in the complex plane) of Equation (3.8) with α & $\beta \neq 1$ lies within the stable region of Equation (3.9), the local linear stability limit deduced for (3.10) need not even be "necessary" at these interior mesh points represented by the region between the two stability boundaries. Such regions are likely to be small, however, if the model equation (3.9) in the above example is appropriately chosen. The local criterion is clearly not sufficient, since the v. Neumann stability condition itself is not sufficient and since the influence of the boundary conditions on computational stability has not yet been investigated. Nevertheless, the local linear stability analysis does appear to provide useful guidance in practical applications,

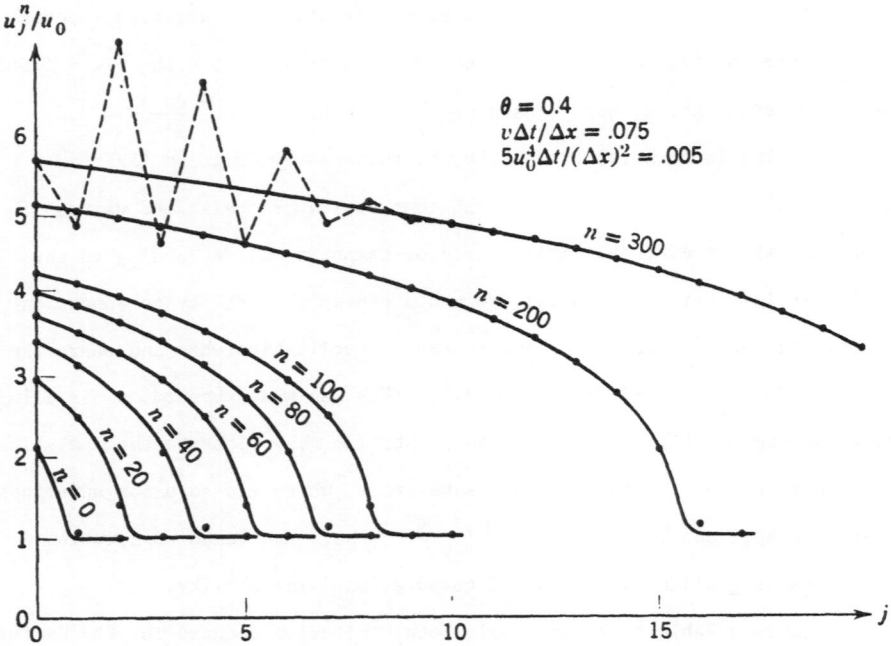

Fig. 1. Running-wave solutions of the non-linear equation $\partial u/\partial t = \partial^2(u^5)/\partial x^2$. The curves show the exact solution, given by equation (8.25) and the dots show the solution of the difference equation (8.27) with $\theta = 0.4$ and with Δt and Δx so chosen that $v\Delta t/\Delta x = 0.075$ and $5u_0^4\Delta t/(\Delta x)^2 = 0.005$. The numbers on the curves are cycle numbers.

especially if the influence of the boundary conditions can be separately investigated, and if the linearized model for the difference relations at the
interior points is properly selected. Such fortunate circumstances are,
however, not to be presumed in complicated equations systems.

3.3 Application to Navier-Stokes Equations

The Navier-Stokes system is quasi-linear due to (i) the variable convective velocity, and (ii) the variable density and energy (and hence the variable
diffusivity). The system is further complicated in the case of small diffusivity (or large Reynolds number), especially in multi-dimensional flow problems, where many viscous terms occur. If the standard procedure of local
linearization is followed, the resulting linearized equations are very long.
The v. Neumann stability analysis for such equations inevitably leads to un-
wieldy algebraic expressions, so that the explicit limit on Δt at each mesh
point can only be obtained at much more labor than that required for Equation
(3.10). It is then impractical to consider checking the stability limit at
many mesh points, even if infrequently. It appears imperative to look for
simple but meaningful model equations, such as Equation (3.9) in the previous example. The search for such useful models is considerably complicated
by the change of the asymptotic behavior of the Navier-Stokes equations system
in different regions of the field of computation, as outlined in the previous
chapter. Near solid boundaries, or where viscous effects are important, the
region is locally parabolic or elliptic. Far away from solid boundaries, the
direct viscous effect is negligible and the flow region is primarily hyperbolic.
It is unfortunate that a difference algorithm, when applied to practical dif-

ferential equations of different types, will lead to difference equations
with quite different stability behavior.

Tables I and II illustrate the application of a few common difference
algorithms to the simple wave equation and to the simple diffusion equation
respectively. An algorithm often yields a stable difference equation for
the diffusion equation (such as the forward time-centered space algorithm,
Scheme 1 in Table II), while it provides an unstable difference equation for
the simple wave equation (Scheme 3 in Table I). Friedrich's modification,
which renders the wave problem stable (Scheme 4 in Table I), on the other
hand, leads to an unstable diffusion problem (Scheme 2 in Table II). The
centered-time and centered-space algorithm (given in the tables) is another
example which illustrates the same point; there are many further examples
like these. Such schemes are not useful for integrating the Navier-Stokes
equations.

There exist many schemes which are stable for both types of equations,
but where Δt is subject to different restrictions in different regions.
Usually $c \frac{\Delta t}{\Delta x} < 1$ is required for the wave equation, and $\frac{\Delta t}{\Delta x^2} \leq$ some fractional
constant g for the diffusion equation; such is true for the forward-time,
backward-space algorithm, Schemes 2 and 1 respectively in the two tables.
The condition $c \frac{\Delta t}{\Delta x} \leq 1$ is known as the Courant-Friedrich-Levy (CFL) condition
of zone of dependence, and is to be satisfied generally for difference forms
of wave equations.[1] The CFL condition states that the zone of dependence
of the difference formulation must include the zone of dependence of the
differential equation. When such a scheme (stable for both types) is used in
integrating the Navier-Stokes equations, computational stability might be

expected if Δt is locally chosen to be the more restrictive of the wave and diffusion limits:[2]

$$\Delta t < \mathop{\mathrm{Inf}}_{j} \left[\frac{\Delta x}{c} , \gamma \frac{\Delta x^2}{\nu} \right] \quad . \qquad (3.12)$$

Here c is related to the local wave speed, ν is the local kinematic viscosity coefficient, and γ is some constant less than unity. The precise values of c and γ may be determined from the v. Neumann stability analysis of the linearized Navier-Stokes equations, after dropping the viscous terms or the dynamic terms respectively. The most restrictive of these local limits over all mesh points may then be taken as the Δt for the next time increment. In actual computation, it is often necessary to reduce this most restrictive limit on Δt further by introducing an empirical safety factor which may have to be rather small.

It might be thought that such a safety factor is needed because of the unknown effect of the boundary conditions. But actual computation often indicates that instability is initiated from the interior. Thus, this instability appears to be, at least in part, due to the fact that the pure diffusion and/or the pure wave equations are rather poor models for the interior points of the linearized Navier-Stokes equations. It is true that, in linearized form, the Navier-Stokes equations may be visualized as the superposition of a wave and a diffusion equation. The stability limit, however, is not generally the superposed diffusion and wave limits. This is because the determination of the eigenvalues of a linear equation, which requires finding the roots of a polynomial equation with constant coefficients, is not a linear problem. A small perturbation on a coefficient of the characteristic polynomial often leads to an inproportionately large change in the largest eigenvalue (or the spectral radius), depending on the specific difference algorithm.

To illustrate the situation, consider the one-dimensional Burgers' equation with constant c and ν:

$$\frac{\partial u}{\partial t} + c \frac{\partial u}{\partial x} = \nu \frac{\partial^2 u}{\partial x^2} \quad . \tag{3.13}$$

When c and ν are taken as the local values at a mesh point, (3.13) serves as a linearized model of the Navier-Stokes equation in one space dimension, possessing the essential characteristic of changing type (of partial differential equation) in different regions. If Equation (3.13) is discretized with forward-time and backward-space differences for the convective term, and a centered-space difference for the diffusion term, the v. Neumann stability limit is

$$\Delta t \leq \left(c/\Delta x + 2\nu/\Delta x^2 \right)^{-1} \quad , \tag{3.14}$$

which is almost half of the hyperbolic limit $\Delta x/c$ or the diffusion limit $\Delta x^2/2\nu$, if they are approximately equal. Thus the safety factor to be applied to condition (3.12) should be about 1/2 (or less) for stable computation of the interior points alone.

The situation is even more critical for multidimensional flow problems. Consider the following two models for 2-D problems:

$$\frac{\partial u}{\partial t} + c \left(\frac{\partial u}{\partial x} + \frac{\partial u}{\partial y} \right) = \nu \left(\frac{\partial^2 u}{\partial x^2} + \frac{\partial^2 u}{\partial y^2} \right)$$

and

$$\frac{\partial u}{\partial t} + c \frac{\partial u}{\partial x} = \nu \left(\frac{\partial^2 u}{\partial x^2} + \frac{\partial^2 u}{\partial y^2} \right) \quad , \tag{3.15}$$

with the convective term $\nu \frac{\partial u}{\partial y}$ represented by $c \frac{\partial u}{\partial y}$ and zero respectively. The stability limits for the two cases, assuming $\Delta x = \Delta y$, are

$$\Delta t \leq \frac{1}{2} \; (c/\Delta x \; + \; 2\nu/\Delta x^2)^{-1}$$

$$\text{and} \qquad \Delta t \leq \quad (c/\Delta x \; + \; 4\nu/\Delta x^2)^{-1} \qquad\qquad (3.16)$$

respectively. This means that the safety factor to be applied to condition (3.12) should be ∿1/4 or 1/3 for 2D flow problems, and should be even smaller for 3D flow problems. Thus, although it is simple and convenient, the stability condition (3.12) based on superposing the wave and the diffusion parts of the Navier-Stokes equations is not very useful.

The local stability condition based on the linearized Burgers' equation was found to be quite satisfactory for the integration of not only the non-linear Burgers' equation without a safety factor, but also of the full Navier-Stokes equations with properly treated boundary conditions.[3] The one-dimensional Burgers' model should be applied locally to the flow along the streamline through a mesh point. This yields stability limits of the form of Equations (3.14) and (3.16), in which c should be interpreted as the local signal speed $|q| + |a|$, where q is the stream velocity. Both the local speed of sound a and the kinematic viscosity coefficient ν should be evaluated at the local temperature or energy. Δx should be evaluated along the streamline in some manner, and may well be taken, for example, as the smaller of (Δx , Δy) for 2-D problems. For a given choice of the difference algorithm for dis-cretization, the expression for the local stability limit on Δt will depend on how the various local quantities are evaluated from the explicitly calcu-lated variables. It is advisable to choose a simple (though perhaps less accurate) form of such an expression which is convenient for the explicit determination of the limit on Δt at each point. This calculation is to be carried out at many points and at many time intervals for an estimate of the most restrictive limit on (the smallest value of) Δt for the next time interval. It may also be convenient to check the local linearized stability limit once every few time steps rather than every time step, and to adjust the magnitude of Δt adopted for the next few steps accordingly.

3.4 Treatment on the Boundary

When the appropriate local linearized stability limit is obeyed, computational instability at the interior points can usually be avoided, although oscillations of fairly large (but bounded) amplitudes are often present in the calculated results. These oscillations originate from the boundaries, both interior and exterior, and do not represent computational instability in the previously discussed sense of boundedness of the solution. Such bounded oscillations are often referred to as Nonlinear Instability, which is basically a different phenomenon more directly related to the question of accuracy, and which probably cannot be clarified by the heuristic local linear stability treatment[4] discussed in the previous section.

Genuine unstable computation can result when certain boundary treatments are applied to a given difference algorithm. For such cases, the local linearized analysis can often foretell the impending computational instability. Consider the integration of the inviscid gas dynamic equation (2.7) with the Leap-Frog scheme. (Scheme 5 in Table I):

$$U_j^{n+1} - U_j^{n-1} = -A_j \frac{\Delta t}{\Delta x} \left[U_{j+1}^n - U_{j-1}^n \right] \quad , \tag{3.17}$$

which is second order accurate in both time and space and is always stable at all of the interior points for any value of $\Delta t/\Delta x$. To initiate the integration, both U_j^0 and U_j^1 should be available at all $j = 0,1,2,...J$, and boundary

conditions must be provided at boundaries j = 0 and j = J. Note that both
the initial value U_j^1 and the boundary data at j = J are not specified by the
initial data of the differential problem for the propagation of a small wave
in an unbounded flow field. These data are extraneous and are required by
the use of a "higher-order accurate" difference algorithm in which a first
order differential coefficient is replaced by a second order difference quotient.

The extraneous initial data U_j^1 are usually obtained from a Taylor series
expansion about t = 0, where the necessary higher order temporal derivatives
are evaluated from the initial data U_j^0 through use of the differential equation
and its time derivatives. But, it is not obvious how the extraneous boundary
data at j = J should be defined. One natural way is to extrapolate along x,
assuming that $\frac{\partial}{\partial x} U$ is small:

$$U_J^n = U_{J-1}^n \tag{3.18}$$

This is not a bad physical approximation. Computationally it leads to the
difference relation

$$U_{J-1}^{n+1} - U_{J-1}^{n-1} = -A_{J-1} \frac{\Delta t}{\Delta x} \left[U_{J-1}^n - U_{J-2}^n \right] \tag{3.19}$$

for advancing the mesh value at the point J-1 immediately preceding the
boundary point J. Being different from equation (3.17), equation (3.19) need
not remain stable on the boundary although the computation in the interior is
stable. In view of the presumption of the local linear analysis that the
stability at a given mesh point is independent of its neighbors, we apply
the v. Neumann stability analysis locally to this difference equation. With
A_{J-1} taken as a constant and U taken as a scalar unknown, the difference re-

lation (3.19) is found to be locally always unstable with the amplification factor $|\lambda| = |U_{J-1}^{n+1}/U_{J-1}^{n}| > 1$. This is so because the v. Neumann analysis leads to the algebraic relation

$$\lambda - \frac{1}{\lambda} = -A_{J-1} \frac{\Delta t}{\Delta x} \left[(1-\cos k\Delta x) + i \sin k\Delta x \right] = -2(f_r + if_i) \quad , \quad (3.20)$$

where k is the wave number under consideration and $2f_r$ and $2f_i$ are the real and the imaginary parts of the right hand side. Thus

$$\lambda = -(f_r + if_i) \pm \left[1 + (f_r + if_i)^2 \right]^{1/2} . \quad (3.21)$$

For some choice of k, f_i will be zero and $|\lambda|$ will be greater than unity regardless of the magnitude of $A_{J-1} \frac{\Delta t}{\Delta x}$. Actual computation confirms the instability, i.e., that $|U_j^n|$ diverges as n. If $A_{J-1} \frac{\Delta t}{\Delta x}$ should be taken as unity, and if the initial data satisfies $U_j^n = (-1)^{j+n}$ for n=0 and 1 and all j = 0, 1...J-1, the solution of the difference equation can actually be shown to continue as

$$U_j^n = (-1)^{j+n} + (-1)^j F(j + n)$$

with $\qquad F(j + n < J) = 0 \qquad$ and

$$F(n + J) = (-1)^{n-1} 2n \quad . \quad (3.22)$$

Higher order accurate extrapolation formulas based on the assumption that higher order derivatives equal to zero (instead of Equation (3.18) will only change f_r and f_i, and will still lead to computational instability in the same manner.

Careful examination of the local stability analysis suggests that stable computation will result if the extraneous boundary value U_J^n is obtained as:

$$U_J^n = \frac{1}{2} (U_{J-1}^{n+1} + U_{J-1}^{n-1}) \qquad (3.23)$$

i.e., U_{J-1}^n in the first order extrapolation formula (3.18) is replaced by the average of its temporal neighbors. Then the difference relation on the boundary is

$$U_{J-1}^{n+1} = \frac{1-\alpha}{1+\alpha} U_{J-1}^{n-1} + \frac{2\alpha}{1+\alpha} U_{J-2}^n \qquad (3.24)$$

where $\alpha = \frac{1}{2} A_{J-1} \frac{\Delta t}{\Delta x}$, with $\alpha > 0$. Thus

$$| U_{J-1}^{n+1} | \leq \max. (|U_{J-1}^{n-1}|, |U_{J-2}^n|) \qquad , \qquad (3.25)$$

and the advanced values U_{J-1}^{n+1} remain bounded. Alternatively, if the local v. Neumann analysis is followed, then

$$\lambda = \frac{1-\alpha}{1+\alpha} \frac{1}{\lambda} + \frac{2\alpha}{1+\alpha} e^{-ik\Delta x}$$

and
$$|\lambda| \leq |\frac{1-\alpha}{1+\alpha}| \frac{1}{|\lambda|} + \frac{2\alpha}{1+\alpha} \quad .$$

Thus
$$-1 < - \frac{1-\alpha}{1+\alpha} < |\lambda| < 1 \qquad \text{if } \alpha < 1 \qquad ,$$

or
$$-1 < |\lambda| < \frac{\alpha-1}{\alpha+1} < 1 \qquad \text{if } \alpha > 1 \qquad ,$$

and computational stability can be expected.

Thus the local linear stability analysis will help to avoid unfortunate choices of unstable boundary conditions and will sometimes suggest appropriate choices to secure stable computation. It must be cautioned that if a partic-ular choice of the boundary condition fails to represent the physical situation, the computed stable solution need not be a good approximation to the genuine solution of the given physical problem. It is commonly found that oscillations of finite amplitudes appear to be generated at the various boundaries in a computed stable solution, and that such oscillations appear to propagate into the interior of the field of computation (or away from a shock wave or other interior boundary). These oscillations represent error components, superposed on the correct solution of the physical problem, and are likely introduced by the "errors" in the difference treatments of such boundaries. Indeed, there are also nonoscillatory errors caused by the difference treatments on the boundary; and such errors may actually be more serious because of their deceptively smooth appearance in the results of a stable computation. These errors tend to be overlooked, especially in view of the difficulty in securing a stable computation. An important aspect of studying the accuracy of computed results is to determine if the various boundary conditions are appropriate and to estimate the associated errors.

IV. IMPLICIT COMPUTATION AND RATE OF CONVERGENCE

Implicit difference algorithms generally lead to stable difference equations when applied to simple wave and simple diffusion equations, as indicated in Tables I and II. The local linear stability analysis for equation (3.3) illustrates further that stability is "improved" when the fraction θ of the spatial derivative evaluated at the advanced time level (and hence implicitly) is increased from zero to 1/2. The system becomes unconditionally stable when $\theta \geq 1/2$. Implicit difference algorithms are traditionally used in the solution of Laplace or Poisson equations without any problem of computational stability. The implicit difference algorithm, then, appears to be the most desirable from the point of view of avoiding computational instability, especially for complicated problems with mixed behavior. It will be demonstrated that the merit of the implicit schemes is not without reservation, since there may indeed be other difficulties involved which are as serious as computational instability.

With implicit difference algorithms, the difference relation at a given mesh point contains the unknown advanced values of quantities at neighboring mesh points. Thus, it is necessary to treat the system of difference relations at all of the interior mesh points simultaneously; and hence, the solution of the difference formulation based on a totally implicit scheme will require the inversion of matrices of very large dimension. This imposes severe requirements on the memory capacity and on the arithmetic speed of the computer, and also calls for skill in rendering efficient inversion of sparse but large matrices, (inevitably through some iterative procedure). The rate of convergence, or the number of iterations required to solve the systems of equations to a prescribed accuracy, is of great concern. This is because the computational effort required to complete a "sweep" over the field of computation (i.e., to advance the values of the functions at all mesh points for one time step) is generally much larger for the implicit difference formulation than for the explicit difference formulation. It is hoped, however, that in the absence of a stability limit with an implicit difference formulation, the time steps may be taken so many times larger than that allowed by the stability limit of the explicit formulation, as to more than compensate for the much larger computational effort per time step for the implicit formulation. In the following sections, this question will be examined.

4.1. Simple Time Dependent Problem

The advantage of the implicit formulation is best illustrated in the solution of time-dependent heat transfer problems in multispace dimension,

(4.1a), or in the solution of Laplace equations for the steady state problem, (4.1b).

$$\frac{\partial u}{\partial t} = \nu \nabla^2 u \qquad\qquad (4.1a)$$

$$0 = \nabla^2 u \qquad\qquad (4.1b)$$

For such problems, the system of simultaneous difference equations to be solved can be conveniently arranged in the form

$$AU = f \qquad , \qquad\qquad (4.2)$$

where U is the vector unknown representing the temperature at all the N interior mesh points, arranged in some appropriate order. Here f is a known vector of dimension N and A is an N x N tridiagonal matrix, often diagonally dominant. The solution of the system (4.2) for the unknown vector U is equivalent to the inversion of the matrix A, giving U as $U = A^{-1}f$. Computationally, a highly efficient method can be used to solve the system (4.2) with approximately 5N operation counts. This is to be compared with N counts for the solution of an explicit system. (Conventionally, each multiplication and division counts as one operation, while addition, subtraction and other data management operations are neglected. The evaluation of coefficients is ignored here on the presumption that the same amount of computational work is needed in both the implicit and the explicit cases). Thus, the computational effort to advance the solution for one time step with the implicit format is about 5 times as much as that with an explicit format. As illustrated in Table 2, most stable explicit schemes will possess a stability limit (easily verified by v. Neumann analysis) of the type $s = \dfrac{\nu \Delta t}{\Delta x^2} \stackrel{\sim}{<} (1/2, 1/4)$; here 1/4 is for two dimensional problems (see equation 3.16 with c = o.).

With good spatial resolution, i.e., a small Δx, the time step for the explicit scheme will be limited to $\Delta t \lesssim \frac{1}{2\nu}\Delta x^2$, which is indeed very small. Thus, if computation with the implicit formulation should be carried out with a time step larger than $\frac{5}{2\nu}\Delta x^2$ (or even say with $\Delta t = \Delta x$), considerable saving in the computational effort required for the determination of the temperature field U at a later time will result.

Such a benefit is illusory, however, if the determination of the solution at some specific later time is required to possess a specified accuracy. Assume that all variables are properly non-dimensionalized, and that it is required to achieve an accuracy of 10^{-2}. This accuracy is presumed to be solely dependent on the truncation error (i.e., all other errors are suppressed in the formulation and computation). Suppose that the explicit scheme 1 in table II is used, which is first order accurate in time and second order accurate in space (i.e., $e_t = 0(\Delta t, \Delta x^2)$). Then the field of computation, defined by $x = 0$ to 1 and $y = 0$ to 1 for a two dimensional problem from time $t = 0$ to 1, should be divided into at least 10 equal parts in both the x and the y directions, i.e., $\Delta x = \Delta y = 1/10$ - preferably say with $\Delta x = \Delta y = 1/20$ to allow some margin of safety. The stability limit will require a Δt (with time non-dimensionalized by the square of characteristic length divided by the diffusivity) as small as $\frac{1}{4}\Delta x^2 \approx 10^{-3}$ if $\Delta x = 1/20$, or $1/4 \times 10^{-2}$ if $\Delta x = 10^{-1}$. The relative magnitudes of Δt and Δx are such that the local truncation error $e_t = 0(\Delta t, \Delta x^2)$, and hopefully the computed results will remain consistent with the accuracy requirement. (In this case, the accumulation of the local truncation error will remain of the same order.)

Suppose that scheme (1) in Table II is now modified so that the spatial derivative is replaced by the implicit difference

$$s(\; U_{j+1}^{n+1} \; - \; 2U_j^{n+1} \; + \; U_{j-1}^{n+1} \; \cdots \;)$$

for both the x and y directions, with the same local truncation error $e_t = 0(\Delta t, \Delta x^2)$. This scheme (Laasonen) is unconditionally stable, i.e., Δt can be taken arbitrarily large without suffering computational instability. However, with Δt much larger than Δx^2, the local truncation error is of $0(\Delta t) \gg 0(\Delta x^2)$. Thus with $\Delta x = 1/20$, as in the explicit case, and with Δt taken as $\Delta x/5$, which is 16 times larger than the stability limit of the explicit scheme, the computational effort will be only $\sim 1/3$ of that with the explicit scheme. However, the solution so obtained is less accurate, with $e_t = 0(\Delta t)$ and $\Delta t = \frac{\Delta x}{5} = 10^{-2}$. This is marginally acceptable to the required accuracy 10^{-2}, allowing no room for the accumulation of the local truncation errors. Formally, the above solution from the implicit scheme should be compared with the solution from the explicit scheme taking $\Delta x = 10^{-1}$, with $e_t = 0(\Delta x^2)$ and $\Delta x^2 = 10^{-2}$. The computational effort of this explicit scheme is then actually 80% of the implicit scheme with the same local truncation error. Alternatively, if the implicit scheme is to produce a result with accuracy comparable to the explicit solution computed with $\Delta x = 1/20$ and $\Delta t = 1/4 \; x(\frac{1}{20})^2$, then the time step Δt for the implicit calculation should be taken at most as $\Delta x^2 = \frac{1}{400}$, so that $e_t = 0(\Delta t = \Delta x^2)$. Then the computational effort for the explicit format will again be 80% of that of the implicit scheme of comparable accuracy.

In the above example, the effectiveness of the implicit algorithm is largely nullified by the first order temporal accuracy of the difference scheme. It may be that implicit schemes with second order temporal accuracy will be more effective in reducing the overall computational effort, but such higher order schemes are cumbersome. From this point of view alone, the implicit schemes would certainly appear to be advantageous in the solution of steady state problems via asymptotic temporal approach, since the temporal accuracy is then of little concern. But, as will be discussed in the next section, it is not certain if such large temporal steps are conducive to rapid convergence to the steady state. It should be noted that, in the above example, the solution of the implicit formulation calls only for the inversion of a tridiagonal matrix, which can be implemented most efficiently in $\sim 5N$ operations. For fluid dynamics problems, the matrices resulting from an implicit formulation will be far more complex; and the solution of such matrix equations will be far more time consuming. It, therefore, appears prudent not to expect significant savings in computational effort by the use of implicit difference algorithms without some detailed investigation.

4.2 Iterative Solution of Steady State and Asymptotic Temporal Approach

Most of the fluid flow problems of practical interest are at a steady state or a quasi-steady state in which the temporal variations of the flow variables are negligible. Discretization of such steady state equations will generally lead to implicit difference relations in terms of the steady state values of various physical quantities at all the interior points and at the boundary points. Except for the solution of potential flow problems of incompressible fluids, the differential equations will be non-linear and considerably more complicated than the Laplace equation. The resulting implicit difference relations will give rise to a rather sparse matrix A, when written in the format of equation (4.2). This sparse matrix A will not, however, be

tridiagonal or block-tridiagonal, or of any of the other special forms convenient for the solution of the system of equations. In fact, the nonlinear
terms will first have to be quasi-linearized so that the coefficients in the
matrix A can be evaluated with some assumed approximate values. The system
of linear equations will then be solved iteratively until the solution from
(4.2) agrees with the assumed solution, under certain convergence criteria.

Let superscript n indicate quantities evaluated with the nth iterate
of U and use the system of difference relations (4.2) to calculate the
$(n+1)^{th}$ iterate. Then equation (4.2) becomes

$$A^n U^{n+1} - f^n = 0 \quad , \tag{4.3a}$$

which is indeed the same as

$$A^n (U^{n+1} - U^n) = f^n - A^n U^n \quad . \tag{4.3b}$$

Equation (4.3b) can now be considered as obtained from a time-dependent
equation in which the terms with spatial derivatives are the same as those in
the steady state equation (4.2), but with an added temporal term

$$\lim_{\Delta t \to 0} \Delta t \; A^n \; \frac{U^{n+1} - U^n}{\Delta t} \stackrel{\sim}{=} \Delta t \; A(u) \; \frac{\partial u}{\partial t}$$

and with a forward temporal difference quotient replacing $\frac{\partial u}{\partial t}$.

The iterative solution of a steady state problem based on an implicit
algorithm is then not substantially different from the solution of a time-
dependent problem, albeit the artificial temporal term may not correspond
to the temporal terms in the time-dependent form of the Navier-Stokes
equations. The physical meaning of the individual fictitious temporal terms
can be easily identified when the matrix operator A is written in expanded

form. The iterative index n can be identified with the temporal index n in the time-dependent formulation, although the equivalent time-dependent physical problem may contain artificial sources of mass, momentum and energy. These artificial sources are small, but are distributed over the entire field of computation - in the interior as well as on the boundary - and vanish in the steady state limit.

In the numerical solution of the Navier-Stokes equations in multi-space dimensions, there will be a few thousand mesh points and 4 or 5 unknown quantities at each mesh point. Thus, the dimension N of the vector U will commonly be $0(10^4)$ or larger. To solve equation (4.3a) for the successive approximations to the solution of the nonlinear equation (4.2) at each time (or iterative) step by the standard Gaussian elimination process, requiring $\sim \frac{1}{3} N^3$ operations per step, is out of the question. It is, therefore, imperative to develop highly efficient iterative methods. Thus in equations (4.3), the matrix operator A^n is split into two parts, with B^n operating on U^{n+1} and (A^n-B^n) operating on U^n. This gives

$$B^n U^{n+1} + (A^n-B^n) U^n = f^n \tag{4.4a}$$

or

$$B^n (U^{n+1} - U^n) = f^n - A^n U^n \quad , \tag{4.4b}$$

where B^n should be some easily invertible matrix so that U^{n+1} can be readily found. This U^{n+1} will replace U^n in the next iteration, until finally $U^{n+1} \stackrel{\sim}{=} U^n$ according to some steady state criterion. In this manner, the iterative solution of the quasi-linearized equation (4.3) has incorporated the iterations that were called for by the quasi-linearization of the nonlinear equation.

If B is chosen as the identity matrix I, equation (4.4b) is then identical to the explicit difference equation which would be obtained using a forward-time difference and the spatial difference algorithm of the implicit equation (4.3a). Thus, as in time-dependent explicit schemes, the iterative solution of equation (4.4b) with B = I corresponds to tracing the physical development in time of the flow field from an initial state toward the steady state. The local accumulations of mass, momentum and energy in the cell around each mesh point are precisely as they would be in the explicit scheme for time-dependent flows.

Alternatively, the matrix B may be chosen as diagonal, with its diagonal elements equal to the diagonal elements of A, i.e. $b_{ii} = a_{ii}$ and $b_{ij} = 0$ for $i \neq j$. Such an iterative process is then known as Jacobi iteration. Since $b_{ii} = a_{ii}$ are not identically unity, the temporal terms may be larger (or smaller) than the accumulation term in the physical, time-dependent flow. The excess (or deficiency) of a particular quantity U may be attributed to the presence of a source (or sink) of that quantity at the mesh point under consideration. These artificial sources (sinks) will tend to zero when the asymptotic steady state is approached.

If the matrix B is chosen to be the main tridiagonal elements of A, i.e. $b_{ij} = a_{ij}$ for $|i-j| \leq 1$ and $b_{ij} = 0$ for $|i-j| > 1$; then the artificial temporal terms will contain spatial derivatives. They then represent doublets and quadruplets around the mesh point. The situation is quite complicated algebraicially and physically, but it is very natural physically how a steady state may be reached via such time-dependent states provided that all of these sources, and doublets, etc., properly vanish in the steady state limit. In practice, the choice of B is dictated by the desire to reduce the computational effort required to obtain the steady solution, irrespective of its physical correspondence to some temporal flow field. The purpose here is to show that the asymptotic temporal approach and the iterative solution of the implicit formulation to obtain steady state results are fundamentally similar. The iterative method does take much more computational effort per iteration or per time step. But it permits the use of a much wider variety of temporal artifices to produce a very rapid convergence to the steady state, possibly resulting in less overall computational effort. It is possible, of course, that for some choices of B, there may not be any steady state solutions; or that the steady state solutions reached through such computational artifices may be different from the physical solution which one wished to obtain.

4.3 Iterative Methods

One of the most popular choices for the matrix B is the lower-triangular part of A i.e. $b_{jk} = 0$ if $k > j$ and $b_{jk} = a_{jk}$ if $k \leq j$. This is the Gauss-Seidel iteration or successive relaxation procedure. The $(n+1)^{th}$ iterate is given by

$$U_j^{n+1} = \frac{1}{a_{jj}^n} \left(f_j^n - \sum_{k=1}^{j-1} a_{jk}^n U_k^{n+1} - \sum_{k=j+1}^{N} a_{jk}^n U_k^n \right) \quad , \qquad (4.5a)$$

from which the successive scalar components of U^{n+1} can be explicitly calculated in the order of increasing j, where the latest available mesh values are used throughout. This semi-explicit solution of U^{n+1} can be given in matrix form as:

$$U^{n+1} = U^n + (B^n)^{-1} (f^n - A^n U^n) \quad , \qquad (4.5b)$$

where $(f^n - A^n U^n)$ is the residue and $(B^n)^{-1}$ is the inverse of the matrix B^n.

If the vector calculated from equations (4.5) is taken as a provisional solution, and if the new iterate U^{n+1} is evaluated as some weighted average of U^n and this provisional value, with weights $(1-\beta)$ and β respectively, then,

$$U^{n+1} = (1-\beta)\ U^n + \beta[U^n + (\ B^n)^{-1}(f^n - A^n U^n)] \tag{4.6a}$$

which is the same as

$$U^n+1 = U^n + \beta(B^n)^{-1}\ (f^n - A^n U^n) \tag{4.6b}$$

or $\qquad U^{n+1} = U^n + (B^n/\beta)^{-1}(f^n - A^n U^n)$

β is often called the acceleration (or relaxation) parameter. Equation (4.6b) suggests that β may be interpreted alternatively as a multiplier of either the residue in equation (4.5), or of the operator B^n operating on the artificial temporal sources in equation (4.4b). An appropriate choice of β is expected to effect a faster convergence of the iterative sequence, with the process referred to as successive over (or under) relaxation when $\beta > 1$ (or $\beta < 1$). For the integration of the Laplace equations in a rectangular domain, the optimal relaxation parameter β^* for the fastest convergence can be evaluated (as a function of the mesh spacing) and is usually around 1.8 - 1.5. For more complicated situations, the selection will have to be empirical and the optimal choice need not even be an over-relaxation. Unfortunate choices of β can lead to diverging sequences, even for Laplace equations (i.e. beyond $2 \geq \beta \geq 0$)..

Each Gauss-Seidel iterative step requires N^2 operations. This is to be compared with the count of $N^3/3$ for a Gauss elimination solution for the

quasi-linear steady state. The iterative solution would, therefore, be
advantageous if it converges within N/3 iterations, since the nonlinear
iterations for the solution of equation (4.3) would then be avoided. Now
with $N = 0(10^3)$, it is hoped that by proper choice of the relaxation parameter β, much fewer iterations than N/3 may be needed to reach a steady
state. In principle, if the steady state is defined by $||U^{n+1} - U^n||/||U^n||$
$< 10^{-m}$, the number of iterative steps required for convergence can be esti-
mated by m/R, where R is the rate of convergence with $R \overset{\sim}{=} \log_{10}(\frac{1}{\rho})$, and
ρ is the geometric mean of the spectral radii of the matrices $(B^n)^{-1} A^n$ at
successive iterative steps n. Such an estimate of R is not possible in practice because of the complexities of the matrix A and its dependence on the
solution U^n.

For the integration of some form of the hydrodynamic equations, it is not
uncommon that hundreds of such iterations are needed. This is partly due to
the nonlinearity of the equations system and partly due to non-optimal choices
of the relaxation parameter. It is also true that such iterations often fail
to converge, despite a wide range of choices of the acceleration parameter.
Now, if the physical state of the flow is steady or quasi-steady, the asymptotic temporal approach using the correct time-dependent equations (B=I) may
be expected to converge on purely intuitive grounds, provided that the difference system is stable and consistent with the time-dependent Navier-Stokes
equations. But when the implicit iterative method is used, its convergence
to the steady state cannot be presumed on physical grounds, since artificial
sources of mass, momentum and energy are introduced purely algebraically.
The particular temporal variations of these sources need not provide any steady

state, although in cases without such external artificial sources, nature
has demonstrated that a steady state will eventually be reached. It might
even be legitimate to question whether a steady state so reached would be
the same as one reached under zero external sources, since the time integrals
of the artificial sources may appreciably alter the integrals of motion of
the system. It is regrettable that no useful answer can be derived physically.

Mathematically speaking, the matrix B in equation (4.4b) can be quite
arbitrarily chosen, and can even be selected differently for different steps.
Convergence to a steady state is assured provided that

$$\lim_{n \to \infty} (B^n)^{-1}(B^{n-1})^{-1} (B^{n-2})^{-1} \ldots (B^1)^{-1}(B^0)^{-1} = 0 \qquad (4.7a)$$

and
$$\lim_{n \to \infty} [(B^n)^{-1}A^n][B^{n-1})^{-1} A^{n-1})] \ldots [(B^0)^{-1}(A^0)] = 0 \quad . \qquad (4.7b)$$

These relations can be secured if the spectral radii of all $(B^n)^{-1}$ and all
$(B^n)^{-1}A^n$ are less than unity. If the form of B chosen should be the same
for all iterations, (4.7b) is not really much different from the local
linearized stability criterion of v. Neumann, with $B^{-1}A$ replacing the amplification matrix G. (See Chapter III, Section 3.1 and 3.2). The essential
difference lies, then, in the freedom of choice of the form of the matrices
B^n at different iterative steps $\dot{n}$. It is not clear whether condition (4.7a)
implies the physical requirement of conservation of the integrals of motion.
It is also not practical to find the spectral radii or bounds on the eigenvalues of these complicated matrices. There is no counterpart of the local
linear stability analysis to provide some idea of the rate of convergence in
a complicated problem. There is only the practical alternative of trying it
out on the computer.

In practice, the possible choices of the form of B are severely limited
to those which are easily invertible. It is difficult to find such a choice
that shows significant improvement over the optimal overrelaxation process,
if inferences from the study of the solution of Laplace equations are a
reliable guide. A further substantial reduction of operational counts per
iterative step can be derived, however, from cyclic processes built upon the
Gauss-Seidel iterative procedure. If the field of computation constitutes
p columns of q elements per row, with $p \cdot q \overset{\sim}{=} N$, the matrix B may be chosen as
block-lower-triangular so that each of the q blocks consists only of the p
(or q) elements in each column (row). Then $B^{n-q+1} B^{n-q+2} \ldots B^n$ can be taken
as the lower triangular matrix in successive blocks, with zero elements everywhere else. This is the line Gauss-Seidel process operating on successive
columns (or rows). Such a line process can be accelerated by employing some
proper acceleration parameter.

The line processes along columns and rows (or diagonals or other convenient directions) may be employed in succession, such as the sequence of
operators $(B^{n-q-p+1} \ldots B^{n-q})(B^{n-q+1} \ldots B^n)$ and its cyclic repetition. A
set of acceleration parameters may be employed with the cyclic column-row
sequence. This is known as the alternating direction method, or the method of
Peaceman and Rachford[6] (who first demonstrated the power of such cyclic
iterative methods for the solution of Laplace equations). Such line methods
derive the benefit of reduced computational work from the basic fact that the
operational count of the Gauss-Seidel process is proportional to the "square"
of the vector length of the unknown. Thus the operational count of a complete
cycle is, with $p = q = N^{1/2}$ for example,

$$p \cdot q^2 + q \cdot p^2 = (p+q)pq \sim 2N^{3/2} \quad , \tag{6.8}$$

compared to N^2 for the point Gauss-Seidel process. This means a decrease in the number of operations per sweep by the factor $2/N^{1/2}$, significant to an order of magnitude with $N = O(10^3)$. The extension of such a cyclic process to problems in three space dimensions with $p \sim q \sim r \sim N^{1/3}$ is obvious, in which case, the total operational count per cycle is $\sim 3N^{4/3}$, and the factor of operational count reduction will be $3/N^{2/3}$. The advantage enjoyed by the alternating direction iteration (ADI) or any such cyclic line iteration process over the point Gauss-Seidel relaxation process is clear. Success in reducing the overall computational effort in the solution of steady flow problems with such schemes requires, in addition, an appropriate choice of the acceleration parameters, suitable for the type of problems at hand and the class of prescribed boundary data. This is where the uncertainty resides.

For the solution of Laplace equations, the optimal acceleration parameters and the maximum rates of convergence of these processes can be explicitly determined. The ADI process is certainly the most efficient. This is likely to be true for the integration of purely elliptical equations, especially those with the Laplace operator as the leading term. For more complicated equations, including the equations of hydrodynamics, success depends on the ability to select appropriate acceleration parameters for the problem at hand and on a proper implementation of the boundary conditions. For hyperbolic problems with discontinuous solutions as interior boundaries, success with implicit methods is yet to be demonstrated.

4.4 Fractional Time and other Alternating Direction Methods

An alternating direction iterative method, known as the time splitting or fractional time step method, has been developed extensively in the Soviet Union by Yanenko, Marchuk, etc.[5] The key idea is to split the operator into a sum of implicit difference operators, each of which should lead to an easily invertible tridiagonal matrix. The successive split operations in a complete cycle serve as a "weak" approximation to the original operator. They prefer the unconditional computational stability and formal second-order accuracy of the Crank-Nicholson algorithm, as illustrated in Equation (3.6) when θ takes the value $\frac{1}{2}$. Second-order accuracy is needed, since a first order-accurate scheme can hardly meet the accuracy requirement of practical problems with the currently available computing machine. The development of this method and its relative merits when applied to gas dynamic applications is presented below.

Consider first the equation

$$\frac{\partial \phi}{\partial t} + L\phi = 0 \quad , \tag{4.9}$$

where L is a linear spatial differential operator, explicitly independent of time t. Discretizing with the Crank-Nicholson algorithm (which is second-order accurate in both time and space) gives

$$\frac{\phi^{n+1} - \phi^n}{\Delta t} + L \left(\frac{\phi^{n+1} + \phi^n}{2} \right) = 0 \quad . \tag{4.10}$$

Letting I be the identity operator, we have

$$(I + \frac{\Delta t}{2} L) \phi^{n+1} = (I - \frac{\Delta t}{2} L) \phi^n$$

or

$$\phi^{n+1} = (I + \frac{\Delta t}{2} L)^{-1} (I - \frac{\Delta t}{2} L) \phi^n = C \phi^n \quad .$$

For the simple heat diffusion equation $L_x \sim -\sigma \frac{\partial^2}{\partial x^2}$, the matrix $(I + \frac{\Delta t}{2} L_x)$ is tridiagonal, and the spectral radius of the matrix C can be obtained as

$$\rho(C) = \frac{1-S\psi}{1+S\psi} \quad \text{with} \quad \psi = 4 \sin^2 (\frac{\pi}{2} \frac{j}{j+1})$$

where $\quad 0 \leq x = j\Delta x \leq (J+1)\Delta x$.

Thus

$$||\phi^{n+1}|| \leq \frac{1-S\psi}{1+S\psi} ||\phi^n|| \leq \cdots \leq (\frac{1-S\psi}{1+S\psi})^n ||\phi^0|| \quad ,$$

which establishes the boundedness and unconditional stability. Indeed, with $\Delta t/\Delta x$ taken as constant, the computational error is bounded as $||e|| \leq ||e_o|| + 0(\Delta x^2)$. When the C.F.L. condition $\Delta t/\Delta x \leq 1$ for wave equations is satisfied, this scheme is expected to work for both the diffusion and the wave equations, and is hoped to work for Navier-Stokes type equations, at least in the 1-D case. (As shown in section 3.3, it does not necessarily follow that an algorithm successful for the wave and diffusion equations individually will automatically succeed for the Navier-Stokes.)

Consider now the heat diffusion problem in three space dimensions

$$\frac{\partial \phi}{\partial t} - \sigma (\frac{\partial^2}{\partial x^2} + \frac{\partial^2}{\partial y^2} + \frac{\partial^2}{\partial z^2}) \phi = 0,$$

and take $L = L_x + L_y + L_z$ or $L_1 + L_2 + L_3$. While the operators $(I + \frac{\Delta t}{2} L_x)$ can be easily inverted, the combined matrix $[I + \frac{\Delta t}{2} (L_x + L_y + L_z)]$ is no longer tridiagonal and, though highly sparse, cannot be simply inverted. So the equation is integrated in three successive steps for the time interval $t_n \leq t \leq t_{n+1}$, these fractional steps being formally designated as $t_{n+1/3}$, $t_{n+2/3}$ and $t_{n+3/3} = t_{n+1}$. For the step at $t_{n + \alpha/3}$, the Crank Nicholson algorithm

$$\phi^{n+\alpha/3} = (I + \frac{\Delta t}{6} L_\alpha)^{-1} (I - \frac{\Delta t}{6} L_\alpha) \phi^{n + \frac{\alpha-1}{3}}$$

is used, giving for the complete cycle

$$\phi^{n+1} = \prod_{\alpha=1}^{3} (I + \frac{\Delta t}{6} L_\alpha)^{-1} (I - \frac{\Delta t}{6} L_\alpha) \phi^n$$

$$= \left\{ I - \Delta t\, L + \frac{\Delta t^2}{2} [L^2 + \sum_{\alpha=1}^{3} \sum_{\beta=\alpha+1}^{3} (L_\alpha L_\beta - L_\beta L_\alpha) + \ldots] + 0(\Delta t^3) \right\} \phi^n$$

and $\quad \phi^{n+1} = \left\{ (I + \frac{\Delta t}{2} L)^{-1} (I - \frac{\Delta t}{2} L) + 0 (\Delta t^3) \right\} \phi^n$

if $L_\alpha L_\beta$ is commutable.

Thus the split difference scheme will be second order accurate if the split operators are commutable; otherwise, it is only first order accurate. When such commutativity of the split operators for different dimensions (x,y, & z) does not hold, the split scheme of only first order accuracy can be arranged to yield second-order accurate results in two cycles if the second cycle is repeated in the opposite order. For the two consecutive cycles, i.e.

$$\phi^{n+1} = \prod_{\alpha=1}^{3} (I + \frac{\Delta t}{6} L_\alpha)^{-1} (I - \frac{\Delta t}{6} L_\alpha) \phi^n$$

and

$$\phi^{n+2} = \prod_{\alpha=3}^{1} (I + \frac{\Delta t}{6} L_\alpha)^{-1} (I - \frac{\Delta t}{6} L_\alpha) \phi^{n+1},$$

the two non-commutative terms cancel and the second order accuracy is resumed for non-commutative operators L_α. This statement will be true even if the L_α involve differential operators with varying coefficients or if they depend on ϕ (as for gas dynamic quasi-linear equations), so long as such coefficients are smooth and are treated properly.

For unconditional stability, it is required that the operator L and the split operators L_1, L_2 & L_3 be semi-positive definite, that is, that the inner product $(L\phi, \phi)$ be ≥ 0 for any arbitrary function and be defined over the entire field of computation. This condition is crucial in securing unconditional computational stability. Taking the norm of ϕ^{n+1}, and defining the norm of an operator as the natural norm induced by any vector norm, gives

$$||\phi^{n+1}||^2 = \frac{[(I + \frac{\Delta t}{2} L)^{-1} (I - \frac{\Delta t}{2} L) \phi^n, (I + \frac{\Delta t}{2} L)^{-1} (I - \frac{\Delta t}{2} L) \phi^n]}{(\phi^n, \phi^n)} \cdot ||\phi^n||^2$$

Defining

$$(I + \frac{\Delta t}{2} L)^{-1} \phi^n = \xi^n,$$

then

$$||\phi^{n+1}||^2 = \frac{||(I - \frac{\Delta t}{2} L) \xi^n||^2}{||(I + \frac{\Delta t}{2} L) \xi^n||^2} \, ||\phi^n||^2 = \Lambda^2 ||\phi^n||^2.$$

It follows that

$$||(I - \frac{\Delta t}{2} L) \xi^n||^2 = [(I - \frac{\Delta t}{2} L) \xi^n, (I - \frac{\Delta t}{2} L) \xi^n]$$

$$= ||\xi^n||^2 - \Delta t \, [L(\xi)^n, \xi^n] + \frac{\Delta t^2}{4} ||L(\xi^n)||^2$$

$$||(I + \frac{\Delta t}{2} L) \xi^n||^2 = [(I + \frac{\Delta t}{2} L) \xi^n, (I + \frac{\Delta t}{2} L) \xi^n]$$

$$= ||\xi^n||^2 + \Delta t \, [L(\xi^n), \xi^n] + \frac{\Delta t^2}{4} ||L(\xi^n)||^2 \, .$$

Here Λ^2 corresponds to the square of the spectral radius $\rho(C)$ for the simple 1-D heat diffusion problem with the Crank-Nicholson algorithm. Since both

$||\xi||^2$ and $||L(\xi)||^2$ are positive, and since $\Delta t > 0$, the amplification factor Λ^2 will be $\geq$ or $<$ unity depending on whether $(L\ (\xi),\ \xi) \geq$ or < 0.

The successive application of split operators at each step leads to

$$||\phi^{n+1}||^2 = \Lambda_1^2 \Lambda_2^2 \Lambda_3^2 ||\phi^n||^2.$$

The conditions for Λ_1^2, Λ_2^2 & Λ_3^2 to be less than or equal to unity are the same as those required for the semi-positive definiteness of the split operators L_1, L_2 & L_3; i.e. $(L_\alpha \phi, \phi) \geq 0$ for $\alpha = 1,2,3$. This semi-positive definiteness is a sufficient (but not necessary) condition for unconditional stability of a complete computational cycle.

Now if this restriction of semi-positive definiteness is enforced, the applicability of the split scheme will be practically limited to the simple diffusion equation or the Laplace equation in a rectangular domain with Dirichelet boundary conditions. It seems intuitively logical that the method should be applicable to a wider class of circumstances than those for which proofs have been given. This situation is really no better off than that encountered in the question of stability for explicit schemes. Indeed, there is not even a necessary criterion for computational stability of this split method, comparable to the von Neumann stability criterion for explicit schemes.

In the theoretical treatment of gas dynamic flows by Marchuk and Yanenko[5] the semi-positive definiteness condition is satisfied by imposing special "periodic" boundary conditions on the problem; in which case it is clear that

$$(L\phi, \phi) = 0$$

so that $\Lambda^2 = \Lambda_1^2 = \Lambda_2^2 = \Lambda_3^2 = 1$, and the norm is preserved:

$$||\phi^{n+1}|| = ||\phi^n|| = \ldots = ||\phi^\circ||$$

This appears to be an excellent feature for initial value problems. But
it also implies, for example, that any error in the initial data (if it is a
guess) will not decrease (in the mean square norm) at later times. Therefore,
the splitting scheme should not be used to obtain steady-state solutions with
periodic boundary conditions, because the results will never be better (within
the integral norm) than the initial guess.

For treating practical problems, the physical significance of this
stability requirement $(L\phi,\phi) \geq 0$ needs to be more carefully examined.
Post-multiplying the equation

$$\frac{\partial \phi}{\partial t} + L\phi = 0$$

by ϕ, and summing (or integrating) over the entire field of computation gives

$$\frac{\partial}{\partial t} ||\phi||^2 + (L\phi,\phi) = 0,$$

i.e.

$$\frac{\partial}{\partial t} ||\phi||^2 = - (L\phi,\phi).$$

If L is semi-positive definite, i.e. $(L\phi,\phi) \geq 0$, then $-\frac{\partial}{\partial t} ||\phi||^2 \leq 0$.
This of course implies the boundedness of the solution   at all times and
guarantees a decreasing sequence of $||\phi||^2$. Now the gas dynamic equations
(in primary physical variables) are conservation laws for mass, momentum, and
energy. $L\phi$ is the net flux of these conserved quantities out of unit physical
volume. If ϕ, and hence $L\phi$, are periodic over a parallelopiped in physical
space in order to secure $(L\phi,\phi) = 0$, then the outflux and the influx across
the boundary of computation exactly balance. Thus, with ϕ identified as mass,
momentum and energy, this condition excludes the loss of these quantities

throughout the entire field of computation. This means that the computed results should not be expected to show body forces acting on some immersed body, or heat transfer to or from the body. Thus any lift, drag and heat transfer that may be presented in the computed results must originate from some computational artifices, and are physically meaningless.

If now the flow field is computed with periodic boundary conditions in the transverse plane, then $\Lambda_2^2 = \Lambda_3^2 = 1$. A deficit in the out-flux $L\phi$ (when there is a body drag or an energy sink to the body) and a positive ϕ will render $(L_1\phi,\phi) \leq 0$, and hence $\Lambda_1^2 > 1$. Thus $\Lambda^2 = \Lambda_1^2 > 1$, and the computation will be "unstable." To secure computational stability under such circumstances, it is necessary to modify the boundary conditions in the transverse plane, so that Λ_2^2 & Λ_3^2 are sufficiently smaller than unity to render $\Lambda_1^2 \Lambda_2^2 \Lambda_3^2 < 1$. Guidance is badly needed here for handling these boundary conditions properly to secure computational stability with the split schemes. And even if computational stability is achieved by some means, there is little idea how the calculated results of such important quantities as body drag, lift, and heat transfer will compare with the physical situation.

The fractional time step method outlined above was not meant to be applied to steady state problems, because each of the interative solutions ($\phi^{n+1/3}$, $\phi^{n+2/3}$ and ϕ^{n+1}) satisfy different equations, none of which approximate the steady state equations. Even if the exact steady state solution of a given problem is used as the initial data, the fractional time step method will generate solutions that will not quite settle down to any sort of a steady state limit. This situation can be remedied by retaining the terms that were dropped in the fractional step method, and evaluating them with the previous (or otherwise known) iterate. One possible method is:

$\phi^{n+\alpha/3} = (I + \frac{\Delta t}{6} L_\alpha)^{-1} \{I + \frac{\Delta t}{6} L_\alpha - \frac{\Delta t}{3} \sum_\alpha L_\alpha\} \phi^{n+\frac{\alpha-1}{3}}$, which would indeed be the same as the alternating direction iterative solution of the implicit formulation of the steady state problem with B = $(I + \frac{\Delta t}{6} L_\alpha)$ and   A = $\frac{\Delta t}{3} \sum_\alpha L_\alpha$ given as equation (4.4a). This method is then similar to the Douglas and Gunn [7] extension of the Peaceman-Rachford Alternating Direction method for the solution of steady state problems. With these additional terms, it is not possible to conjecture what the stability behavior and the convergence rate of the difference formulation will be. Some experience of the research group at Langley Research Center, NASA (the author is grateful for this private communication) indicates that the overall computational effort in using such schemes (for the numerical integration of the Navier-Stokes Equations for some mixed supersonic-subsonic flow fields) is much larger than that experienced by the author on similar problems with explicit formulation.

While no generalization is implied, the fundamental reasons expounded in this and the previous sections, coupled with some practical experience, serve as an appropriate caution against being overly optimistic about the advantage of such implicit methods.

The split operator L_x can be split further as $L_x = L_{xc} + L_{xv}$ (for example), where L_{xc} is the convective part and L_{xv} the viscous part of L_x. In this manner, each momentum equation is split into 6 parts, and each of the 6 parts give rise to either a wave operator or a diffusion operator. The question of rendering a stable computation for each step is somewhat simplified in the interior. There will be difficulties in formulating the boundary conditions and in achieving higher order accuracy, especially for time dependent problems. Future developments in such split schemes can be important, however, for computing flows in 3-space dimensions.

V. ACCURACY AND CONSERVATIVE FORMULATION

The physical conservation laws of mass, momentum and energy are
established for arbitrary macroscopic volumes of a homogeneous fluid. By
reducing the volume to a macroscopically small "point," but a microscopically
large domain (to justify the continuum model) the Navier-Stokes partial
differential equations are derived. They are used as convenient mathematical
relations governing smooth point functions in the flow field. Now, to facilitate the numerical integration of the partial differential equations system,
the Navier-Stokes equations are discretized into a system of difference
equations for finite elements of a spatial domain. Such difference equations
may as well be obtained directly from consideration of the fundamental physical
laws for such finite discrete spatial domains (with the help of interpolation
formulas). It is, however, more common that discretization is effected by
replacing a differential coefficient with a difference quotient according to
a Taylor series truncated to some order of accuracy. The errors associated
with the interpolation formula or the truncated Taylor series are called
truncation errors, some of which are given as e_t in Tables I and II. The
mathematical requirement of consistency means simply that the truncation error
will vanish as $\Delta t, \Delta x \to 0$.

The conservation laws for each finite spatial element are properly
approximated to some formal order of accuracy (given by the truncation error)
by the difference equations deduced in either manner mentioned above. However,
when the difference forms of such conservation laws are summed over a large but
arbitrary collection of such finite spatial elements, the conservation laws

may be seriously violated. This is because the small higher-order errors will accumulate when summed over the very large number of the small discrete elements which make up the finite domain of computation. Now for an appropriate description of a physical problem to the accuracy of say $0 \, (\Delta x^2)$, it is essential that such conservation laws should be accurate to $0(\Delta x^2)$ over not only the differential elements, but also over finite volumes. If the truncation errors of the conservation laws in finite space are to be of $0(\Delta x^2)$, the errors must not accumulate when neighboring mesh cells are summed up. If the truncation errors are allowed to so accumulate, the difference formulation used should be higher-order accurate, so that the accumulation of such small higher-order truncation errors over arbitrary mesh combinations throughout the field of computation will not exceed $0 \, (\Delta x^2)$. But the difference form of Navier-Stokes equations, uniformly accurate to better than $0 \, (\Delta x^2)$, is extremely cumbersome to construct and execute. Thus, with the limited spatial resolution currently available, it is imperative to pervent or limit the accumulation of truncation errors.

It is highly commendable to (a posteriori) verify the extent to which the computed results conserve mass, momentum and energy over the entire field of computation. This is not an alternative to requiring no accumulation of the truncation errors. The truncation errors are generally representative of dipoles or quadruples, rather than of simple sources or sinks. They distort the local flow field much more than they cause apparent deviations in the overall mass, momentum and energy balances. The consequence of such dipoles and the like is, indeed, familiar to aerodynamicists. A circular cylinder in a uniform incompressible flow can be represented by a doublet. A thin airfoil or a thin wing in a subsonic or supersonic flow can be represented by some

distribution of sources and sinks or dipole pairs within the framework of the
linearized theory known as the method of singularities. If a series of tiny
little vanes or thin sheets is not to be tolerated in the test section of a
windtunnel, then the distributed dipoles arising from the truncation errors of
every computational cell must be correspondingly suppressed (if not completely
eliminated) in numerical solutions. . Such suppression can be achieved with some
close attention to the formulation of the difference problem.

5.1 Conservative Difference Formulation

The conservation relations are written in divergence from as Equations
(1.1) to (1.3) for the density ρ, the momentum ρu_i, and the energy density e
per unit volume. These five quantities are the scalar components of the vector
function V in Equation (1.6), etc., and will be considered as the "Primary
Dependent Variables" in terms of which the physical laws are stated and the
practical results desired. The conservation laws provide the integrals of
motion when proper initial and boundary data are specified over a specific
but arbitrary volume. When neighboring volumes are summed, the contributions
on their common boundary cancel identically, so that the integrated conservation
laws retain the same form. This is the crucial property that enables the
integral theorems of Stokes and Green to cast the conservation principles into
field descriptions in terms of different variables (Dia. 1). An adequate
approximation of the conservation laws in difference form should preferably
retain this property, at least to the order of accuracy required. Such a
summable property is implicit in the mathematical abstractions of continuity
and differentiability of the functions in question. Thus, the differential
formulations in terms of various dependent and independent variables are all
equivalent, although the forms of the partial differential equations may be

CONSERVATION LAWS OF SOURCE-FREE FLUID FLOW FOR ARBITRARY VOLUMES

BOUNDED BY SURFACE S WITH FLUXES F CROSSING BOUNDARIES

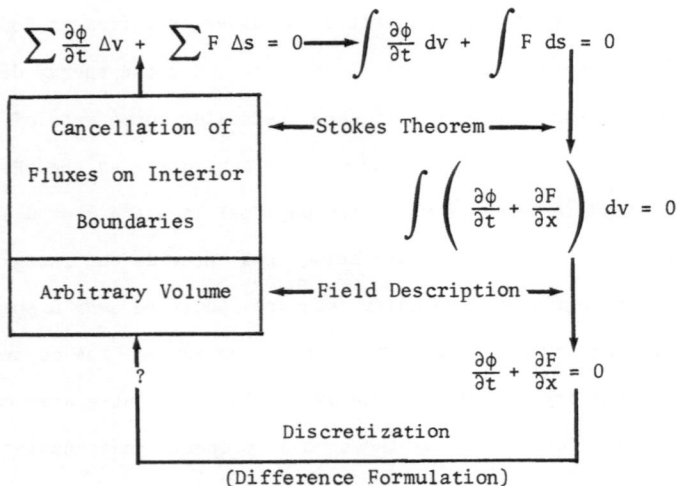

DIAGRAM 1

much different. This is not the case for the difference approximations of the conservation laws that may be formally "derived" from the varieties of forms of equivalent partial differential equations. This is because the difference functions are discrete or at least not differentiable beyond a certain order when the discrete values are joined by locally smooth functions. A summable difference formulation - in the sense that when cells in the field of computation are summed, the fluxes in the physical space (x_j) of the primary dependent variables cancel identically along their common boundary - will be called a "Conservative Difference Formulation".[3,10] The computational space need not be the physical space, and the dependent variables computed need not be the primary ones. Nevertheless, the fluxes in physical space and of the primary variables are still those that are required to be summable for the conservative difference formulation.

For illustrative purposes, consider the discretization of the continuity relation from the integrated conservation law expressed in the primary variables ρ, ρu, and ρv, in the two dimensional physical space (x,y) divided into uniform rectangular cells $\Delta x \, \Delta y$. $\rho_{j,k}$ is the average density of the fluid in the cell $j\Delta x$, $k\Delta y$. The net increase of mass in the cell during Δt is $(\rho_{j,k}^{n+1} - \rho_{j,k}^{n})\Delta x \cdot \Delta y$. The mass fluxes of ρU and ρV should be evaluated on the boundary, but ρU and ρV are known only as the average momenta of the fluid in the cells. Thus the boundary fluxes are evaluated through (second order accurate) linear interpolation, as the arithematic average of the mean momentum in neighboring cells. If increasing j and k are taken as the positive directions, the conservation of mass is stated as:

$$(\rho_{j,k}^{n+1} - \rho_{j,k}^{n})\Delta x \Delta y + \frac{\Delta t \Delta y}{2} \left\{ \left[(\rho U)_{j+1,k}^{n} + (\rho U)_{j,k}^{n} \right] - \left[(\rho U)_{j,k}^{n} + (\rho U)_{j-1,k}^{n} \right] \right\}$$

$$+ \frac{\Delta t \Delta x}{2} \left\{ \left[(\rho V)_{j,k+1}^{n} + (\rho V)_{j,k}^{n} \right] - \left[(\rho V)_{j,k}^{n} + (\rho V)_{j,k-1}^{n} \right] \right\} = 0$$

$$(5.1)$$

For the neighboring cell $(j-1)\Delta x.k\Delta y$, the difference form of mass continuity

relation can be obtained from (5.1) by replacing j by j-1. The two cells

have a common boundary at $(j-\frac{1}{2})\Delta x.k\Delta y$. The outflux from the cell $(j-1,k)$

crossing this common boundary is $\frac{1}{2}[(\rho U)^n_{j,k} + (\rho U)^n_{j-1,k}]$, which is identically

the same as in the influx to the cell (j,k). Thus, when the two mass

continuity equations (5.1) for the cells (j,k) and $(j-1,k)$ are added, the flux

terms across the common boundary cancel out. The resulting difference equation

is identical to the one obtained when the conservation law is applied directly

to the combined cells, and is accurate to $O(\Delta x^2)$. The addition of other

neighboring cells behaves in the same manner. Similar results will be obtained

for the momentum and the energy relations; thus a conservative difference

formulation accurate to $O(\Delta x^2)$ is formulated. It is easily verified that the

same difference formulation will be obtained with the forward-time, centered-

space difference algorithm applied to the differential equations system (1.1)

to (1.3) written in divergence form. Indeed, the first order accurate

algorithm of backward or forward spatial differences will also yield a conser-

vative difference formulation, but of first order accuracy of $O(\Delta x)$, provided

that the differential equation is discretized in divergence form and that the

physical space is divided uniformly.

If the continuity equation should be written in expanded form for discre-

tization, such as $u\frac{\partial\rho}{\partial x} + \rho\frac{\partial u}{\partial x}$ for the net mass flux in the x-direction, the

centered-space difference algorithm can represent the net x-flux as

$$\frac{\Delta t\Delta y}{2}\left[U_j(\rho_{j+1} - \rho_{j-1}) + \rho_j(U_{j+1} - U_{j-1})\right] \tag{5.2a}$$

or as $\dfrac{\Delta t\Delta y}{4}\left[(U_{j+1} + U_{j-1})(\rho_{j+1} - \rho_{j-1}) + (\rho_{j+1} + \rho_{j-1})(U_{j+1} - U_{j-1})\right]_{(5.2b)}$

The influx to the cell (j,k) from the cell (j-1,k), crossing the boundary at $(j - \frac{1}{2})$ Δx is, (respectively):

$$\frac{\Delta t \Delta y}{2} \left[U_j \rho_{j-1} + \rho_j U_{j-1} \right] \qquad (5.3a)$$

or $\qquad \frac{\Delta t \Delta y}{4} \left[(U_{j+1} + U_{j-1}) \rho_{j-1} + (\rho_{j+1} + \rho_{j-1}) U_{j-1} \right] \qquad (5.3b)$

The outflux from the cell (j-1,k) into the cell (j,k) crossing the same common boundary, as may be obtained from Equation (5.2a)(5.2b) by putting $j \rightarrow j-1$, is (respectively):

$$\frac{\Delta t \Delta y}{2} \left[U_{j-1} \rho_j + \rho_{j-1} U_j \right] \qquad (5.4a)$$

or $\qquad \frac{\Delta t \Delta y}{4} \left[(U_j + U_{j-2}) \rho_j + (\rho_j + \rho_{j-2}) U_j \right] \qquad (5.4b)$

The outflux (5.4a) is identical to the influx (5.3a), and will cancel when the two cells are summed. Thus the difference algorithm (5.2a) will lead to a conservation difference formulation, without the differential equation being written in divergence form. But the outflux (5.4b) is different from the influx (5.3b). When the two cells are summed up, they do not cancel completely, but produce a net mass source of magnitude proportional to $\Delta x \cdot \Delta y \cdot \Delta t$ along the common boundary. This is formally negligible in a second-order accurate algorithm, but renders the difference formulation from algorithm (5.2b) not summable and not conservative. Even if such errors accumulate randomly over a field of computation with $1/\Delta x^2$ meshes, the accumulated truncation error will be $O(\Delta x)$ rather than $O(\Delta x^2)$. If the first order accurate backward or forward spatial difference algorithm is used for discretizing $u \frac{\partial \rho}{\partial x} + \rho \frac{\partial u}{\partial x}$ the net x-flux will be:

$$\frac{\Delta t \cdot \Delta y}{2} \left[U_j(\rho_j - \rho_{j-1}) + \rho_j(U_j - U_{j-1}) \right] \tag{5.5a}$$

or

$$\frac{\Delta t \cdot \Delta y}{2} \left[U_j(\rho_{j+1} - \rho_j) + \rho_j(U_{j+1} - U_j) \right] \tag{5.5b}$$

Neither of the two will lead to a conservative difference formulation, even to an accuracy of $O(\Delta x)$. The above examples demonstrate that both the centered difference algorithm and the divergence form of the differential equation are conducive to a conservative difference formulation with uniform mesh size in physical space. On the other hand, the difference formulation based on integrated conservation laws, even with only linear interpolation, leads straightforwardly to a conservative difference form of second order accuracy.

Consider now the effect of nonuniform mesh sizes in physical space, with

$$\frac{(\Delta x)_{j+1}}{(\Delta x)_j} = \eta_{j+1/2} \qquad \text{and} \qquad \frac{(\Delta x)_j}{(\Delta x)_{j-1}} = \eta_{j-1/2}$$

when the integrated conservation laws and linear interpolation are used to discretize the continuity relation. The net flux into the cell at $j\Delta x$, during the time interval Δt, is obtained in a straightforward manner (illustrated here only for x-fluxes):

$$+ \Delta t \cdot \Delta y \left[\frac{\eta_{j+1/2}}{1+\eta_{j+1/2}} (\rho U)_j + \frac{1}{1+\eta_{j+1/2}} (\rho U)_{j+1} \right] \tag{5.6}$$

$$- \Delta t \Delta y \left[\frac{\eta_{j-1/2}}{1+\eta_{j-1/2}} (\rho U)_{j-1} + \frac{1}{1+\eta_{j-1/2}} (\rho U)_j \right].$$

The first bracket represents the outflux from cell j, and the second bracket represents the influx to cell j. If, in the first bracket, j is replaced by j-1, then the outflux from the cell at $(\Delta x)_{j-1}$ becomes identical to the influx into the cell at $(\Delta x)_{j}$, across their common boundary . They therefore cancel when the two cells are summed. Thus the algorithm (5.6) will, in physical space, lead to a conservative difference formulation despite the variable spacing. Algorithm (5.6) clearly indicates how the centered spatial difference algorithm should be modified to accommodate variable physical spacing, in order to achieve a conservative difference formulation and second-order accuracy. The choice of this particular weighted average of $(\rho U)_{j+1}$, $(\rho U)_{j}$, and $(\rho U)_{j-1}$ is, however, not obvious from the point of view of discretizing $\frac{\partial}{\partial x}$ (ρu) with second-order accuracy through Taylor series expansions without the consideration of fluxes on the boundary.

Variable mesh sizes in physical space are commonly achieved through some transformation $x = x(\xi)$ of the independent variables, or inversely as $\xi = \xi(x)$. The difference formulation is then derived from the transformed differential equation by discretizing with a uniform mesh spacing $\Delta\xi$ in the transformed ξ-space, according to some difference algorithm. This transformation of the spatial coordinates is often suggested by the desire to bring the boundaries into coordinate lines-such as $\xi \sim x/1+x$, so that $x = \infty$ corresponds to $\xi = 1$-or the use of spherical, cylindrical **or other** convenient body coordinates is dictated by the contour of the solid body present in the flow field. The intuitive process of discretization in the ξ-space is not likely to produce a conservative difference formulation. Even with a uniform mesh spacing $\Delta\xi$, the cancellation of influx and outflux in the transformed space does not guarantee the same in physical space, due to the presence of the metric coefficients.

Consider the mass continuity relation in cylindrical polar coordinates (r,θ,z):

$$\frac{\partial \rho}{\partial t} + \frac{1}{r}\frac{\partial}{\partial r}(r\rho u) + \frac{1}{r}\frac{\partial}{\partial \theta}(\rho v) + \frac{\partial}{\partial z}(\rho w) = 0 , \tag{5.7}$$

where u, v, and w are the radial, azimuthal and axial velocity components.
Even if the mesh spacings Δr, $\Delta\theta$, and Δz are uniform, and the central space
difference algorithm is adopted, there remains the question of how the metric
coefficient r should be treated in discretizing Equation (5.7) to obtain a
conservative difference form. Now the integrated conservation relations
in the physical space with curviliner coordinates stand as:

$$\Delta r.\Delta z.(r_j\Delta\theta).\Delta_t(\rho)$$

$$= \Delta t.\Delta z\Delta_r(\rho ur.\Delta\theta) + \Delta t.\Delta z.\Delta r\Delta_\theta(\rho v)$$

$$+ \Delta t.\Delta r.(r_j\Delta\theta).\Delta_z(\rho w), \tag{5.8}$$

where Δ with subscript r, θ, or z stands for the net flux of the quantity
in the parenthesis in some difference form. The left hand side of Equation
(5.8) represents the net increase of mass in the volume element. If the
flux terms on the right hand side are expressed either in the form (5.1) for
uniform mesh sizes, or in the form (5.6) for nonuniform mesh sizes ,
the difference form of the continuity equation will be conservative (or
summable). Thus, the metric coefficient r arising in the volume element
should be treated as r_j, while the metric coefficient r arising in the
surface element should be treated differently for the influx and the outflux
surfaces, depending upon the specific difference algorithm. It appears,
therefore, that the conservative difference formulation can be more

conveniently obtained by considering the integrated conservation relations
in the physical space-despite the curvilinear coordinate system that may
have to be adopted.

The treatment of the conservation relations of the momentum vector is
considerably more complicated than that of the scalar mass, because of the
stress and inertia terms due to curvature, and because of the need to
consider the appropriate vector components. Complicated as this may be, the
flux terms can be clearly identified, and conservative difference formulations
can be obtained. Often it is desirable, for the purpose of achieving a
simpler difference formulation, to relax the condition of identical cancellation of influx and outflux crossing the same common cell boundary. A more
lenient requirement may be that the influx and outflux crossing the same
boundary differ by a sufficiently small higher order quantity (allowing some
error accumulation), possibly supplemented by identical cancellation of the
fluxes over a group of say four neighboring cells. This may be permissible,
since the ultimate objective of the conservative difference formulation is
to prevent an undue accumulation of truncation errors over finite volumes
which would cause serious deterioration of the accuracy of the computation.

With conservative difference formulation, the accumulated truncation
error E_T of a set of calculations can be estimated (to an order of magnitude)
at any point within the field of computation. Moreover, the error of the
computed results at a point can be separated into two parts (despite the
fact that the difference problem is essentially nonlinear):
(i) the truncation error E_T, and (ii) the error at the point caused by the
errors on the boundary of the field of computation.

5.2 Heuristic Error Estimate and Accuracy

The accuracy question has been little explored. This may be due partly to preoccupation with stability questions, and partly to the difficulty of constructing an upper bound on the error of a computation for the type of initial-boundary value problems found in fluid dynamics. It may be possible that those convergence proofs which naturally include an estimate of the error bounds can be extended from periodic boundary value problems to more realistic boundary conditions. Such a difficult, complicated, and rigorous a priori error estimate generally gives an error bound much too large to be practically meaningful. Heuristic, rough, a posteriori error estimates will often suffice. Indeed, it would be preferable to have an estimate that is simple and generally applicable, though not rigorous and precise. Inview of this, the nonlinear Burgers' equation is conveniently adopted for analysis as a one-dimensional model of the Navier-Stokes equations.[10] As shown in Section (3.2), it is a useful model for stability analysis, being quasi-linear and possessing both wave and diffusion characteristics. It is also convenient for the study of akcuracy because many exact solutions of this equation are known, so that computational errors can be quantitatively evaluated and compared with theoretical estimates.

The Burgers' equation (in dimensionless form) is:

$$\frac{\partial u}{\partial t} + u \frac{\partial u}{\partial x} = \frac{1}{R_e} \frac{\partial^2 u}{\partial x^2} \tag{5.9}$$

having the steady state solution:

$$u(x) = - \alpha \tanh (\alpha R_e x/2)$$

with $u(x=0) = 0, \quad u(x=-1/2) = 1$

and $|u(x=\pm\infty)| = \alpha = 1/\tanh (\alpha R_e/4) .$ $\tag{5.10}$

This steady state solution for the range $-1/2 \leq x \leq 0$ has been calculated as the long-time limit of the temporal problem via several difference algorithms. The quasi-linear term $u \frac{\partial u}{\partial x}$ is always treated in the divergence form $\frac{\partial}{\partial x} (u^2/2)$, where

$$\left(\frac{U^2}{2}\right)_{j+1/2} = \left[\left(U_{j+1}^2 + U_j^2\right) + aU_{j+1} U_j\right]\Big/2(2+a) \tag{5.11}$$

and

$$\Delta\left(\frac{U^2}{2}\right)_j = \frac{1}{\Delta x} \left[\left(\frac{U^2}{2}\right)_{j+1/2} - \left(\frac{U^2}{2}\right)_{j-1/2}\right]$$

Here "a" is a parameter. The simple centered spatial difference corresponds to $a = 0$. The centered-space difference in non-divergence form results when $a = \infty$, in which case,

$$\Delta\left(\frac{U^2}{2}\right)_j = \frac{U_j(U_{j+1} - U_{j-1})}{2\Delta x} \tag{5.12}$$

If it is presumed that an approximate steady-state solution $U(t,x,\Delta t,\Delta x)$ will be reached, departing only slightly from the genuine solution (5.10), then a linearized differential equation for the error can be derived and solved. Linearization permits the separation of the errors, i.e., the truncation errors E_T and the boundary errors E_b, although the difference equations derived from (5.11) are all nonlinear. The linearized differential error analysis implied (and hence the linearization procedure presumes) that the error created over the entire field of computation from a given source is proportional to the magnitude of the source, i.e., the cumulative error is of the same order of magnitude as the local error. Accordingly, for the results of the linearized error estimates to be applicable, it is imperative that the difference formulation of the nonlinear equation (5.9) to be conservative (or summable).

The linearized analysis shows that E_T at any point in the field of computation is proportional to $(Re_{\Delta x})^2$ for the second order accurate conservative difference formulations derived from (5.11). Here $Re_{\Delta x}$ is the Reynolds number, based on the length Δx and the velocity difference between the point $x = 0$ (with maximum velocity gradient) and the point $x = 1$ (with nearly the asymptotic velocity). A quantitative estimate of E_T[10] is:

$$\frac{E_T}{(Re_{\Delta x})^2} = M_o E_o + \frac{M_1 E_1 + (1+3a)M_2 E_2}{2 + a} + \frac{M_3}{2} E_3 , \qquad (5.13)$$

where M_o is the constant defining the steady state criterion

$$Sup \left[U_j^{n+1} - U_j^n \right] < M_o \Delta x^3 ,$$

$M_1 (Re_{\Delta x})^2$ and $M_2 (Re_{\Delta x})^2$ are the coefficients of the truncated quasilinear convective terms, and $M_3 (Re_{\Delta x})^2$ is the coefficient of the truncated viscous terms. M_1, M_2, and M_3 are expected to be of O(1) for reasonable difference algorithms and for reasonably smooth solutions. E_o, E_1, E_2, and E_3 are universal functions of the genuine solution u(x) that vanish on both aries and have their absolute magnitudes less than 0.1 (Fig. 2). For small values of M_o and the parameter "a" in equation (5.11), the truncation errors E_T are expected to be of the order of $(Re_{\Delta x})^2/10$ for second-order accurate schemes. Actual computations with $M_o = O(\Delta x)$ and $\Delta x = 1/20$

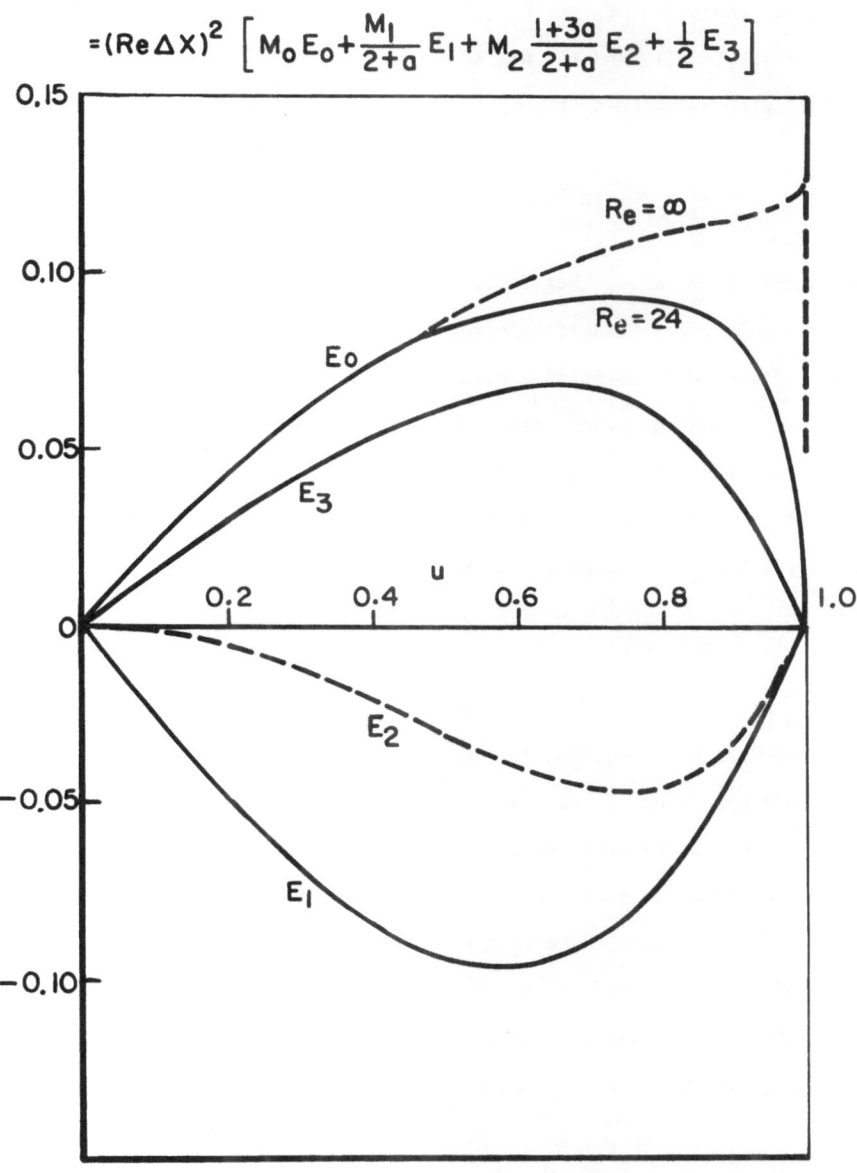

TRUNCATION ERROR $E_T(u)$

$$= (Re\,\Delta x)^2 \left[M_0 E_0 + \frac{M_1}{2+a} E_1 + M_2 \frac{1+3a}{2+a} E_2 + \frac{1}{2} E_3 \right]$$

FIG. 2

for various schemes have verified the quantitative values of Equation (5.13) and the dependence of $\dot{E}_T$ on $(Re_{\Delta x})^2$.

For $Re_{\Delta x} = 0(1)$ and for all of the finite values of $a = 0(1)$ tested, the following estimate of the maximum absolute truncation error is valid:

$$E_T < 3 \times 10^{-2} \ (Re_{\Delta x})^2 \qquad\qquad (5.14)$$

This simple formula is, therefore, recommended as a preliminary estimate of the bound of the truncation errors of a second-order accurate conservative difference formulation. With a non-conservative difference formulation, the truncation errors can accumulate and can thus become considerably larger than the estimate given by Equation (5.14)

The boundary errors in the field due to a fractional error ε_b in the boundary value is given by the linearized analysis as:

$$E_B = \varepsilon_b E_h , \qquad\qquad (5.15)$$

where E_h is a universal function that is unity on the boundary where the erroneous boundary condition is applied, and decays very slowly toward the other boundary, where it vanishes. The decay is so slow that the error retains more than half its value until within the last few tenths of the field of computation near the other boundary, depending upon the magnitude of the Reynolds number. (Note that E_h is plotted against $u(x)$ in Fig. 3. The decay into the field is even slower when $u(x)$ is replaced by x.)

For Neumann boundary conditions, the boundary error is still given by Equation (5.15) but ε_b is evaluated as

BOUNDARY ERROR $E_b(u) = \epsilon_b E_h$

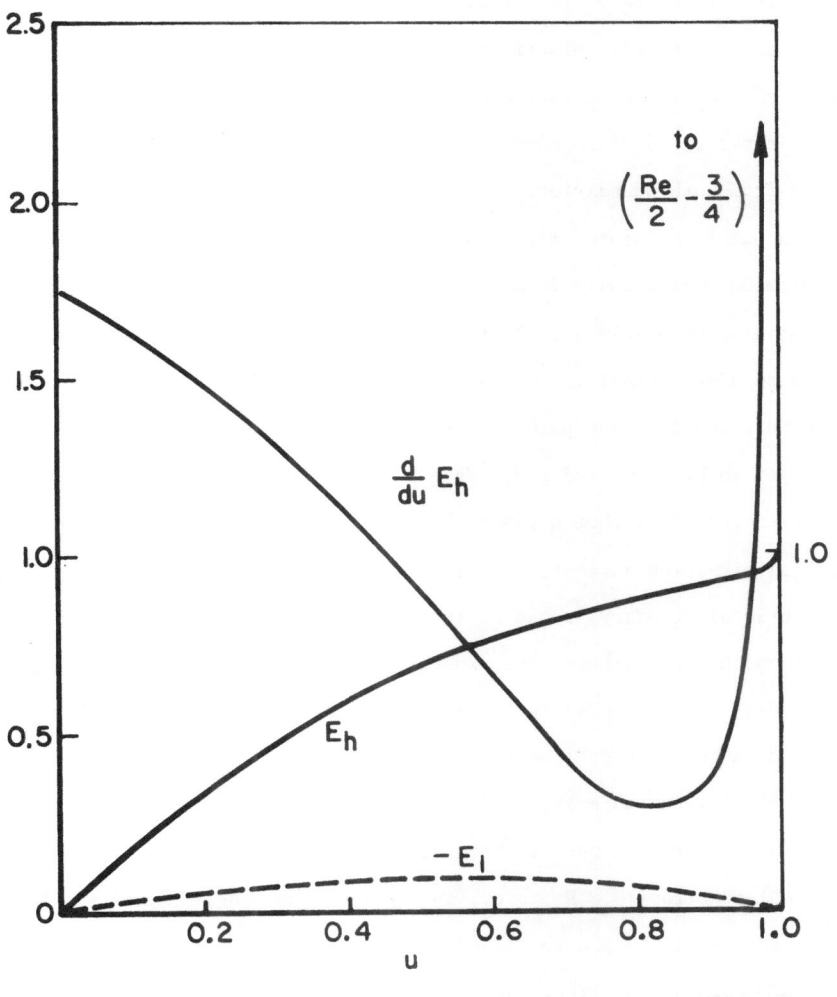

FIG. 3

$$\varepsilon_b \overset{\sim}{=} -2\varepsilon_b'/\alpha Re \ , \tag{5.16}$$

where ε_b' is the fractional error in the spatial derivative on the boundary. Within the framework of linearized error estimates, the superposition of (5.15) and (5.16) with proper coefficients will enable an estimate of the errors caused by a Cauchy-type condition. The boundary error at a given point in the field of computation will be the sum of the decayed boundary errors from both boundaries.

In multidimensional flow problems, it is presumed that the results of the previous model analysis may apply primarily in the direction along streamlines, or nearly so. This leaves the estimate of the contributions of the boundary errors from those portions of the boundary of the field of computation that are primarily parallel to the local streamline directions yet to be accounted for. No helpful suggestions can be made here, except to recommend that a description be rendered as nearly correct as the physical situation suggests. In fact, the treatment of this portion of the computational boundary is one of the two outstanding difficulties that the author and his co-workers have experienced in various problems. (The other outstanding difficulty is the treatment of internal shockwaves, to be explored in the next section.)

The decay characteristics described by the universal function E_h may be used where the one-dimensional model is appropriate. The various universal functions E_o, E_1, etc., and E_h from the model results may be recognized as "influence functions" describing error propagation in the field of computation. They can be empirically established, a posteriori, by introducing a known error at a specific point (on the boundary for the specific boundary error **and at** chosen interior points for truncation errors) and then computing the solution under the modified condition. The difference between the two sets of solutions then gives the influence function E in question. Usually, during the developmental stage of a difference formulation for a given physical flow problem, such information can be derived from preliminary results and can be used for the purpose of a posteriori error estimation. Clearly, the a posteriori determination of such influence functions is desirable to provide additional checks on the behavior of the computational program.

Without referring to any specific computational problem, the following general observations can be inferred from the model study. They are applicable only for a conservative difference formulation in which the truncation errors do not accumulate, so that the truncation and the boundary errors can be treated separately and estimated by Equations (5.14) to (5.16).

(1) The steady state criterion $|U_j^{n+1} - U_j^n| < 0(\Delta x)^4$ is sufficiently accurate for a second-order accurate scheme.

(2) The truncation error E_T is expected to be $(Re_{\Delta x})^n$ for conservative difference formulations of nth order formal accuracy, and the influence functions $E_{1,2}$, etc. are not likely to possess maximum magnitudes much less than 10^{-1}. With $Re_{\Delta x} > 1$ in practical cases, the maximum truncation error is not likely to be reduced appreciably from that of second-order accurate scheme, as may be estimated from (5.14)

(3) Boundary errors cannot be efficiently reduced by reducing the mesh sizes. They decay very slowly, and are generally considerably larger than the truncation errors in practical cases with $Re_{\Delta x} = 0(1)$. The primary

effort required in achieving a reasonably accurate solution of complicated practical problems lies in the sophistication of the treatment of the various boundary conditions. The field of computation and the choice of coordinates should be properly defined to facilitate a more accurate implementation of the boundary conditions.

The general observations made above carry an important message for those interested in obtaining solutions for complicated fluid dynamics problems with reasonable accuracy to suit practical purposes. Much attention should be paid to the formulation of the problem. Attempts to improve the accuracy of the numerical solution of a poorly formulated problem by extending the computation to satisfy a more restrictive steady-state criterion or by refining the mesh (even with the help of much larger and faster computers) can prove to be not only expensive but frustrating.

A similar attempt at model analysis was made for time-dependent flow. It was found that for flows with slow and monotonic temporal variations, the behavior of error propagation in the second-order accurate conservative difference formulation is essentially similar to that described above for steady state problems. For oscillatory flows, conservation in the spatial domain apparently fails to help. Test calculations[3] for some simple damped oscillations as exact solutions of the Burgers' equation indicate the serious effect due to the phase errors of the different oscillatory components caused by the dispersive truncated terms; the computed results become highly inaccurate after one or two cycles. It has been illustrated [8, 9, 11] that fourth-order accurate difference algorithms will substantially improve the accuracy of the computed results beyond a few cycles of oscillations.

It is, however, a tremendous task to compute a complicated equation system like the Navier-Stokes with formal forth-order accuracy.

5.3 Shock Waves and Artificial Viscosity

In all of the previous discussions, the question of "non-smooth" or even "discontinuous" solutions is deferred. In problems of practical interest, shock waves and contact discontinuities are often the prominent features of the flow field. The presence of such discontinuities, (or in general, regions of very large gradient) causes difficulties in the computation.

Discontinuous initial and boundary data are often imposed on purely elliptic or parabolic problems. Such data may cause oscillations in the vicinity of the boundary, but they are never very serious. This is because of the inherent nature of these systems to smooth out any discontinuities in time and in space. The accuracy of the computed results may suffer somewhat (according to the modulus of continuity of the functions involved), but this can often be remedied by using a higher-order accurate difference algorithm. This inherent tendency to smooth out any discontinuity can also be troublesome as in treating flow problems involving an interfacial discontinuity formed by two different fluid media - especially when the interface is not stationary - since an initially sharp discontinuity diffuses in the course of the computation (if not artificially maintained).

For hyperbolic problems, a discontinuity in the initial-boundary data propagates into the field of computation and causes excessive computational disturbances downstream, particularly in its zone of influence. It also produces upstream influences. For quasi-linear gas dynamic problems, a

shock discontinuity can physically arise from a perfectly smooth boundary
due to the coalesence of smooth compression waves. Thus, when this flow
field is computed with an algorithm that works well for smooth fields, quite
severe oscillations can develop approximately at the location where the shock
discontinuity would appear. Such oscillations are fairly large, but do not
necessarily lead to the catastrophic divergence of linear instability. It
may be that the amplitude of such shock-induced oscillations are limited by
non-linear effects; thus the phenomenon may well be called nonlinear-instabi-
lity. But this is certainly not instability in the sense of violating the
requirement of boundedness discussed in Chapters II and III. Even if bounded,
however, such oscillations are highly damaging to the accuracy of the results,
not only in the vicinity of the shock, but over most of the flow field. Since
practical interest is often centered in the vicinity of such a shock discontinuity, much has been done for computing a shock discontinuity.

It is natural to treat a shock front or an interfacial contact discontinuity as an internal boundary, and to compute the smooth solutions on both
sides of the discontinuity separately. The jump conditions across the
discontinuity will connect the two solutions together. This shock-matching
or shock-fitting procedure is easily carried out in one space dimension for
a known discontinuity, i.e., a discontinuous front propagating into a homogeneous medium at rest or in uniform motion. If the shock should be propagating into a non-uniform medium or a homogeneous medium in non-uniform motion,
the shock strength and speed will vary, and the Hugoniot relations across
the shock will have to be supplemented by some additional matching conditions

to be derived form the difference results in the vicinity of the shock front.
Oscillations often appear on one or both sides of the shock discontinuity,
probably as a result of inaccuracies in the location of the shock and in the
values of functions in its vicinity. The oscillations may be alleviated if
the shock location is fixed at a mesh point and if the mesh dividions are
rezoned at every time or iterative step. The computational procedure in terms
of such shock coordinates rapidly becomes complicated.

In two space dimensions and with a curved shock of unknown shape and
location, the computational details of such a shock matching procedure become
more tedious and inaccurate. With fixed mesh points, the shock front is
generally off the mesh points. Thus it becomes difficult to determine the
direction normal to the front, resulting in a highly inaccurate matching
process. The use of curvilinear shock coordinates is convenient, and may
possess other desirable features for treating inviscid steady-state flow
problems with uniform supersonic flow on the upstream side of the shock
front. [12] They are not suitable, however, for a shock wave imbedded in a
non-uniform inviscid flow field, or for viscous and inviscid flow fields
involving more complicated shock configurations, such as shock intersections
and Mach reflections, or transonic shocks that terminate in the flow field.
The tedious shock matching can in principle, be implemented even for such
complicated configurations, but the procedure is too complicated to be
manageable, and the results so obtained are uniformly poor.

To avoid shock matching, v.Neumann and Richtmeyer[13] introduced the
artificial viscosity method for computing shock propagation in an inviscid

flow field. A quadratic viscous pressure term $\rho\alpha^2\Delta x^2\left|\frac{\partial u}{\partial x}\right|\frac{\partial u}{\partial x}$, where α is

a numerical constant chosen conveniently, is added to the differential

equation before discretization. Quadratic dependence on the velocity gradient

is to promote rapid decay of the artificial viscous term away from the shock

front (which possesses a steep velocity gradient). With $\alpha < 1$, typical

results of the calculation for one-dimensional shock propagation into a

uniform field give a sharp shock front, spreading over ~ 2 meshes, and a

calculated shock speed within 0.1% of the correct value. But sizable

oscillations develop downstream over an extended range, without appreciable

damping (spatially and temporally). By increasing α to $\gtrsim 2$, the magnitudes

of the oscillations are reduced, but the shock front spreads wider, over 4 or

more meshes. A reasonably smooth downstream solution is obtained only when

α is so large as to be $O(\Delta x^{-1})$, and the shock front spreads over many meshes.

By then the artificial viscous term is no longer small in the apparently

smooth inviscid region, and the apparently smooth results of computation fail

to be a satisfactory approximate solution near the shock front.

The artificial viscosity method is physically sound, simply implemented,

and easily extended formally to multispace dimensions by including derivatives

in the other spatial dimensions. The large spread of the shock front and the

induced oscillations generally become more objectionable, however, Many

artifices can and have been devised to improve the appearance of the computed

results. The artificial viscous term may be dropped when the gradient of

velocity becomes less than a pre-assigned value, or the downstream oscillations

may be suppressed or eliminated by some smoothing process, or they may be

limited to a permissible range about the mean through some filtering process.

Excellent results can generally be obtained for simple test problems

with known shocks. The merit of such procedures in computing shock propagation into non-uniform flow fields is yet to be demonstrated, particularly
with respect to the accuracy of such smoothed results.

The Lax-Wendroff treatment [14] of a shock wave utilizes the fact that
the Hugoniot relations are simply the conservation laws integrated over the
discontinuity. Thus, with the inviscid equations written in divergence
form for the physically conserved quantities, shock matching can be avoided
because the difference equations for such conserved quantites are, indeed,
the approximate form of the Hugoniot relations. (Note that the divergence
form of the transformed dependent variables, rather than the physically conserved variables, may not result in such approximate Hugoniot relations).
One-dimensional computations show that this method leads to a quite sharp
shock front ($\sim 2\Delta x$) and accurate shock speed. But sizable oscillations are
generated at the shock front, although they are rapidly damped and disappear
within 8 to 10 meshes from the front. This damping is derived from the
dissipative term $\Delta x^3 . r(1-r^2)\frac{\partial^4 u}{\partial x^4}$, with $r = u\Delta t/\Delta x$, which may be visualized as
an artificial viscosity that spreads out the shock front. The quadratic
viscous terms adopted by v.Neumann and Richtmeyer does not appear to provide
as much damping of the shock-induced oscillations as does this linear viscous
term. But the peak amplitude of the shock-induced oscillation near the front
is often larger for the linear than for the quadratic artificial viscous
term. Additional artificial viscous terms are often introduced to reduce
the amplitude of such oscillations.

The introduction of artificial viscous terms into the differential

equation before discretization is fundamentally not much different from
the process of dropping higher order terms in a truncated Taylor series
during discretization. Since such viscous terms contribute to the stability
of the difference formulation, artificial viscosity is very widely employed
for problems without shocks. These artificially-introduced viscous terms are
often substantially larger than the Navier-Stokes viscous stress terms evaluated with the physical viscosity coefficient of the fluid. This is justifiable in the solution of inviscid flow problems (i.e., flow problems
visualized as the asymptotic limit where viscous stress terms are negligable),
as long as the contributions due to the artificial viscous terms are "negligi-
bly small" compared to those from the inviscid terms, and provided that the
somewhat spread-out shock front is visualized as a "sharp" discontinuity.
Such large artificial terms are clearly not tolerable for viscous flow problems, since the effect of the fluid viscosity will be overshadowed by the
effect of the pseudo-viscosity.

There are many numerical solutions of the Navier-Stokes equations--some
with first-order accurate algorithms, some with second-order accurate
algorithms--using large artificial viscous terms, and at large Reynolds
numbers (based upon fluid viscosity) of the order of 10^6. These computed
results are very insensitive to the large fluid Reynolds number[15]. This
is understandable, since the pseudo-viscosity in such calculations is
substantially larger than the real fluid viscosity, and therefore changes
in the fluid Reynolds number will not significantly alter the effective
Reynolds number(based on the total viscosity included in the difference
formulation). If one wishes to quantitatively evaluate the viscous effects,
both the artificial viscous terms introduced into the differential equations

system and the pseudo-viscous terms implicit in the difference form should remain substantially less than the physical fluid viscous term. Thus for viscous flow problems, artificial viscosity terms of the type used by v. Neumann and Richtmeyer should satisfy

$$\alpha^2 \Delta x^2 \; (\frac{\partial u}{\partial x})^2 \; << \; \nu \frac{\partial^2 u}{\partial x^2} \Delta x,$$

or dimensionally

$$\alpha^2 \; \frac{\Delta u \; \Delta x}{\nu} \; = \alpha^2 \; Re_{\Delta x} \; << \; | \; . \tag{5.17}$$

With $Re_{\Delta x}$ generally larger than unity, the constant α must be chosen appreciably less than unity. This severely restricts the usefulness of an artificial viscous term, either for securing computational stability, or for suppressing shock-induced oscillations in viscous flow problems.

For a second-order accurate conservative difference formulation, the errors introduced by the pseudo-viscous terms are included in the truncation error E_T, the absolute upper bound of which may be estimated as $E_T < 3 \times 10^{-2}$ $(Re_{\Delta x})^2$ according to the results based on the Burgers' model equation given in the previous section. Thus $Re_{\Delta x}$ may be as large as 1 or even 2 without having the cumulative truncation errors exceed a few percent. Note that this $Re_{\Delta x}$ is defined in terms of the local change in velocity per mesh when the Burgers' model is fitted to the "local flow field" of large velocity gradient. With a Reynolds number of $0 \; (10^3-10^4)$ based on the viscous flow dimension and the reference velocity in the inviscid flow field, it is possible to provide sufficient number of mesh points over the linear dimension so that the local values of $Re_{\Delta x}$ will be considerably smaller than 10 and $E_T \lesssim$ a few percent except in the region of shock-induced oscillations. If the shock front is visualized as an interior boundary, and the shock-induced oscillation as a form of propagating boundary error, the errors in the results computed with the second-order accurate difference formulation will generally be dominated by boundary errors.

Shock-induced oscillations mar the appearance of the computed solution much more seriously than the less conspicuous sources from the exterior boundary, though they need not cause larger errors. The difficulty is compounded where a shock wave, either incident or emerging, intersects an exterior boundary. In the next section, the relation between boundary treatment and shock-induced oscillations will be explored.

5.4 Shock-Induced Oscillations

Shock-induced oscillations are often considered unavoidable when a shock wave is encountered in computation with a higher-order accurate difference algorithm. While a first-order accurate algorithm does not give rise to such oscillations, the smear of the shock front becomes excessive and the cumulative truncation errors become large. Thus, when a shock wave is encountered in a computation, it is often held to be necessary to choose between these two evils. The following is an attempt to clarify the origin of the spurious oscillations, and to show that a certain class of second-order accurate difference algorithms can, under favorable circumstances, avoid such spurious shock-induced oscillations.

Consider the solution of a linear steady state problem via the time-dependent approach. Let the spatial difference operator be split into two parts, $L_1(T)$ and $L_2(T)$, where T is the shift operator for the spatial indices, i.e., $TU_j = U_{j+1}$, $T^{-1}U_j = U_{j-1}$, and $T^2U_j = T \cdot TU_j = U_{j+2}$, etc. Construct the class of two-step difference algorithms for the time interval $n\Delta t$ to $(n+1)\Delta t$:

$$\widetilde{U}_j^n - U_j^n = L_1(T)U_j^n + L_2(T)U_j^n$$

$$U_j^{n+1} - U_j^n = L_1(T)\widetilde{U}_j^n + L_2(T)U_j^n \tag{5.18}$$

where $\widetilde{U}_j^n$ is a provisional or predicated value of U_j^{n+1}. The second or final step is a corrector step. $L_1(T) + L_2(T)$ is second-order accurate and consistent with the differential operator in the steady state.

Let the boundary conditions applied in the first or provisional step be

$$B(T)\ \widetilde{U}_j^n = 0 \ , \tag{5.19}$$

and let the boundary values of $\widetilde{U}_j^n$ derived from these boundary conditions used in the first step, be used in the second step for the computation of U_j^{n+1} at the correspoinding boundary points. In this manner it is maintained that $U_j^{n+1} - \widetilde{U}_j^n \equiv 0$ at all of the boundary points for every time step. The boundary values at each boundary point may change from step to step and may contain errors implicit in the boundary conditions (5.19). By subtracting the two steps in the difference equations (5.18), the following difference relation is obtained:

$$U_j^{n+1} - \widetilde{U}_j^n = L_1(T)\left(\widetilde{U}_j^n - U_j^n\right). \tag{5.20}$$

In the event that a steady state is approached in the sense that $U_j^{n+1} = U_j^n$, then Equation (5.20) becomes (in the steady state limit):

$$\left[I + L_1(T)\right]\left(\widetilde{U}_j^n - U_j^n\right) = 0 \ . \tag{5.21}$$

Thus $\widetilde{U}_j^n - U_j^n$ is governed by the linear system of difference equations (5.21), and is subject to zero boundary values over the entire boundary. If there are no eigen solutions to this system of equations, it follows that in the steady state limit $U_j^n = \widetilde{U}_j^n = U_j^{n+1}$. The solution in the steady state limit is thus the solution of the correct steady state equation

$$\left[L_1(T) + L_2(T) \right] U_j^n = 0 \qquad (5.22)$$

Now if the boundary values of $\hat{U}_j^n$ and U_j^{n+1} are not kept the same in successive iterations, $\tilde{U}_j^n$ must be eliminated from Equations (5.18). Then, in the limit of the steady state with $U_j^{n+1} = U_j^n$, the solution will be determined by the equation

$$\left[I + L_1(T) \right] \left[L_1(T) + L_2(T) \right] U_j^n = 0 \qquad (5.23)$$

This solution will contain the "correct" steady state solution (5.22), to the extent that the boundary conditions $B(T)U_j^n = 0$ represent a correctly-posed situation. But it will also contain the nontrivial solutions of Equation (5.21) when $\hat{U}_j^n - U_j^n$ is not identically zero, as a result of the slight difference in the boundary values of $\hat{U}_j^n$ and U_j^{n+1}. Naturally, such extraneous solutions are possible sources of shock-induced oscillations, and can indeed be identified in the course of computation as being proportional to the difference between the provisional and the final solutions. From the practical point of view, it is simplest and most desirable to use the identical boundary values from (5.19) to suppress all of the spurious fundamental solutions arising from Equation (5.21).

There are many two-step difference algorithms, but most are not of the class (5.18), except for the Cheng-Allen scheme and Brailovskaya's scheme. For the linearized Burgers' Equation (3.13), the difference forms can be cast into: Cheng-Allen Algorithm[10,16]

$$\begin{cases} L_1(T) = \dfrac{1}{1+2s} \left[\left(-\dfrac{r}{2} + s \right) T + \left(\dfrac{r}{2} + s \right) T^{-1} \right] \\[4mm] L_2(T) = \dfrac{-2s}{1+2s} \end{cases} \qquad (5.24)$$

Brailovskaya Algorithm[17]

$$\begin{cases} L_1(T) = r(T - T^{-1}) \\ L_2(T) = s(T - 2 + T^{-1}) \end{cases} \tag{5.25}$$

where $r = c\Delta t/\Delta x$ and $s = \nu\Delta t/\Delta x^2$. When (5.24) is substituted into Equation (5.23), the general solution U_j is obtained as

$$U_j = \sum c_k \xi_k^{\,j} \qquad k = 1, 2, 3, 4$$

where

$$\begin{cases} \xi_1 = 1 \\ \xi_2 = \dfrac{2s+r}{2s-r} = \dfrac{1+\frac{1}{2}\,Re_{\Delta x}}{1-\frac{1}{2}\,Re_{\Delta x}} \end{cases}$$

$$\xi_{3,4} = \left[-(1+2s) \pm \left\{ (1+2s)^2 + (r^2-s^2) \right\}^{1/2} \right] \Big/ (2s-r) \;. \tag{5.26}$$

ξ_1^j and ξ_2^j are the two proper fundamental solutions of the correct steady state equation $[L_1(T) + L_2(T)]U_j = 0$, because in the limit $Re_{\Delta x} \to 0$, they approach the two fundamental solutions 1 and exp(Rex) of the steady state differential equation, $c\dfrac{\partial u}{\partial x} = \dfrac{1}{Re}\dfrac{\partial^2 u}{\partial x^2}$. ξ_3^j and ξ_4^j are the two extraneous fundamental solutions of the two-step scheme that constitute the errors or "spurious solutions" arising from the solution of the equation

$$\left[I + L_1(T) \right] U_j = 0$$

or of Equation (5.21).

With both r and s > 0, and $\left|\dfrac{2s-r}{2s+1}\right| < 1$, it is found that

$$\xi_3 \sim -\frac{r+2s}{1+2s} < 0$$

$$\xi_4 \sim -\frac{(1+2s)}{2s-r} \gtreqless 0 \text{ as } r \gtreqless 2s.$$

Thus ξ_3^j always represents a mesh-to-mesh oscillation, while ξ_4^j can be either oscillatory or monotonic. The steady state limit of the difference $U^{n+1} - \hat{U}^n$ can be given as:

$$\hat{U}_j - U_j = c_3 \xi_3^j + c_4 \xi_4^{\ j} \tag{5.27}$$

where c_3 and c_4 are determined by the difference in the values of $\hat{U}^n$ and U^{n+1} at $j = 0$ and $j = J$ on the boundary. When the boundary values of $\hat{U}^n$ and U^{n+1} are kept the same at every step, then $c_3 = c_4 = 0$ and no spurious solution will be present in the computed steady state result. Otherwise, oscillations can be expected.

If Brailovskaya's scheme (5.25) is substituted into Equation (5.23), the same proper fundamental solutions $\xi_1^{\ j}$ and $\xi_2^{\ j}$ are obtained, but the pair of extraneous solutions $\xi_3^{\ j}$ and $\xi_4^{\ j}$ are given somewhat differently as $\xi_{3,4} = [1 \pm (1 + 4r^2)^{1/2}]/2r$, with $\xi_4 < 0$ always. The overall situation is much the same, however.

It may be pertinent to repeat here that the spurious solutions will be suppressed so long as the same values of $\hat{U}^n$ and U^{n+1} are used on the boundary at every step. Such boundary values can be determined by the approximate boundary conditions $B(T)U_j^n = 0$, and may contain errors. In this event, they may cause errors in the constants c_1 and c_2 in the steady state solution

$$U_j = c_1 \xi_1^{\ j} + c_2 \xi_2^{\ j} . \tag{5.28}$$

There will not be any catastrophe if the boundary values are not excessively in error and if the mesh size of the steady state solution is not too coarse, so that the inequality

$$Re_{\Delta x} < 2 . \tag{5.29}$$

is maintained.

This last restriction $Re_{\Delta x} < 2$ has little to do with suppressing the spurious fundamental solutions, $\xi_3^{\,j}$ and $\xi_4^{\,j}$, but is rather to keep $\xi_2^{\,j}$ from becoming oscillatory and failing to be a valid approximation to the fundamental solution exp $(Rej\Delta x)$ of the differential problem. It is clear from Equation (5.26) that when $Re_{\Delta x} > 2$, the appropriate form of $\xi_2^{\,j}$ is

$$\xi_2^{\,j} = (-1)^j \left[\frac{1 + 2/Re_{\Delta x}}{1 - 2/re_{\Delta x}} \right]^j , \qquad (5.29)$$

which is oscillatory and rapidly amplifying with increasing j, and hence fails to serve as any meaningful approximation to exp $(Rej\Delta x)$. Thus, to obtain a valid steady state solution without spurious oscillations based on algorithms (5.24) or (5.25), not only should identical boundary values be used at the provisional and the final steps, but also the mesh size must be sufficiently refined so that $Re_{\Delta x} < 2$. Sample calculations for steady state solutions of the linearized Burgers' equation (3.13) verified the abrupt change in the behavior from a smooth to a violently oscillatory limiting solution when $Re_{\Delta x}$ increases beyond the critical value of 2.

For linear problems with variable coefficients, the various fundamental solutions of the difference equations cannot be displayed. It is nevertheless expected that the spurious solutions will be suppressed if the same operators $L_1(T)$ and $L_2(T)$ and the same boundary values are used for the successive iterative steps in each time interval. Regarding the proper fundamental solutions of $[L_1(T) + L_2(T)]U_j = 0$, it is known that one of them must be unity to satisfy the consistency requirement. The other will become oscillatory for too large a $Re_{\Delta x}$. Whether the critical value of $Re_{\Delta x}$ will be 2, or how it may vary with x, is uncertain. For nonlinear problems with sufficiently smooth solutions, the complete suppression of spurious fundamental solutions in the first variation of the nonlinear difference operator at each time step may be expected. This is because the spurious fundamental solutions contained in the computed results of the nonlinear equations will have been reduced to higher order small quantities in Δt by the stratagem described above. Such higher order small quantities in Δt are of little significance in the steady state limit. Thus, the outstanding problem for eliminating shock-induced oscillations is to satisfy the requirement of sufficiently small mesh size Δx corresponding to the restriction of $Re_{\Delta x} < 2$ for the linearized Burgers' equation. It is anticipated that, for nonlinear problems, there may not be such a sharp value for the critical $Re_{\Delta x}$. The transition from a smooth to an oscillatory steady state solution may take place gradually over some range of values of $Re_{\Delta x}$. This has been verified in actual computation. It is hoped that the following heuristic model will give a general idea of where this critical range of $Re_{\Delta x}$ may be.

When a second-order accurate conservative difference algorithm of the class (5.18) is used for the integration of the Navier-Stokes equations, and when the stratagem just described is followed in the treatment of the boundary conditions, the shock wave (if present in the computation) is not regarded as a discontinuity, but as a "smooth" region with large gradient, spread out by the pseudo-viscosity. This shock transition region usually spreads out over two or more mesh points to connect the smooth, asymptotically uniform flow fields both up and downstream of the shock region. The transition profile as calculated is not intended to be accurate. Its primary function is to accomplish a smooth connection, hopefully without inducing oscillations propagating into the smooth flow field in its neighborhood. Thus, the transition profile, joining a scalar function u with asymptotic values $u_\infty = \pm \alpha$ in the up and downstream regions respectively, might as well be computed approximately, based on the nonlinear Burgers' equation as a model for the local flow field. This means that the local profile might be approximated by the steady solution (5.10), with x = 0 and u = 0 located at the point of maximum slope in the transition profile actually computed with the full Navier-Stokes equations. Thus, the computed maximum value of $\frac{\partial u}{\partial x}$ (properly nondimensionalized in the transition region) will define the effective Reynolds number of the transition region.

$$\left(\frac{\partial u}{\partial x}\right)_{max\ computed} = Re/2 \tag{5.30}$$

In this manner, the poorly defined thickness of the transition region is avoided. The parameter α can be taken as unity when the reference velocity is taken as the change in the velocity (or the particular scalar quantity in dimensionless form) from the point of maximum gradient to the asymptotic value. If the computed transition profile is approximately symmetric with respect to the inflection point, this reference velocity will be half the jump across the shock.

Letting the asymptotic values of u across the shock transition region be U_1 and U_2, then assuming we have $U_1 > U_2$,

$$\left(\frac{\partial U}{\partial x}\right)_{max} = \frac{U_1-U_2}{2} \quad \left(\frac{\partial u}{\partial x}\right)_{max} = \frac{U_1-U_2}{4} \ Re$$

$$\left(\Delta U\right)_{max} = \left(\frac{\partial U}{\partial x}\right)_{max} \Delta x = \frac{U_1-U_2}{4} \ Re_{\Delta x} \tag{5.31}$$

Now it is pressumed that the critical value of this $Re_{\Delta x}$ is essentially the same as if the computation were done with the same algorithm, but based on the Burgers' equation so that oscillation-free computed results in the transition region can be effected with $Re_{\Delta x} < 2$. When expressed as an a posteriori criterion in terms of quantities directly available in the computation, according to (5.31) this condition becomes

$$\frac{(\Delta U)_{max}}{U_1 - U_2} < \frac{1}{2} \; ; \qquad\qquad (5.32)$$

i.e, "the maximum change permissible in U per mesh, $(\Delta U)_{max}$, in order to avoid large shock-induced oscillations in the computed results, is one half of the jump $|U_1 - U_2|$ across the discontinuity."

This statement implies that we cannot expect to obtain an oscillation-free shock front containing less than two meshes from a computational solution following the given strategem. Moreover, within the linearized framework, the criterion (5.32) might be equally applicable to any physical scalar variable sustaining a "jump" across some large gradient region, not necessarily a discontinuous front, even though $Re_{\Delta x}$ was defined in terms of flow velocity and viscosity provided that Burgers' model remains appropriate. Criterion (5.32) is explicitly independent of viscosity.

Condition (5.32) stands, however, only as an a posteriori criterion for achieving an oscillation-free shock solution. This is because $(\Delta U)_{max} = \left(\frac{\partial U}{\partial x}\right)_{max} \Delta x$ becomes known only after the completion of the computations; by then, there is no need of a criterion to find out if the computed solution is oscillation-free! Such an a posteriori criterion can, however, be of some help in practice, since $(\Delta U)_{max}$ can be estimated long before the computed solution reaches a satisfactory "steady state". Oscillations will be present in the "transient states" of the computation, whether or not the steady state limit will contain shock-induced oscillations. If the criterion should be satisfied at some transient stage, we may expect an oscillation-free steady state solution with further temporal steps. Otherwise, smaller mesh sizes may be needed.

It is more convenient if this criterion is put into some a priori form, even if it is then less precise (as it must be). Note that the magnitude

$|U_1-U_2|$ depends on the shock strength, the shock orientation relative to the coordinate axes (in a multidimensional problem), and the coordinate direction under consideration. If it is possible to estimate $|U_1-U_2|$, then

$$Re_{\Delta x} \stackrel{\sim}{=} |U_1-U_2|\Delta x/2\nu < 2$$

may be used directly as an a priori limit. This Reynolds number $Re_{\Delta x}$ must not be confused with $Re_{\Delta x,\infty}$, based on the uniform supersonic flow velocity U_∞ far upstream of the flow field, i.e., $Re_{\Delta x,\infty} = U_\infty \Delta x/\nu$. In terms of this $Re_{\Delta x,\infty}$, the criterion becomes

$$Re_{\Delta x,\infty} = \frac{U_\infty \Delta x}{\nu} < \frac{U_\infty}{U_1} \frac{4}{1-U_2/U_1} \tag{5.33}$$

which can be useful a priori if there is some idea as to the shock strength U_2/U_1 and as to the ratio U_∞/U_1 of the reference velocity U_∞ far upstream to the velocity U_1 into which the shock wave is propagating. For complicated flow problems, however, such quantities are usually among the unknowns. Thus, the limit on $Re_{\Delta x,\infty}$ given by (5.33) will have to be based on some rough estimate, or on the "transient states" of the computed solution.

The previous heuristic development is equally applicable to any flow region containing a large gradient other than a shock front. In particular, oscillations originating from boundaries of the field of computation can be similarly alleviated. It is to be emphasized, however, that if the oscillatory extraneous fundamental solutions like ξ_3^j and ξ_4^j are not suppressed by the strategem described above, these extraneous oscillatory solutions will propagate into the neighboring smooth flow fields, even if the mesh size is reduced much below that required by (5.33). At least one of these solutions will be amplifying away from the boundaries of the transition region, while propagating into the neighboring smooth regions on either side. On the other hand, if much too coarse a mesh size is used in the computation, large amplitude oscillations will result despite the fact that the strategem described above is followed, since one of the proper fundamental solutions of the difference equation fails to be a valid approximation to that of the differential problem. To produce an oscillation-free computational solution of a flow problem involving shock waves, it is recommended not only that some form of the two step algorithm (5.18) be used with identical boundary values applied to both iterative steps during a time interval, but also that the mesh size Δx be kept sufficiently small according to the condition (5.33). This recommendation is based on the results of analysis of a simple linear model for the numerical solution of the much more complicated and nonlinear gas dynamic equations. It is recommended in the same spirit that the local linear stability analysis of von Neumann be used to help in achieving computational stability. The practical merit of this recommendation is yet to be examined in greater detail by the computational community.

The previous development has guided the author quite successfully in his early attempts at integration of the Navier-Stokes equations for some complicated flow problems, such as the near wake flow behind a flat base with a sharp corner in the supersonic flow[16], and the hypersonic flow over the sharp leading edge of a highly-cooled flat plate.[18] The flow situations encountered in these examples are just too complicated to provide any meaningful quantitative tests of the validity of the above-mentioned criterion and the accuracy of the computed results. In the following, a simple case will be described which may serve to support and to illustrate that, despite the heuristic arguments for their application to the integration of the Navier-Stokes equations, the strategem and the criterion outlined above are indeed useful in practice.

The Cheng-Allen two-step algorithm, as a member of the class (5.18), is used to integrate the complete Navier-Stokes equations for the propagation of a planar shock wave into a uniform supersonic flow at Mach No. 2, with the shock front inclined at an angle $\beta = 41.84°$ to the uniform inflow.[19] The gas density ρ_1, velocity u_1, energy e_1, and pressure p_1 are taken to be unity in dimensionless form. The theoretical values of these variables downstream of the shock, given by the Hugoniot relations, agree with the values computed at $Re_{\Delta x, \infty} = 10$ to better than 0.1%. The critical Reynolds number per mesh is $(Re_{\Delta x, \infty})_c = 4/(1-0.837) = 24.5$. No oscillations are found, and the shock front is sharp and straight. It is verified that the a posteriori criteria (5.32) are satisfied for the density ρ, the x-velocity component u, the y-velocity component v, the energy e, and the pressure p across the shock. (Figure 4 and Table 3)

When the computation is repeated at $Re_{\Delta x, \infty} = 50$, exceeding the critical value $(Re_{\Delta x, \infty})_c = 24.5$ for the same flow configuration, substantial oscillations are present immediately downstream of the shock. The a posteriori criteria (5.32) for all of the physical variables are found violated. The peak amplitude of the oscillation is about 10%, but such oscillations are essentially damped out a few meshes downstream of the shock. The downstream asymptotic values are reached well within the field of computation; the results obtained from the computation at $Re_{\Delta x, \infty} = 50$ are correct to within 0.3% of the Hugoniot values. (Figure 5)

The smooth incident shock computed at $Re_{\Delta x, \infty} = 10$ was then allowed to be reflected from an inviscid wall. For the reflected shock, the critical Reynolds number is $(Re_{\Delta x, \infty})_c = 4/(0.837 - 0.646) = 21$, which exceeds the $Re_{\Delta x, \infty} = 10$ used in the computation. A smooth, straight reflected shock is obtained. All of the computed downstream asymptotic values agree with the theoretical values to better than 0.1%, and there are no oscillations.

	FLOW	REGION		
	2		**3**	
	NUMERICAL	THEORETICAL	NUMERICAL	THEORETICAL
ρ	1.573	1.575	2.460	2.457
u	0.837	0.837	0.646	0.648
v	-0.181	-0.181	0.000	0.000
e	1.214	1.213	1.466	1.463
p	1.910	1.910	3.600	3.597
θ	-12.23°	-12.23°	12.23°	12.23°
β	41.84°	41.84°	46.23°	46.14°

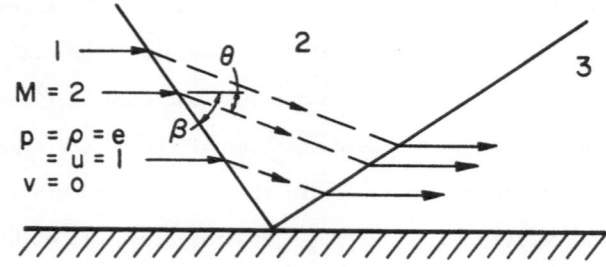

TABLE III
INVISCID SHOCK-REFLECTION CALCULATION

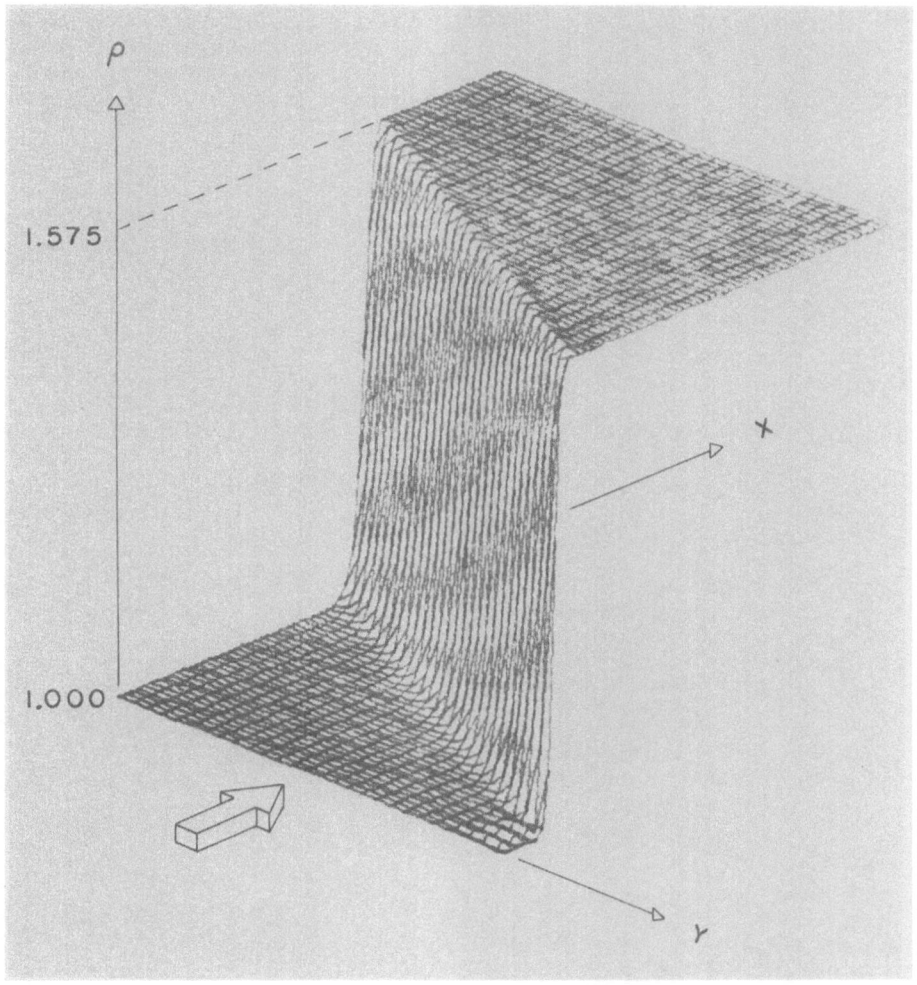

STEADY STATE DENSITY PROFILES

COMPUTED AT $Re_{\Delta x, \infty} = 10$

FIG. 4

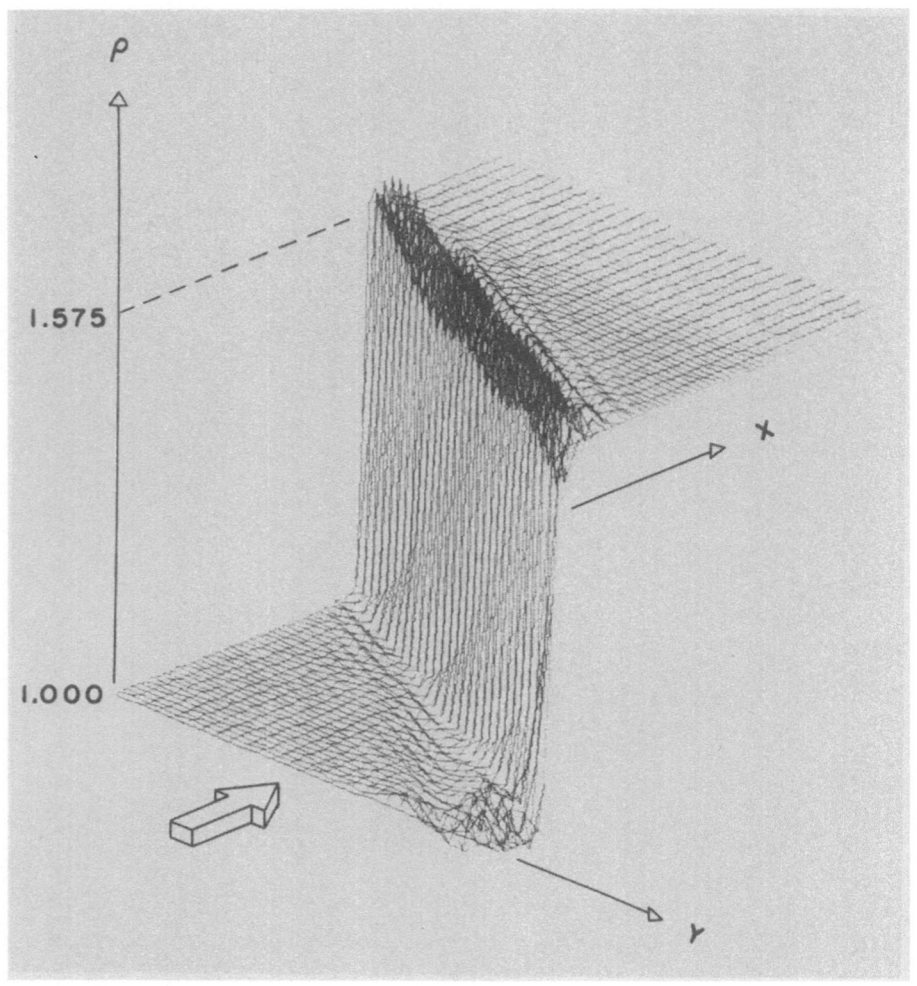

STEADY STATE DENSITY PROFILES

COMPUTED AT $Re_{\Delta x, \infty} = 50$

FIG. 5

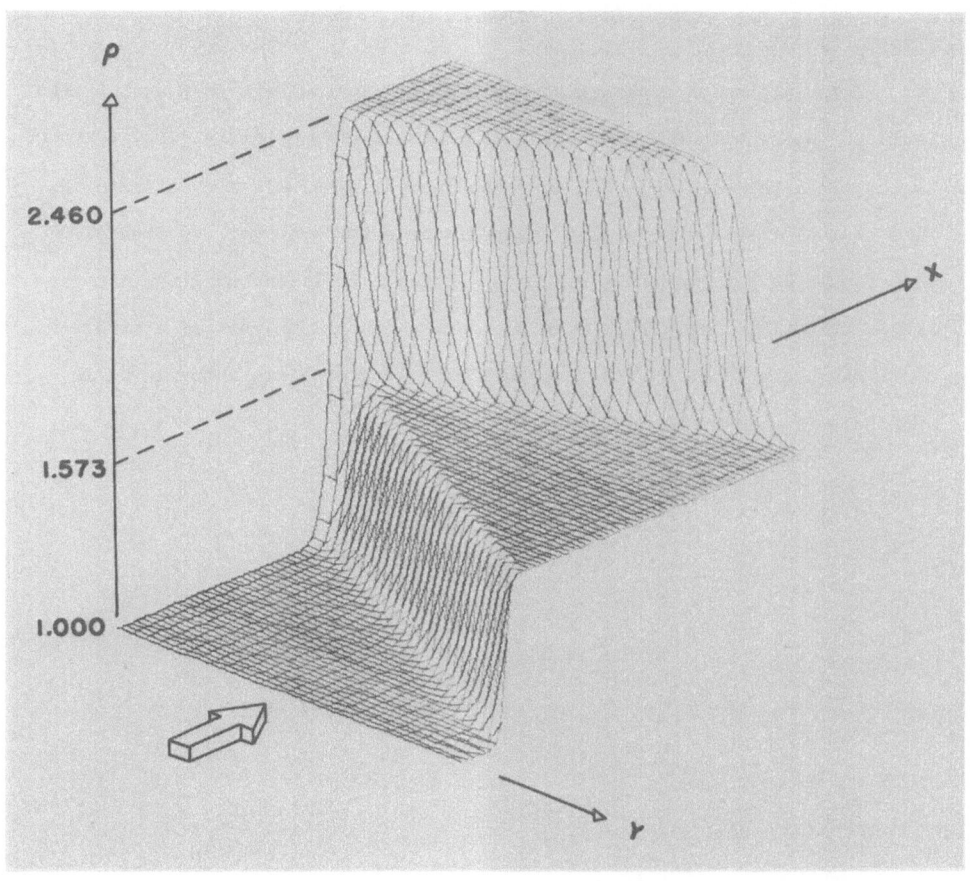

STEADY STATE DENSITY PROFILES

COMPUTED AT $Re_{\Delta x, \infty} = 10$

FIG. 6

(Figure 6)

Computations at intermediate values of $Re_{\Delta x}$ indicate that oscillations begin to appear with $Re_{\Delta x}$ exceeding 10 to 15, increase fairly rapidly around the critical value of 20 - 30, and keep increasing slowly with a larger $Re_{\Delta x}$. This gradual rather than abrupt change in behavior with $Re_{\Delta x}$ is probably what should be expected in a nonlinear system. It is encouraging that the simple criterion obtained from an elementary linear analysis of a simple model may prove to be useful in complicated flow problems encountered in practice.

VI. CURRENT STATUS AND FUTURE PROSPECT

The various problems associated with the numerical integration of the
Navier-Stokes equations have been reviewed in the previous chapters, as to
the mathematical origin of the problems and the basis of various current
techniques for dealing with them. This approach was chosen in preference
to a review (in the form of a glossary) of various solutions from the
literature to facilitate the presentation of an overall view of the problem.

In the days of mechanical desk calculators or card programmed calculators
(CPC), the numerical integration of the hydrodynamic equations was attempted.
The primary concern at that time was the limitation on the computational speed
offered by these machines. While the question of computational stability was
known to mathematicians[1], it was not of much concern to the practitioners.
The dawn of high-speed electronic computers in the mid-1940's changed all that,
demonstrating how often an apparently straightforward computation will lead
to unbounded meaningless results. This problem of stability deserves to be
the first and the most pressing one presented by high speed computation,
because unless the stability question is successfully resolved, no results
of any kind can be obtained. Since the mid-1940's, this stability question
has been studied very extensively, both mathematically and empirically. As
described in Chapter III, much has been learned and understood since then.
But when complicated sets of partial differential equations such as those of
gas dynamics are to be integrated, computational stability remains a formidable problem. As in the older days, so much work is still needed to achieve
a stable computation that one often hesitates to ask any further questions
about any reasonably looking computed solution. For those interested in the
use of computational methods for practical purposes, computational stability
is not the single major problem in obtaining a numerical solution of a

partial differential equation system, but is only a first step in achieving
a solution of value.

With the help of suitable model studies and appropriate choices of
difference algorithms, computational stability can generally be obtained
after some hard work as may be tested in actual machine computation. Now it
is the time to be concerned with obtaining not only some qualitatively correct
solutions, but also quantitatively accurate answers and some estimate of the
error bounds on the computed solution. In applications, the primary purpose
of a computed solution is to seek some reasonably accurate quantitative
estimate of some physical quantities in the flow field. Of course, the accuracy requirements for different applications may vary greatly. Whether or not
a solution is sufficiently accurate for a specific application can only be
judged under criteria dictated by considerations external to the mathematical
analysis. But such a judgment can be made only when the computed solution
is accompanied by some error bound, if not by a strict error estimate. The
error bounds on a computed solution are no less important than the error bars
of a set of experimental data, if such computed solutions are to be practically
useful. With this in mind, the preliminary developments on computational
accuracy given in Chapters IV and V are quite important in practice, although
may not be widely appreciated. Most of the solutions available in the published literature were probably obtained primarily to demonstrate qualitatively what can be done, rather than to solve specific problems in application.
Little attention has been paid to the accuracy of those computed results. In
the few examples described below, we hope to illustrate that, with proper
attention to certain details, quantitatively accurate and useful results can
be obtained for some complicated fluid flow problems.

6.1 Hydrodynamics

The flow of an incompressible viscous fluid in two space dimensions probably represents a simple, non-trivial form of the Navier-Stokes equations. It is most often treated in the stream function-vorticity formulation. The mass continuity equation in two space dimensions (x,y)

$$\frac{\partial u}{\partial x} + \frac{\partial v}{\partial y} = 0 \qquad (6.1$$

can be satisfied by a scalar stream function Ψ defined by

$$u = \frac{\partial \Psi}{\partial y} \qquad \text{and} \qquad v = -\frac{\partial \Psi}{\partial x} \quad , \qquad (6.2)$$

while the vorticity component ω normal to the x-y surface is

$$\nabla^2 \Psi = \frac{\partial^2 \Psi}{\partial x^2} + \frac{\partial^2 \Psi}{\partial y^2} = -\omega(x,y) \quad . \qquad (6.3)$$

The curl of the momentum equation reduces to the vorticity transport equation

$$\frac{\partial \omega}{\partial t} + \frac{\partial \Psi}{\partial y} \cdot \frac{\partial \omega}{\partial x} - \frac{\partial \Psi}{\partial x} \cdot \frac{\partial \omega}{\partial y} = \nu \nabla^2 \omega \qquad (6.4)$$

The divergence of the momentum equation gives the $\nabla^2 p$ in terms of Ψ and ω. Thus the static pressure p can presumably be found independently after the stream function Ψ and vorticity ω have been determined. The solution of a hydrodynamic problem is therefore posed as the simultaneous solution of two elliptic problems for Ψ and ω (represented by Equations 6.3 and 6.4) subject to Dirichlet and/or Neumann boundary conditions on a closed boundary. The physical boundary conditions depend on the particular problem.

A simple case is the decay of a vortex in a closed rectangular box, in which case $u = v = 0$ on the boundary (taken as $x = 0$, $y = 0$, $x = 1$, $y = 1$). This set of physical boundary conditions has to be translated into boundary conditions of Ψ and ω. By definition, $\Psi = 0$ may be assigned on the boundary. Equation (6.3) then serves to determine $\Psi(x,y)$ completely, when $\omega(x,y)$ is given over the field. The remaining physical boundary conditions are

$$\frac{\partial \Psi}{\partial y} = u = 0 \quad \text{on} \quad x = 0 \quad , \quad x = 1$$

$$-\frac{\partial \Psi}{\partial x} = v = 0 \quad \quad y = 0 \quad , \quad y = 1 \tag{6.5}$$

A practical question arises concerning how (6.5) may be expressed as boundary conditions on ω in the solution of Equation (6.4). In practice, this question is by-passed by first solving Equation (6.3) for the advanced values of $\Psi(x,y)$, and then estimating the boundary values of ω from the most recently available advanced values of Ψ near the boundary. This can be done with or without the conditions (6.5) taken into consideration. In principle, the boundary conditions (6.5) should at least be checked a posteriori. There is clearly an error τ_B in the boundary values of ω of the order of Δt, Δx, and/or Δy depending on the formal order of accuracy of the algorithm how the boundary values of ω are calculated from the values of Ψ near the boundary.

Now if Equation (6.4) is intergrated over the volume (x = 0 to 1 and y = 0 to 1) and over the time period (t = 0 to t) of the integration, the total decay of the vorticity is

$$\int_V [\omega_o(t = 0) - \omega(t)] dV$$

$$= \nu \int_o^t dt \int_V \nabla^2 \omega \ dV$$

$$= \nu \int_o^t dt \int_S (\vec{\nabla}.\omega) \cdot \vec{dn} \tag{6.6}$$

i.e., it is proportional to the total outflux of the gradient of vorticity $\vec{\nabla}\cdot\omega$ on the boundary. (The three dimensional analog is obvious.) Thus the non-random cumulative error on the total decay of the vorticity in the box will be of the order $NJ\tau_B$, where N is the number of time steps intergrated and J is the number of spatial meshes in a linear dimension. $\vec{\nabla}\cdot\omega$ on the boundary is assumed to be of the same order of the error in the boundary vorticity τ_B itself (although it is likely much larger). The use of the integral formula has implied that the accumulation of truncation errors over all interior points in the difference calculation has been neglected. Even so, the total decay of the vorticity at later times depend very importantly on how accurately the boundary vorticity is formulated in the computation, and on whether and how the errors associated with such a formulation will accumulate in space (along the boundary) and in time. The question involves more than the local truncation error of the difference formulation of the vorticity boundary condition, since the correct physical boundary condition Equation 6.5 - which represents some integrated condition on the vorticity field rather than the local values of the vorticity - was ignored.

The use of the stream function and the vorticity as the dependent variables is the fundamental reason for the difficulty in implementing the boundary conditions. It also causes considerable complications in rendering a conservative formulation in order to prevent the accumulation of the truncation errors over the interior points. If the physical variables u and v are used as the dependent variables in the difference formulation, the difficulty with the boundary condition is eliminated for the above example, and the conservation of the difference formulation can be readily implemented. The advantage of the vorticity-stream function formulation in reducing the number of partial differential equations may be outweighed by this difficulty alone. The determination of pressure field often brings more serious problems.

For hydrodynamic problems with inflow and outflow boundaries in the field of computation, the boundary treatment in the difference formulation poses a difficulty of a different nature. This is because the physical boundary conditions are prescribed very far up and downstream of the field of computation. The vorticity-stream function formulation does not aggravate this situation much further, and therefore may be preferred for the numerical integration of the hydrodynamic equations. Poisson-type equations can be efficiently solved in different ways. There are many such solutions in the literature. Most of such results cannot be analyzed for an error estimate, primarily because of the non-conservative form of the difference formulation which permits the accumulation of the local truncation errors. Experimental data is generally not available to provide a quantitative estimate of the error in the computed results. Most of such computations serve to demonstrate over and again the feasibility of computing some "reasonable" approximate solutions for different flow fields, but fail to demonstrate their quantitative value. A numerical study of the steady flow of a uniform stream over a sphere will be presented below to illustrate the point.[10] (A sphere was preferred to a circular cylinder since the sphere data are not subjected to the experimental uncertainties as to the two dimensionality of the test configuration).

The flow field of a uniform stream over a sphere is conveniently described by using the spherical polar coordinates. To extend the outer boundary of the field of computation as far downstream as possible (to facilitate the implementation of the boundary conditions), $z = \ln r$ is used in place of the physical radius r. Three different sets of numerical integration have been made by different authors at common Reynolds numbers of 40 and 100.[20,21,22] There is also a set of experimental data by Taneda[23] of some characteristic quantities of the recirculatory wake flow field at these and other Reynolds numbers. Such measured values of wake length and locations of the separation point and the vorticity centers provide for comparisons of the detailed flow field in the most sensitive region, in addition to the overall drag coefficient acting on the sphere.

Jenson,[20] and Hamielec, et al[21] used similar difference relaxation procedures and the same downstream boundary conditions approximating uniform out flow, (Table 4). Both cases were carefully executed and examined very carefully

TABLE IV

Outflow boundary conditions. Sphere wake calculations based on Navier-Stokes equations.

Re_D	Authors	Location radii	Vorticity ω	Vorticity $\Delta\omega_{Oseen}$	Stream function	Velocity $q-q_{Oseen}$	Mesh size $\Delta\theta, \Delta z$
40	Jensen	3	Extrap.	5×10^{-1}	$\psi = \psi_\infty$	5×10^{-1}	$6°, \frac{1}{20}$
	Hamielec, Hoffman, and Ross	7 14	0 0	2×10^{-1} $1/3 \times 10^{-2}$	$\psi = \psi_\infty$ $\psi = \psi_\infty$	2×10^{-1} 10^{-1}	$6°, \frac{1}{20}$
	Rimon and Cheng	20	0	10^{-3}	$\frac{\partial\psi}{\partial x} = 0$	10^{-2}	
100	Hamielec, Hoffman, and Ross	7	0	5×10^{-2}	$\psi = \psi_\infty$	2×10^{-1}	$3°, \frac{1}{40}$
	Rimon and Cheng	20	0	10^{-3}	$\frac{\partial\psi}{\partial x} = 0$	10^{-2}	$6°, \frac{1}{20}$

numerically, making sure that the steady state results obtained are essentially independent of further reduction in the mesh spacing from $\Delta\theta = 6°$ and $z = 1/20$. They obtained a drag coefficient C_D in agreement with that expected from the experimentally well-established "standard drag curve". However, the details of the two solutions were much different. For example, the two results for the vorticity on the wake side of the sphere surface differ by a factor of 2 to 3 for the case with $Re_D = 40$, (Fig. 7a). The streamline patterns in the recirculatory wake are visibly different, although qualitatively similar. Such differences in the detailed results clearly demonstrate the importance of the cumulative effects of the truncation errors due to the non-conservative nature of the difference algorithms which are equivalent to their relaxation procedures, despite the good agreement in such overall results like the drag coefficient C_D. Jenson's results depart considerably further from Taneda's wake data than do the results of Hamielec, et al at $Re_D = 40$, (Fig. 8). Hamielec, et al also calculated the case $Re_D = 100$. They found it necessary to refine the mesh to $\Delta\theta = 3°$ and $\Delta z = 1/40$ to secure a reasonable steady state, and also had to introduce some fine adjustments in order to reproduce the experimental value of the drag coefficient C_D at $Re_D = 100$.

Rimon and Cheng[22] employed the Gauss-Seidel, over-relaxation procedure, and succeeded in developing a conservative difference form that is still reasonably simple despite the contracted curvilinear coordinates and stream function-vorticity formulation. The same mesh size $\Delta\theta = 6°$ and $\Delta z = 1/20$, as used by the previous authors, was employed. The conservative nature of the difference formulation permits an estimate of the upper bound of the cumulated truncation error via Equation (5.14), in which the $Re_{\Delta x}$ should be replaced by $Re_{\Delta z}$ for this calculation in terms of $\Delta\theta$ and Δz. The magnitudes of $Re_{\Delta z}$ for the two cases with $Re_D = 40$ and 100 can be estimated from the computed solution, based upon the velocity gradient in the region near the isolated rear stagnation point in the

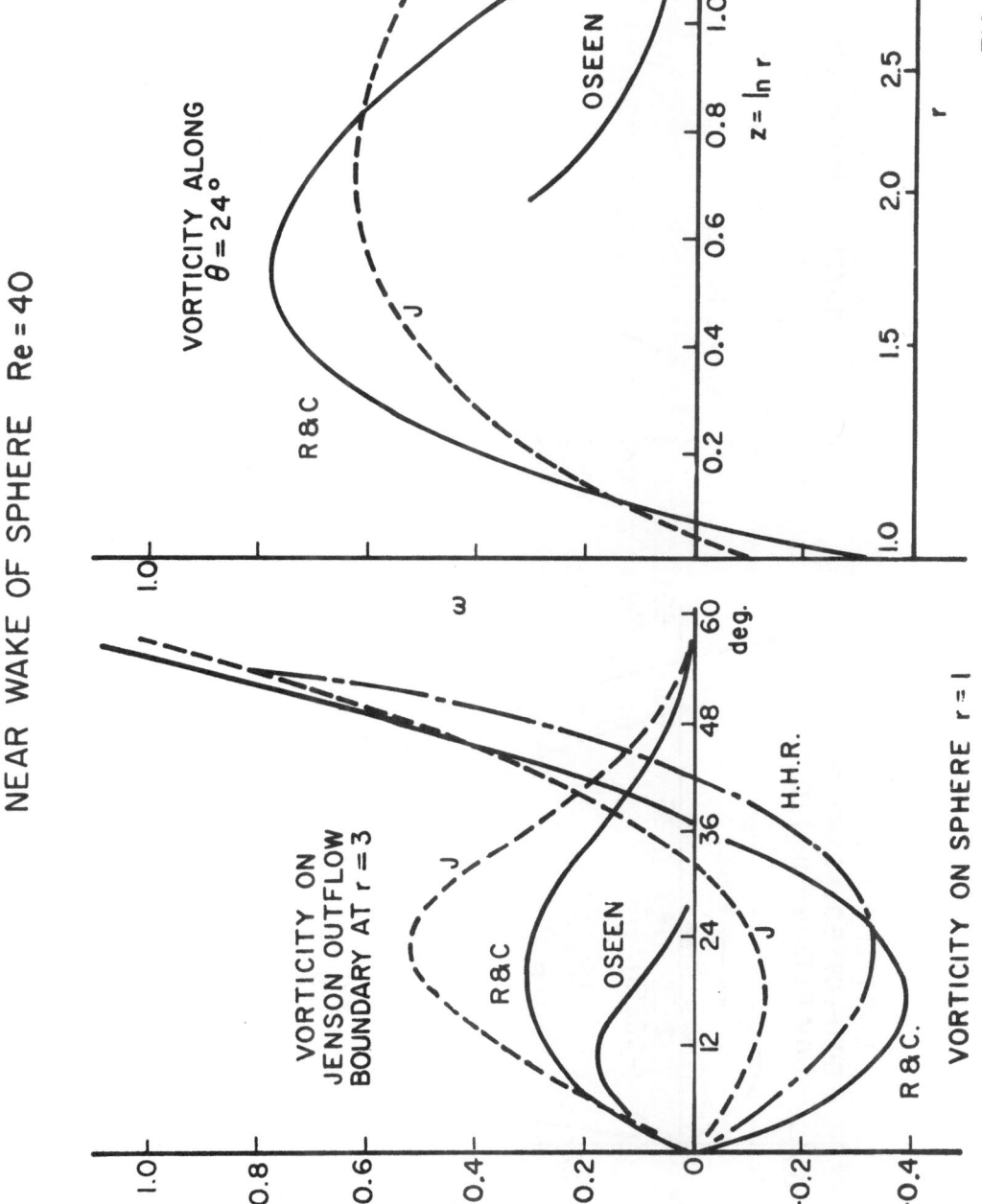

NEAR WAKE OF SPHERE   Re = 40

VORTICITY ALONG
θ = 24°

OSEEN

J

R&C

$z = \ln r$

ω

r

FIG. 7a

VORTICITY ON
JENSON OUTFLOW
BOUNDARY AT r = 3

R&C

J

OSEEN

H.H.R.

J

R&C.

VORTICITY ON SPHERE r = 1

deg.

ω

NEAR WAKE OF SPHERE Re = 100

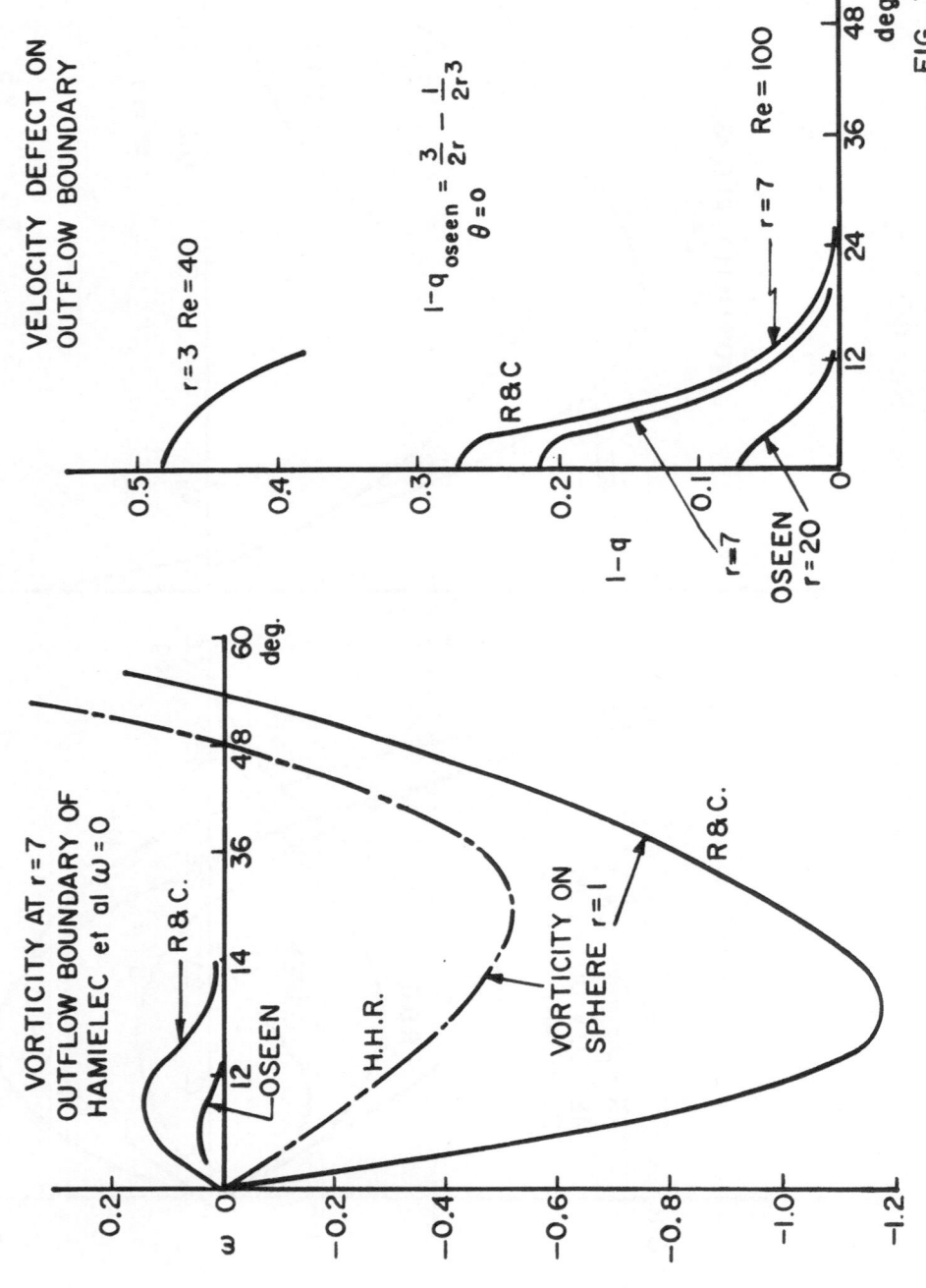

FIG. 7b

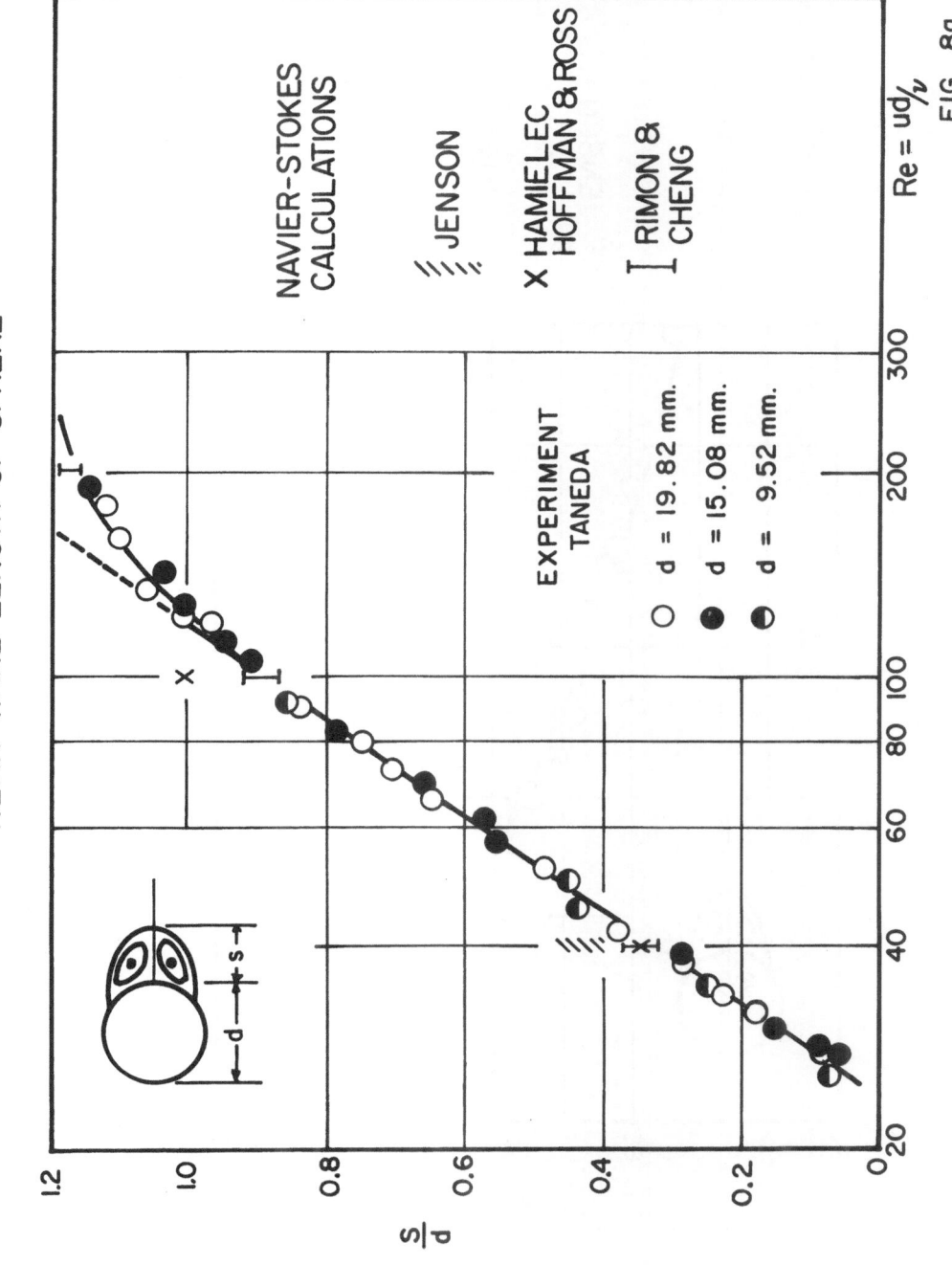

NEAR WAKE LENGTH OF SPHERE

NAVIER-STOKES CALCULATIONS

JENSON

X HAMIELEC HOFFMAN & ROSS

RIMON & CHENG

EXPERIMENT TANEDA

○ d = 19.82 mm.

● d = 15.08 mm.

◑ d = 9.52 mm.

$Re = \dfrac{ud}{\nu}$

FIG. 8a

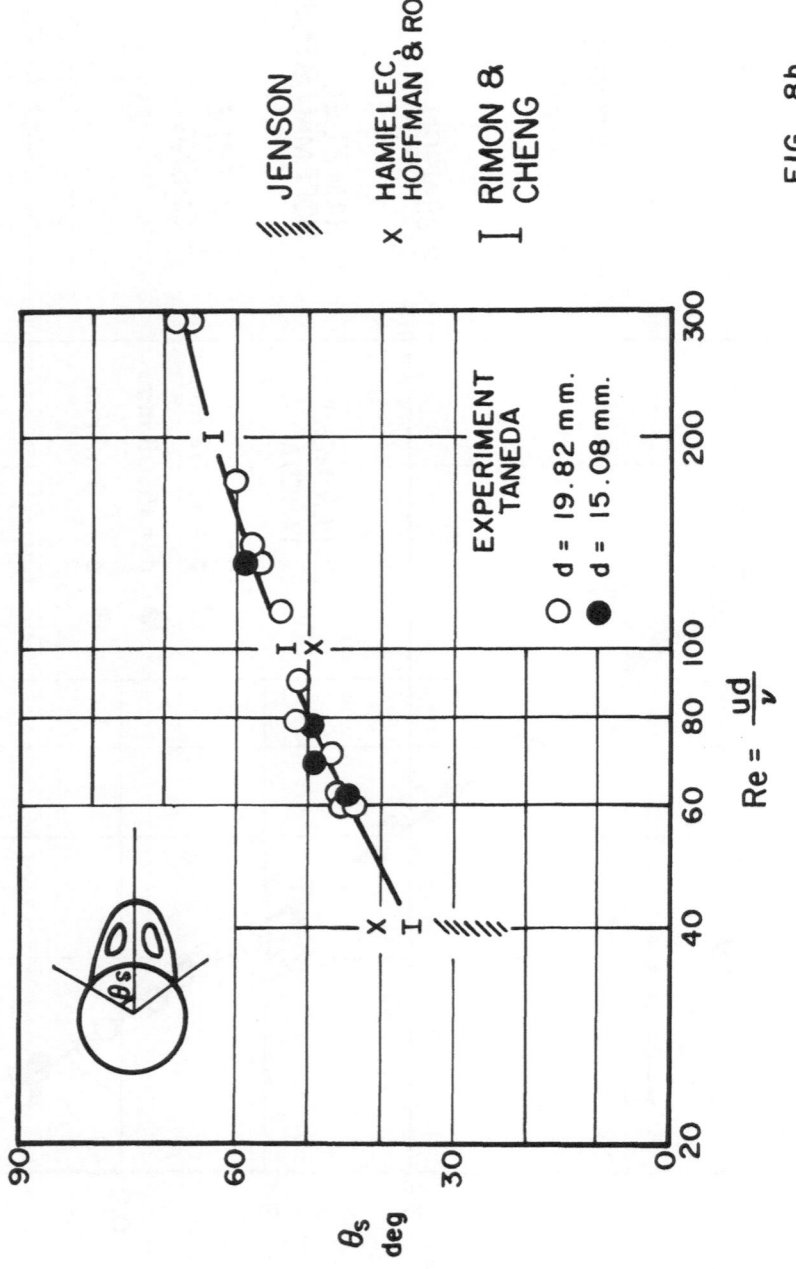

EXPERIMENT
TANEDA

○ d = 19.82 mm.
● d = 15.08 mm.

θ_s
deg

$Re = \dfrac{ud}{v}$

////// JENSON

× HAMIELEC,
HOFFMAN & ROSS

I RIMON &
CHENG

FIG. 8b

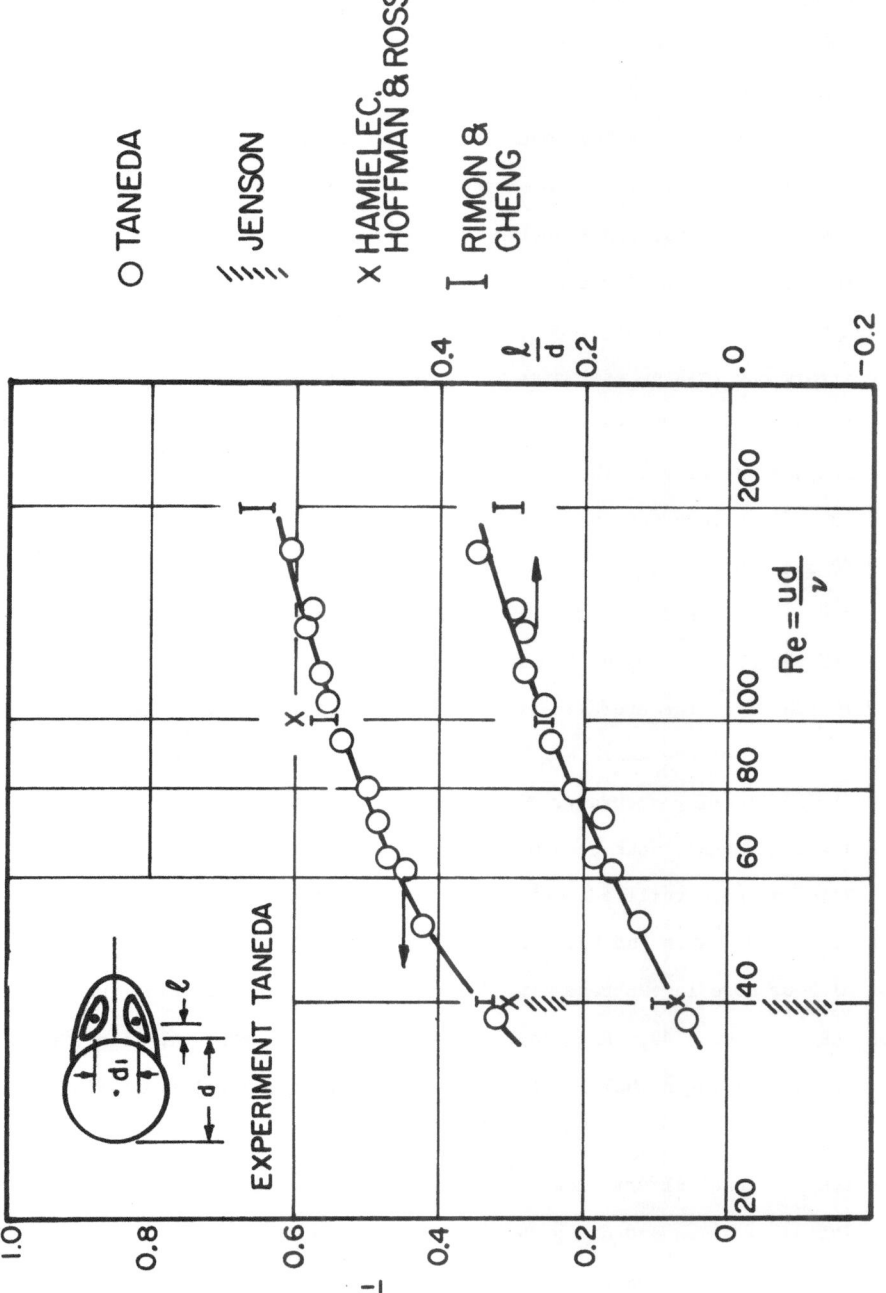

FIG. 8c

wake according to Equation (5.31) with the reference velocity of this mesh Reynolds number based on the velocity difference across the large gradient region. These magnitudes are less than $\frac{1}{2}$ and 1 respectively. Accordingly, the absolute upper bounds of the accumulated truncation errors are $3 \times 10^{-2} \times Re_{\Delta z}^2 \sim 1$ and 3% respectively. The extrapolation condition at the downstream boundary gives the largest contribution to the boundary error. (Both $\frac{\partial \psi}{\partial x} = 0$ and $\omega = 0$ on the out flow boundary commit a fractional error as much as 100%. They are not expected to err in sign, however). The absolute upper bound of the boundary errors may then be estimated with Equation (5.16), where $\varepsilon_b' = 1$, and Re is based on the maximum velocity in the wake region and the length from the rear stagnation point to the out flow boundary in the z-θ plane of computation. This is more than two sphere diameters. The effective Reynolds numbers are then 80 and 200 respectively. Accordingly, the bounds on the boundary errors are estimated as $2\varepsilon_b'/Re$ of 2.5% and 1% respectively. By adding the estimates of the absolute upper bounds of the truncation errors and the boundary errors for each case, the overall estimates of the absolute error bounds are about 3.5% and 4% for the cases $Re_D = 40$ and 100 respectively. This is quite satisfactory engineering accuracy. Thus, the computed results were expected to agree well with Taneda's wake data even for certain details of the wake flow fields. This has been verified, (Fig. 8). The computed vorticity fields in the near wake region of Rimon and Cheng and of Hamielec, et al, however, differ by a factor of 2 or more in the case with $Re_D = 100$, while they differ by much less for the case $Re_D = 40$, (Fig. 7b). This again demonstrates the significance of larger mesh Reynolds numbers on the accumulation of the local truncation errors.

The computational effort expanded in the solution of this problem following the formulation of Rimon and Cheng was not excessive at the time it was done, and is rather small in terms of present-day computing machines. $61 \times 31 = 1891$

mesh points were used. A steady state solution was obtained in about an hour computation on the IBM 7094, with the potential flow field taken as the initial data. In terms of CDC 6600 machine time, the solution would take less than 10 minutes. Fortran language was used without paying much attention to programming efficiency. The computational time can be appreciably reduced if an approximation more accurate than the inviscid flow field should be used as the initial data, and if more attention should be paid to programming efficiency. It is therefore believed that with conscientious effort in constructing the difference formulation, useful quantitative results may be obtained from numerical solution of the Navier-Stokes equations for hydrodynamic flow problems, somewhat more complicated than the sphere problem dealt with here.

The extension of such calculations to steady flows in three dimensional space and to higher flow Reynolds numbers will, however, be more complicated. It will need not only substantially more computer time, but also some analysis in order to gain understanding of certain intricacies in the truly 3-D problems such as the computations in the vicinity of the "separation lines". With greatly increased capability of high speed computers in the foreseeable future to provide the much needed speed and resolution, it is hopeful that good quantitative results even for these practical steady flow problems in three spatial dimensions may be obtained.

Time-dependent hydrodynamic problems in three space dimensions are considerably more difficult and demanding. This is especially true if hydrodynamic turbulence is the subject of investigation. The high frequency components of the turbulent fluctuations can doubtfully be treated with a reasonable accuracy, despite the giant stride in the capability of computing machines foreseen in the future. It appears that some phenomenological theory for the high frequency turbulent components will be needed, while the low frequency components can perhaps be satisfactorily handled by computational methods. This statement is meant to apply whether the integration takes place in the physical space for the physical variables or in the Fourier space for the Fourier components of the physical variables. Much work is needed in any case.[24,25,26]

6.2 Supersonic Gas Dynamics

The gas dynamic equations are basically the same as the hydrodynamic equations, except for the variations in the gas density and in the diffusivities, and for the addition of the equation of energy balance (1.3). The outstanding feature of supersonic flow fields is the presence of shock waves, either generated from within the field, or incident on the flow field from without. Most of the practical problems that call for a numerical treatment of the Navier-Stokes equations involve the generation of shock waves due to the interaction of the inviscid and viscous streams. The computation of a shock wave of unknown strength and location presents considerable difficulty, as was discussed in the previous chapter. The shock-induced oscillations in its neighboring flow field are detrimental to the appearance of the computational solution. Such solutions are often presented after some artificial averaging or filtering procedure, and can therefore be of qualitative value only. Indeed, those solutions relatively free from this criticism owe their success to avoiding the serious consequences of a shock standing in an important part of the flow field. By carefully selecting the field of computation for the problems to be investigated, the consequences of shock-induced oscillations are minimized.

Allen and Cheng[16] treated the near wake flow imbedded in a supersonic stream turning over a sharp shoulder of a flat base with a "recompression shock" generated from the turning of the supersonic stream caused by the closing of the recirculatory wake. In the steady state solution of this problem, the small oscillations caused by the recompression shock appreciably distort the computed results only in the far downstream portion of the rejoined wake flow field near the downstream boundary. Although the oscillations of the flow properties in the flow field are equivalent to those induced by an oscillation of the shock front of only $1/4\Delta x$, they remain as one of the two largest sources of computational errors. It is conjectured that the likely source of the small oscillation is the inaccurate extraneous difference treatment where the shock emerges from the downstream boundary of the field of computation. The conservative difference form of the class (5.18) was used, and the criterion (5.35) was satisfied (although without a substantial margin). Unfortunately, comparable experimental data are not available, and the extension of this calculation to the range of practical Reynolds numbers of 10^3 - 10^4, and for a somewhat more complicated geometrical configuration, was beyond available means then (computation time and storage capacity).

Ross and Cheng[27] studied the question of the permissible ranges of Reynolds numbers and Mach numbers such that the computational solutions with the previous formulation will possess an absolute upper bound on the error of no more than 10%. They limited the number of mesh points to 2100, used an "optimal" ratio of $\Delta x/\Delta y$, and modified their boundary treatment in a non-essential (but simplifying) way. The computational effort was limited to 10-15 minutes of computing time on the IBM 360-91 (equivalent roughly to 20-30 minutes on the CDC 6600, or 4 to 10 hours on the IBM 7044, originally used by Allen). When other restrictions of purely a fluid mechanical nature are superposed, it was established that the range of validity of the computational formulation can be extended to $M \overset{\sim}{=} 4$ and Reynolds numbers of $\sim 1\text{-}2 \times 10^3$ (based on half width of the base.) To extend this computation to the practical range of interest would require substantial refinement in the mesh size, with a corresponding increase in the computational effort. The storage limitation of the computer did not seem restrictive, but rather the computer time and cost that was prohibitive. It may be that an absolute upper bound of 10% is too restrictive, since the maximum fractional error in the solution is likely to be substantially less than the

absolute upper bound. A substantial decrease in the estimate of the computational effort will follow a modest reduction of the accuracy requirement - if the
method of error estimate described in Chapter V should be granted in the absence
of any direct comparison with reliable, appropriate experimental data.

Carter[28] chooses to integrate the Navier-Stokes equation for a steady
supersonic viscous flow over a compression ramp or corner with an imbedded
separated region. The compression waves will eventually coalesce into a shock
wave. Carter kept the upper boundary of the field of computation sufficiently
close to the viscous region so that the waves generated from the viscous layer
may be treated (without serious error) as isentropic waves, and utilized the
simple wave extrapolation condition on the upper boundary. This stratagem, as
was used in the treatment of the near wake problem,[16,27] serves to eliminate
the major part of the undesirable wave reflections from the upper boundary.
By restricting the field of computation to such a narrow strip, and by using a
highly refined mesh with Brailovskaya's difference algorithm (a member of the
class (5.18)), results which compare favorably with experimental data can be
obtained in the comparable Reynolds and Mach number ranges. This difference
formulation is probably not quite conservative, due to the use of the "curved"
body coordinates. But the curvature is sufficiently small (or otherwise localized) so that the accumulation of truncation errors may not be excessive. While
an estimate of the error bounds has not yet been made, the evidence seems to
indicate that this calculation may have come very close to directly generating
some useful practical results. Admittedly, the computational effort in this
calculation seemed to be excessive from the academic point of view (two or more
hours of CDC 6600 per case), but it does not appear prohibitive from the view of
engineering development. Moreover, there is substantial room for improvement if
an error estimate can be made. The 4th generation computers that will be operational shortly promise a further substantial increase in speed of computation

and in storage capability. When successfully implemented, this may render the computational effort of less concern in solving such practical problems.

An academic program has been devoted to developing techniques for handling the difficulty of computing shock waves in a complicated flow field. The results reported in Section 5.4 demonstrate some progress in this direction. There are still tremendous difficulties ahead in cases where a shock wave interacts with other incident waves or when the criterion of (5.33) becomes much too restrictive. Nevertheless, even in their present unsatisfactory state, computational results can be useful in fluid dynamics research to supplement experimental and other efforts. The following treatment of the hypersonic leading edge problem may illustrate the situation.

Over the leading edge of an infinitely thin flat plate, placed in a hypersonic or supersonic stream at zero incidence, a shock wave will develop due to viscous effects in the vicinity of the plate. In this region, the hypersonic strong interaction theory, based on boundary layer type arguments of various forms, fails to provide even a qualitatively adequate description of the flow field. It is not clear to what extent the flow situation will have to be described by the kinetic theory of gases. We feel that the continuum theory, when appropriately modified for the slip effects, should provide a good approximate description of the flow field very near the leading edge and fair smoothly into the results of the hypersonic strong interaction theory further downstream. Thus, asymptotic approximations were introduced only in the formulation of the surface slip conditions, and the full Navier-Stokes equations system was integrated numerically.

Physically, a rather strong oblique shock wave develops rapidly from the leading edge, and produces, in the downstream gas, a high pressure and temperature, both proportional to $M^2\sin^2\theta$, where θ is the local inclination of the shock front to the incoming uniform stream. It is very clear then that any small oscillations in the shock front will produce, in the downstream, corresponding oscillations of significant magnitudes when the upstream flow Mach number is ∿ 20. It is therefore critical to eliminate or suppress shock oscillations from the computation.

The conservative two-step algorithm of Cheng and Allen was used with a 40 × 30 mesh in the physical space x-y and with y = 0 describing the plate surface.[18] The leading-edge shock emerges from the downstream outflow boundary. The downstream outflow boundary is treated by second-order accurate extrapolation along the shock direction where the shock emerges, along the plate on the plate surface, and along directions linearly interpolated in between. Slip conditions are derived from two stream molecular distribution function with the molecules emitted from the plate diffusedly and accommodated fully to the plate conditions. The various computed profiles at constant y = (i-1)Δy from the plate are given in Fig. 9. Around the plate leading edge point a fairly large oscillation is generated by the discontinuous boundary, but dies out rapidly away from the plate leading edge. A minor localized oscillation developed farther downstream at about 2-4 Δy possibly due to some inappropriate difference treatment of the boundary conditions on the plate. This localized oscillation (in the thermodynamic variables) imposes no significant error on the solution.

The absolute upper bound of the errors in the smooth part of the computed solution due to the downstream extrapolation boundary condition is evaluated according to Equation (5.16) to be < 7%. The absolute upper bound of the truncation error is estimated from Equation (5.14) to be also less that 7%. With both the round-off error and the error due to the steady state criterion both less than 1%, the absolute upper bound of the error in the computed solution, away from the immediate vicinity of the leading edge and the out flow boundary is about 16%.[29]

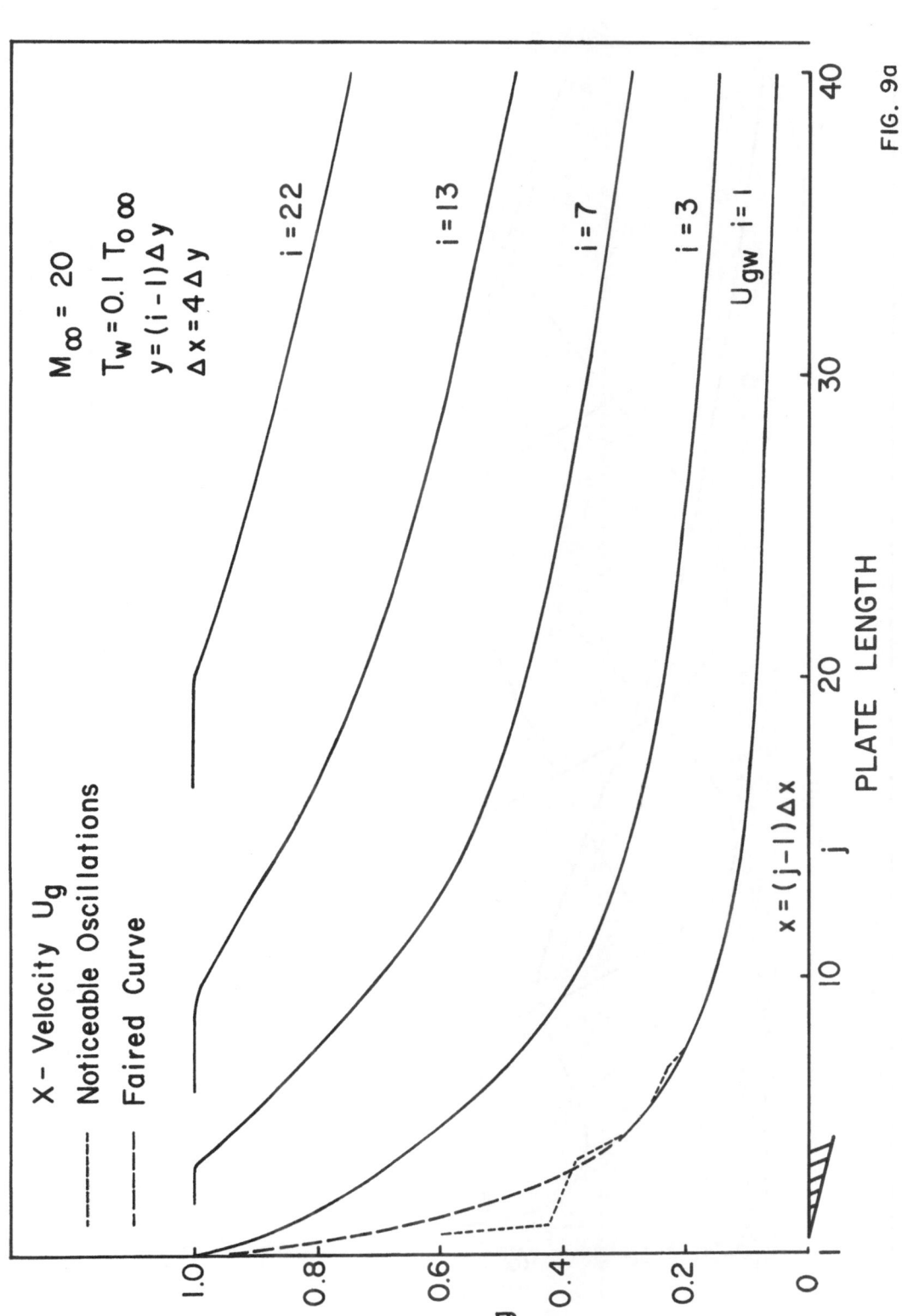

FIG. 9a

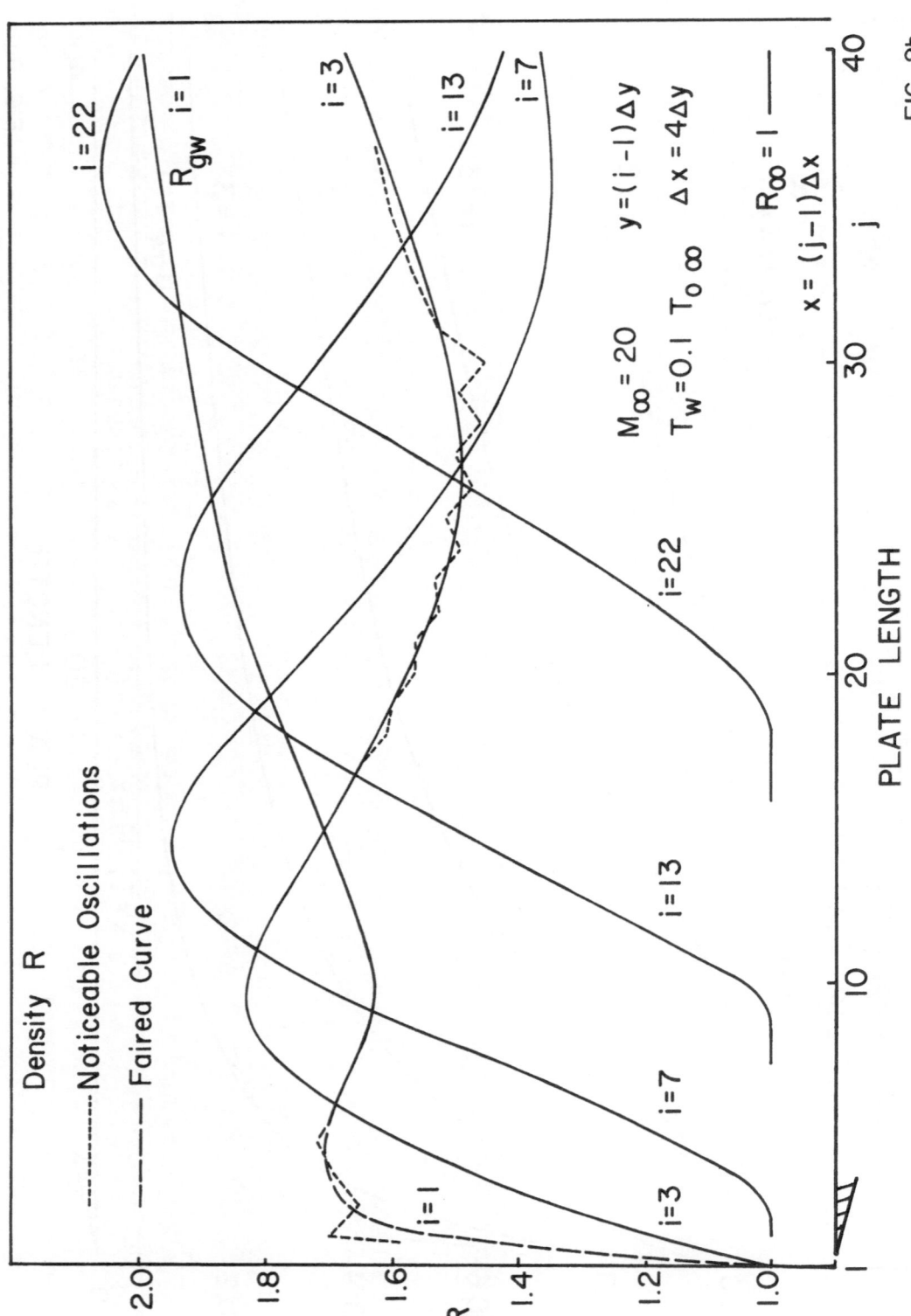

FIG. 9b

There is a collection of measured data from two different hypersonic wind tunnels at Cornell Aeronautical Laboratories and at Princeton University. The test conditions do not correspond precisely but encompassing the case computed and shown above, the hypersonic tunnels and the instrumentations were developed through many years. The model testing was tedious and difficult. The surface pressure data, reduced to dimensionless parameters, agree in general trend; but quantitatively the data from the two sources differ by a factor of two or more. Figure 10 shows how the computed results compromise the two sets of data. Examination of the details help to identify the region of validity of each set of data and the cause of divergence between the two. Figures 11a,b show how the computed results of slip velocity and slip temperature of the gas along the plate surface compare with available experimental data. Figures 12a,b show the computed and measured dimensionless parameters of surface friction and heat transfer rate. The agreement is good and is without any adjustable parameters. Thus computational methods with reasonable accuracy can provide results directly useful in engineering development.

The above computational solution also contributes to the fundamentals of continuum fluid dynamics. It provided a positive indication that the continuum description of fluid flow by Navier-Stokes equations system with appropriate slip conditions is valid for the transition regime, and that the proper surface slip conditions should be consistent with diffused emission of fully accommodated molecules from the gas-solid interface.

In the above review, there are important omissions of many interesting and significant results in the development of computational methods relevant to aerospace applications. They are omitted here to facilitate the presentation of the major themes, hopefully with as little digression as possible. While computational stability remains a problem, difficulties can generally be overcome with some hard work. Stability problems should not be permitted

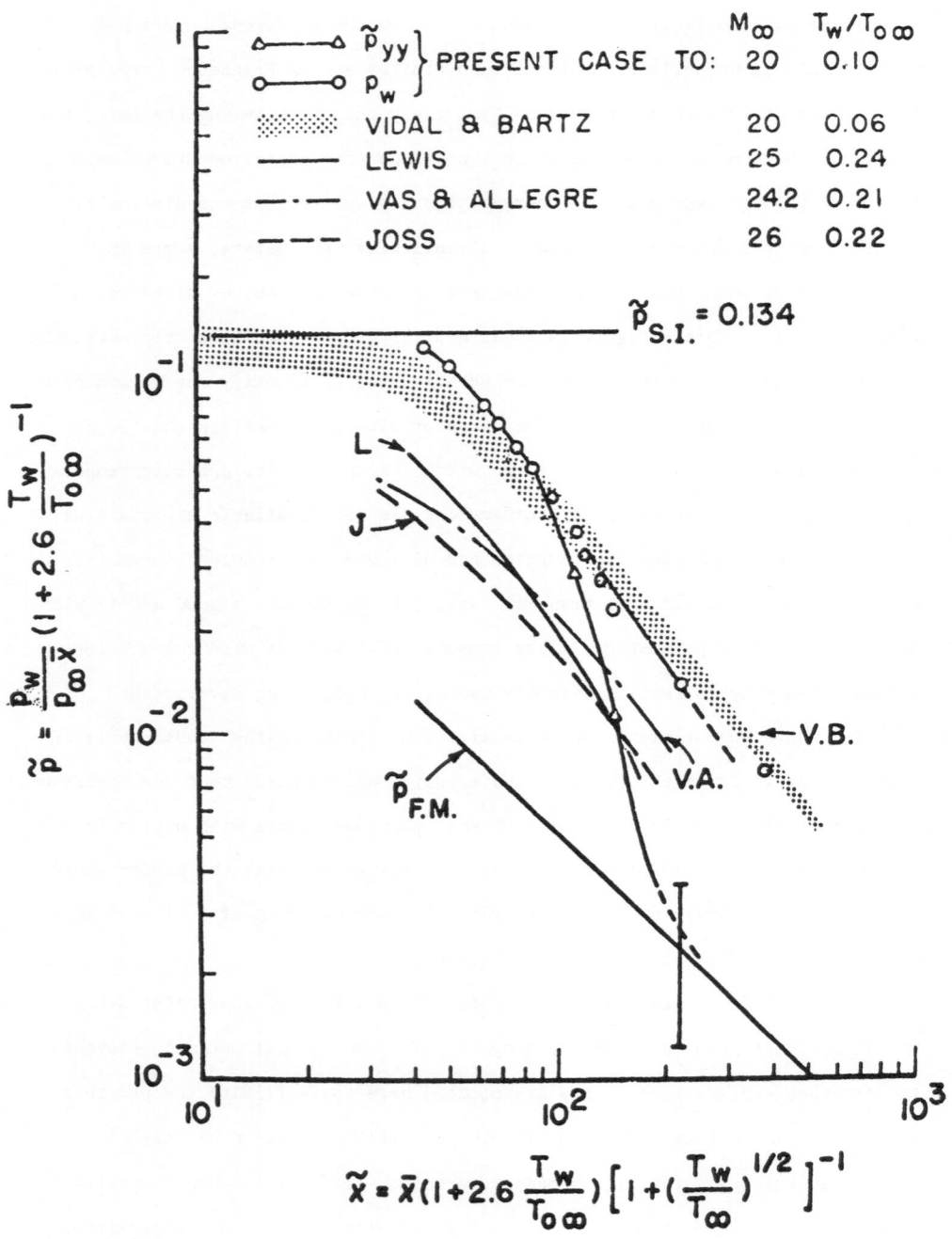

FIG. 10

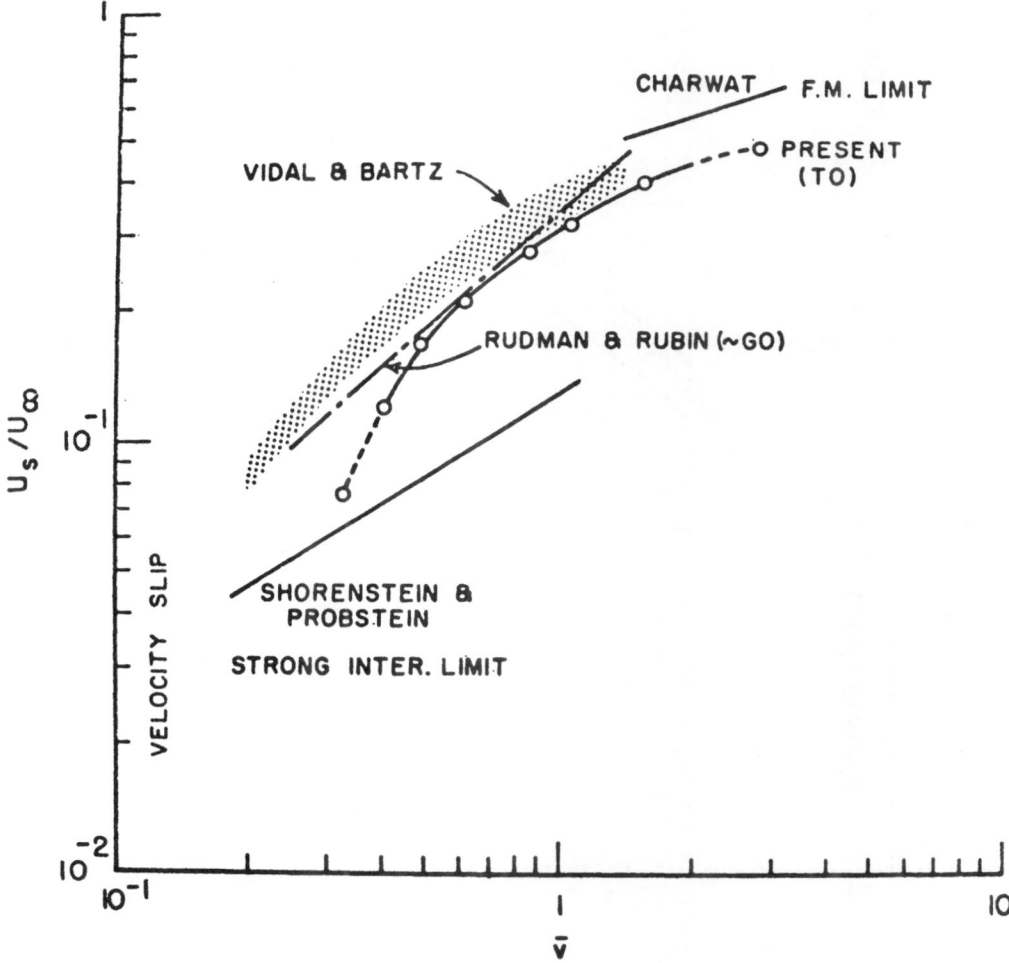

FIG. IIa

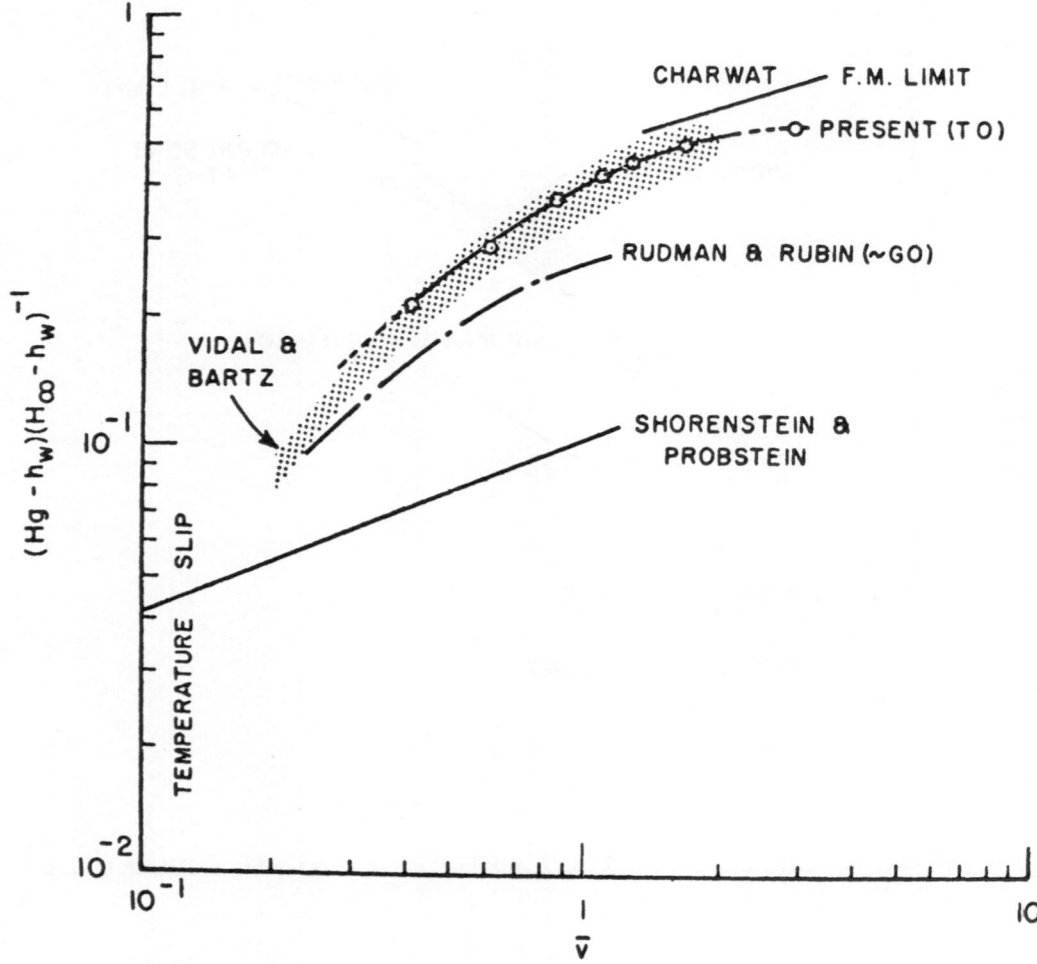

FIG. IIb

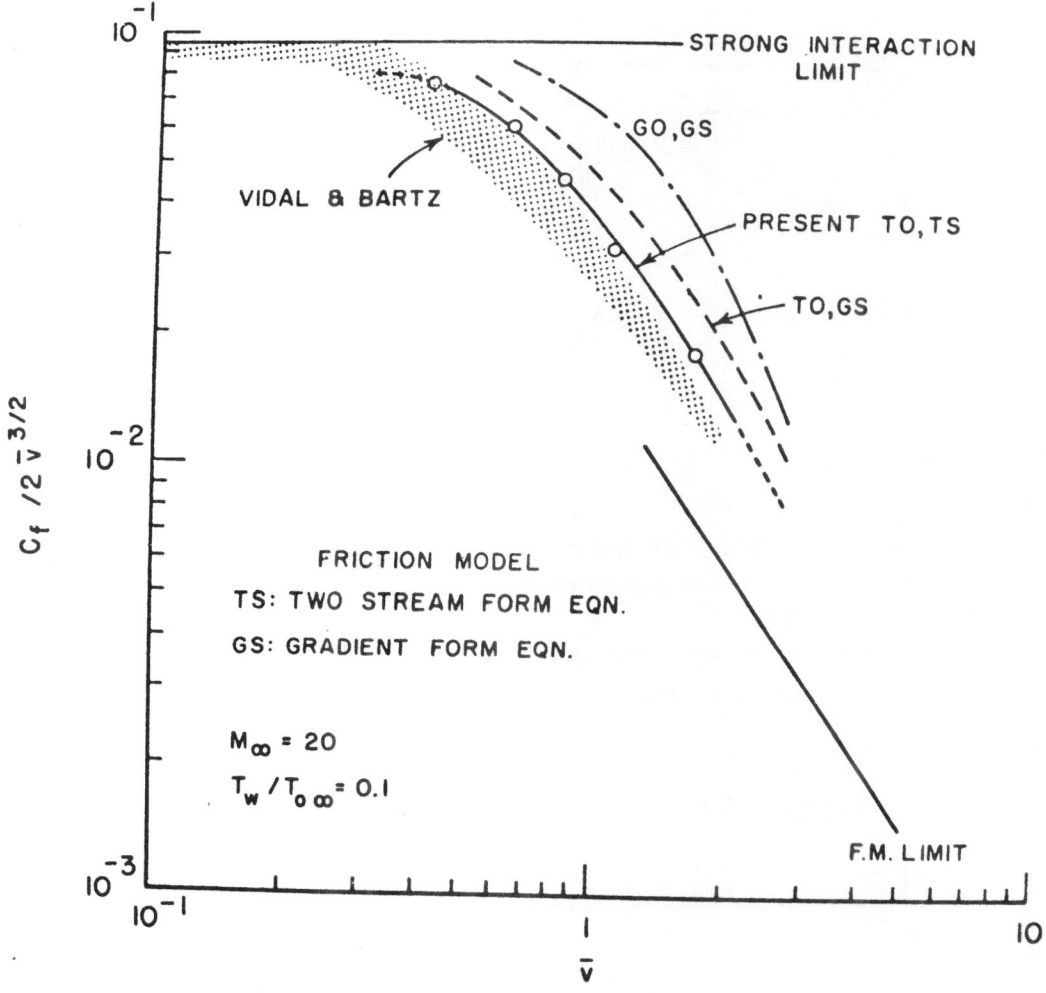

FIG. 12a

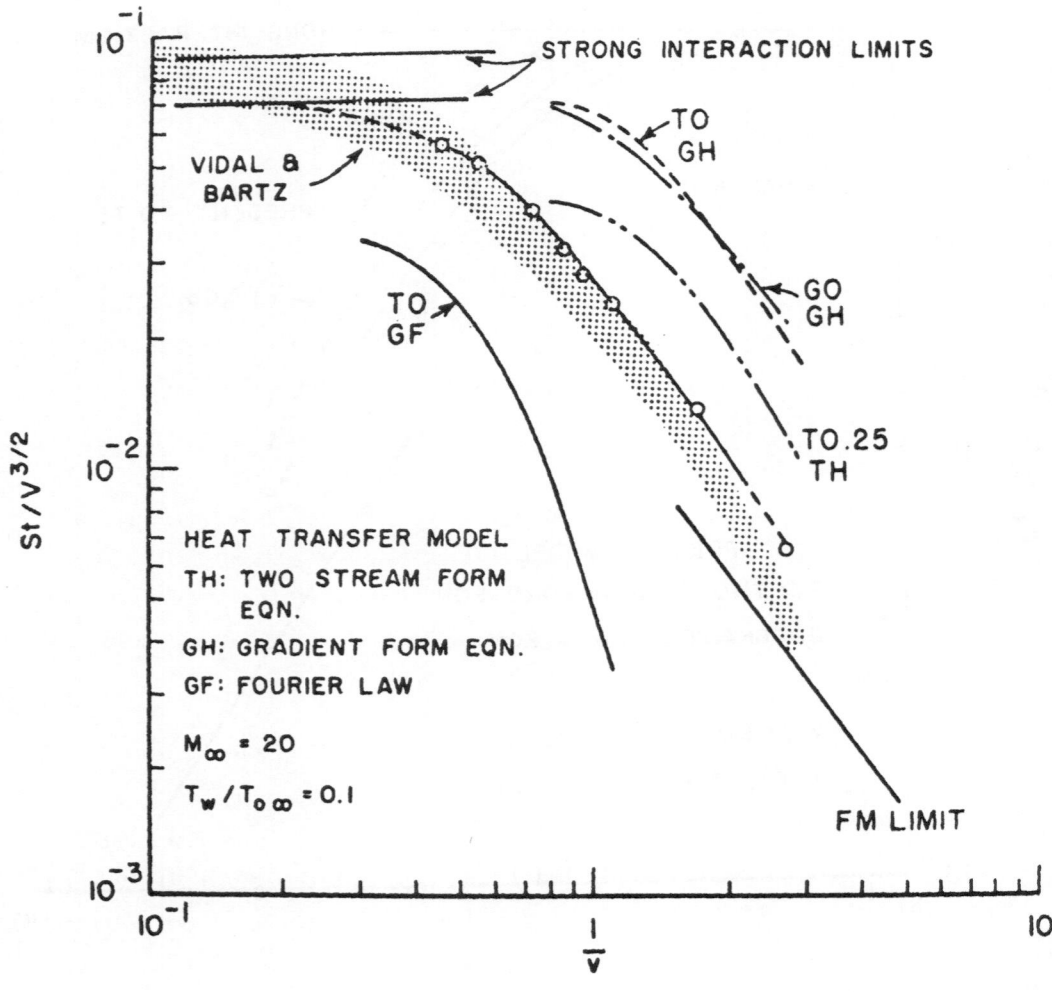

FIG. 12 b

to draw attention away from the need for reasonably accurate computed results. Stable and smooth computational results are encouraging, but can be very deceiving. From the applications point of view, the question of accuracy is crucial. Accordingly, the approach described above to secure "accurate" formulation is of fundamental importance (crude as it is). How such crude criteria may be used and incorporated is demonstrated in this chapter. Much development in this direction is needed. Some fundamental aspects should be understood, and practical methods developed to deal with the various situations. Such problems will not fade away because of the dramatic advances in computer capabilities. Indeed, there are serious problems that will be encountered in the efficient use of the fourth generation computers, if any meaningful speed advantages are to be reaped. Therefore, a few words on the prospects of the coming fourth generation computer will serve to bring to conclusion the present review.

6.3 Future Prospects with the Fourth Generation Computers

It has been a constant allusion that faster and bigger computers will provide the solution to many difficulties associated with the numerical integration of partial differential equations. Such larger and faster computers are needed, but they do not provide the brute force to resolve all computational difficulties without conscientious efforts. Certain aspects of the problems must be fundamentally understood before being satisfactorily dealt with, such as the questions of stability and accuracy. Moreover, the development of computer hardware has reached the point that order of magnitude improvement in the speed of information processing through miniaturization cannot be expected, as in the past. Fourth generation computers promise to bring about a large improvement in speed through "Parallelism" which is very much dependent on the sophistication of software

and on the nature of specific problems to be solved. These computers will
bring complicated problems to the users, as well as to the manufacturer of the
machines.

"Parallelism" is effected primarily in two different ways. Burrough
Company's ILLIAC IV speeds up the arithematic process by using 64 arithematic
units, receiving the same instruction from a common command module to
simultaneously process 64 sets of raw data. Thus arithematic results can be
"effectively" obtained 64 times faster. This is often referred to as a
"single Instruction Multiple Processor" machine (SIMP). Control Data Corporation's STAR (the STring ARay processor) employs the assembly line (or "pipe
line") technique in which a string of data is "continuously" fed into the
"pipe line" to be processed by a standing instruction. In this manner, the
arithematic unit does not become idle when the instructions are being fetched,
decoded, and installed in place to direct the computation, or when the newly
computed data is being sent out of the arithematic unit, or the raw data is
being brought into the arithematic unit. This is often referred to as the
pipe-line machine. Both the ILLIAC IV and the STAR machines possess virtual
memory capacity, i.e., the machines will manage automatically the data
stored in the external memory units used to extend the storage capacity of the
machine. Texas Instrument Corporation's ASC machine (Advanced Science Computer)
incorporates both the multi-processor and the pipe-line concept, but possesses
no virtual memory capability. All of these machines are about to be (or have
already been) delivered by the various manufactuers and are to become operational shortly.

ILLIAC IV is most efficient when the 64 arithematic processors can be
fully utilized. Any vacant processors are simply idling (doing no useful
work) when an operation is performed on less than 64 sets of data. Thus the
demonstration of the speed of ILLIAC IV vs currently available computers is

often in terms of the inversion of a 64 x 64 matrix. STAR is most efficient
when a large amount of raw data (a long string of data) is to be processed
through the same operation (i.e. single instruction), so that the "filling
time" of the pipe becomes negligible. The machine will provide a 64-fold
increase in the effective speed since each word in the STAR contains 64 bits of
binary information. The ASC machine possesses intermediate behavior. Each
manufacturer has developed powerful and intricate software to implement and enhance the advantages of the hardware. But they are all subject to the inherent
limitations of a SIMP or a pipe-line machine.

For any one of these machines, a huge amount of data must be stored, arranged, and retrieved from storage facilities. This must be done efficiently,
commensurate with the processing speed of the machine. Such can be done for
the data in core memory, directly accessible to the central processing unit
(CPU), but it cannot be done for the data stored in external memories. The
speed of a search operation or of a data transmission through the interface to
the CPU is orders of magnitude slower than the arithematic speed of the machine.
If the CPU asks for data in the external storage too frequently, the CPU would
be doing little useful computation, instead, it would be spending time trans-
mitting the data in and out of the external memory units under its virtual
memory operation. If the user should prefer to deprive the machine of its
virtual memory capability, then the user-programmer must assume the re-
sponsibility of managing the data across the interface. An alternative solution
to this problem is to expand the core memory of the CPU of the computer to
match its processing speed. This is unfortunately a very expensive proposition.
There are also other problems of data management in the CPU (though probably
not so serious as the ones just mentioned) more intimately related to the
specific characteristics (hardware and software) of each computer. These
are problems which the user cannot do much about. On the other hand, these

machines present problems to the users which the manufacturer of the machines cannot alleviate.

Currently available computers are serial machines that process and advance the data at one point after another. Simultaneous solution of unknowns at many points, as is required by implicit algorithms, is handled through special procedures such as matrix inversion. If a program designed for the serial machine should be run on parallel computers, no speed advantage will result. (Indeed there will be some loss.) The 64 parallel processors of ILLIAC IV will only have one processor doing useful work. The STring-ARray processor of STAR will operate in its scalar mode (versus the "vector mode" for string array processing). There is not, and will not be, software that will translate an existing serial program into a reasonably efficient "parallel program" for a specific parallel machine. Such a translation is not a matter of translating one language into another; it is a matter of changing the logic in solving a problem. It asks essentially for a new formulation of a specific problem, if one wishes to exploit the speed advantage offered by a specific machine. The user is asked to start anew, for each problem and for each specific parallel computer. The user must pay considerable attention not only to the formulation of a problem for solution, but also to the storage of the data in the external memory in order to match the demand of the data, according to the formulation of the problem.

In writing such a program for use with a specific parallel computer, it is not a simple matter to take advantage of a successful serial program used with the current serial machines. It may indeed be doubtful that the existence of a successful serial program offers any advantage in constructing a good parallel program. Without further elaboration, it may be noted, even for simple problems, that:

1. An efficient serial algorithm need not lead to an efficient

parallel algorithm, while an inefficient serial algorithm may lead to an efficient parallel algorithm.

2. An algorithm that is apparently serial and was constructed for use with a serial computer may possess a great deal of hidden parallelism which can be exploited to suit the particular mode of operation of a specific parallel computer.

3. A parallel program may behave quite differently in the difference solution of a partial differential equation than the corresponding serial program. Here "behavior"refers to the stability of the computation, the rate of convergence to the desired solution, and the accuracy of the solution.

The last note is particularly important. It asks the user to gain as much understanding as possible of the various fundamental problems of difference methods, such as stability and accuracy. With a better understanding, it is hoped that the years of tedious and painful trial-and-error learning that occurred in the development of difference techniques for the serial machines will not be repeated, or at least may be greatly reduced.

For many important practical problems the solution of the Navier-Stokes equations in three spatial dimensions will be required. Even for the steady state solution of such problems, computation for a reasonably accurate solution will require the speed and storage capacity promised by these parallel computers. Complicated boundary conditions do not lend themselves to efficient parallel treatments, and interfere with efficient organization for the parallel computations of fluid flow problems. These problems are in addition to the fundamental difficulties noted above. It is hoped that what has been learned from the serial machines may benefit the development of computational programs which can reap the promised speed advantage of the parallel computer. For this purpose, it is especially important to gain some fundamental understanding of the complicated computational difficulties particular to fluid dynamics. Such understanding cannot be expected from computer scientists, who have a full share of difficulties associated with the operation of the parallel computers in general. Those wishing to solve complicated flow problems with the Navier-Stokes equations must learn how to resolve such difficulties for themselves. The task ahead is formidable. The potential reward is also immense.

ACKNOWLEDGEMENT

A portion of the material included in this manuscript is part of the, as yet, unpublished results of research supported by the Office of Naval Research, U. S. Navy, under Contract Number N00014-67-A-0151-0028.

The author is grateful for the support and also for the permission to report such information.

GENERAL TEXTS FOR BACKGROUND MATERIALS

1. Forsythe, G. E. and Wasow, W. R., Finite Difference Methods for Partial Differential Equations, John Wiley and Sons, Inc. (1960).

2. Isaacson, E. and Keller, H. B., Analysis of Numerical Methods, John Wiley and Sons, Inc. (1966).

3. Richtmeyer, R. D. and Morton, K. W., Difference Methods for Initial Value Problems, second edition, Interscience Publishers, John Wiley and Sons, Inc. (1967).

4. Morse, P. M. and Feshback, H., Methods of Theoretical Physics, McGraw-Hill Book Co. (1953).

5. Goldstein, S. (Editor), Modern Development in Fluid Dynamics, Oxford Clarendon Press (1938).

6. Howarth, L. (Editor), Modern Development in Fluid Dynamics High Speed Flow, Oxford Clarendon Press (1953).

Background materials in Mathematics and Physics in Chapters 1 - 4 are contained in the above general texts and others; and they are not referenced specifically. The physical interpretations of the various results in the present review represent the opinion of the present reviewer whose view may or may not be in full agreement with opinions of the authors of these references. The present review is to put forth and to illustrate, with simple examples, some physical perspective of these mathematical results to help in the numerical integration of the complicated system of equations of fluid flows. It is inevitable that, at places, such inferences may appear far-fetched. The following are the specific references relevant to these physical discussions in the order they appear in the manuscript.

SPECIFIC REFERENCES

1. Courant, R., Friedricks, K. O., and Levy, H., Uber die Partiellen Differenzen gleichungen der Mathemetischen physik, Mathematics Annual, Vol. 100 (1928).

2. Fromm, J. E., "The Time Dependent Flow of An Incompressible Fluid" Methods in Computational Physics, Vol. 3, Academic Press, N. Y. (1964).

3. Cheng, S. I., "The Numerical Integration of Navier-Stokes Equations," J. of American Institute of Aeronautical and Astronautics, Vol. 8, No. 12, (1970).

4. Hirt, C. W., "Heutristic Stability Theory for Finite Difference Equations," J. of Computational Physics, Vol. 2, (1968).

5. Yanenko, N. N., The Method of Fractional Steps, English translation by M. Holt, Springer Verlag (1971).

6. Peaceman, D. W. and Rachford, H. H., Jr., "The Numerical Solution of Parabolic and Elliptic Differential Equations," Journ, Soc. Industrial and Applied Mathematics, Vol. 3, (1955).

7. Douglas, J. and Gunn, J., "A General Formulation of Alternating Direction Methods," Numerical Mathematics, Vol. 6, (1964).

8. Kreiss, H. O. and Oliger, J., "Comparison of Accurate Methods for the Integration of Hyperbolic Equations," Tellus XXIV 3 (1972).

9. Fromm, J. E., "A Numerical Study of Buoyancy Driven Flows in Room Enclosures," Proceedings of the 2nd International Conference on Numerical Methods in Fluid Dynamics, Berkeley, California, Sept. (1970), Springer Verlag.

10. Cheng, S. I., "Accuracy of Difference Formulation of Navier-Stokes Equations" The Physics of Fluids, Supplement II, Dec. 1969.

11. Fromm, J. E., "Practical Importance of Convective Difference Approximations of Reduced Dispersion," The Physics of Fluids, Supplement II, Dec. 1969.

12. Moretti, G. and Bleich, G., "Three Dimensional Flow around Blunt Bodies," Jour. of American Institute of Aeronautics and Astronautics, Vol. 5, No. 9, (1967).

13. v. Neumann, J. and Richtmeyer, R. D., "A Method for the Numerical Calculations of Hydrodynamic Shocks," Jour. of Applied Physics, Vol. 21, (1950).

14. Lax, P. and Wendroff, B., "Systems of Conservations Laws," Communications on Pure and Applied Mathematics, Vol. 13, (1960).

15. Thoman, D. C. and Szewczyk, A. A., "Time Dependent Viscous Flow over a Circular Cylinder," The Physics of Fluids, Vol. 12, (1969).

16. Allen, J. and Cheng, S. I., "Numerical Solutions of the Compressible Navier-Stokes Equation for the Laminar Near Wake," The Physics of Fluids, Vol. 13, No. 1, (1970).

17. Brailovskaya, I. Y., "A Difference Scheme for the Numerical Solution of the Two-Dimensional Unsteady Navier-Stokes Equations for a Compressible Gas," Soviet Physics Doklady, Vol. 10, No. 2, Aug., (1965).

18. Cheng, S. I. and Chen, J. H., "Finite Difference Treatment of Strong Shock over a Sharp Leading Edge with Navier-Stokes Equations," Proc. of 3rd International Conference on Numberical Methods in Fluid Mechanics,

Vol. II, held in Paris, France, July 1972, Springer-Verlag, Berlin.

19. Messina, N. A. and Cheng, S. I., "A Study of the Computation of Regular Shock Reflection with Navier-Stokes Equations," presented at the Symposium on Application of Computers to Fluid Dynamics Analysis and Design, PolyTech. Institute of Brooklyn, Jan. 1973.

20. Jenson, V. G., "Viscous Flow Round a Sphere at Low Reynolds Number," Proc. of Roy. Soc. London, A249, (1959).

21. Hamielec, A. E., Hoffman, T. W., and Ross, L. L., "Numerical Solution of the Navier-Stokes Equations for Flow Past Non-Solid Spheres," Journal of American Institute of Chemical Engineers, Vol. 13, No. 2, March, 1967.

22. Rimon, Y. and Cheng, S. I., "Numerical Solution of a Uniform Flow over a Sphere at Intermediate Reynolds Numbers," The Physics of Fluids, Vol. 12, No. 5, (1969).

23. Taneda, S., "Studies on Wake Vortices, Experimental Investigation of the Wake Behind a Sphere at Low Reynolds Numbers," Journal of the Physical Society of Japan, Vol. 1, (1956).

24. Chorin, A. J., "Computational Aspects of the Turbulence Problem," Proc. of the Second International Conference on Numerical Methods in Fluid Dynamics, held at Berkeley, California, U.S.A., Sept. 1970.

25. Daly, B. J. and Harlow, F. H., "Inclusion of Turbulence Effects in Numerical Fluid Dynamics," Proc. of the Second International Conference on Numerical Methods in Fluid Dynamics, held at Berkeley, California, U.S.A. Sept. 1970.

26. Orzag, S., "Numerical Simulation of Turbulence," (Fourier or Spectral Method), Proc. of Symposium on Statistical Models and Turbulence, held at San Diego, California, U.S.A., July 1971.

27. Ross, B. B. and Cheng, S. I., A Numerical Solution of the Planar Supersonic Near Wake with its Error Analysis," Proc. of the Second International Conference on Numerical Methods in Fluid Dynamics, held at Berkeley, California, Sept. 1970.

28. Carter, J. E., "The Navier-Stokes Equations for the Supersonic Laminar Flow over a Two-Dimensional Compression Corner," NASA TR R385, July 1972.

29. Cheng, S. I. and Chen, J. H., "Slips, Friction, and Heat Transfer Laws in Merged Regimes," to appear in The Physics of Fluids.

VARIATIONAL PRINCIPLES IN FLUID MECHANICS and FINITE ELEMENT APPLICATIONS

B. FRAEIJS de VEUBEKE

Aeronautics Laboratory - University of Liège - Belgium

TABLE OF CONTENTS

1. EULERIAN AND LAGRANGIAN COORDINATES

The fundamental theorems of classical mechanics deal with closed systems, that is fixed sets of material particles which are followed in their motion. In most problems of fluid mechanics, one is rather interested in the phenomena that occur in a fixed region of space traversed by the particles. This modification of point of view requires some essential transformations to the classical variational formulations of Hamiltonian dynamics.

By $a = (a_1, a_2, a_3)$ we denote generalized lagrangian or "material" coordinates identifying a particle. They may, but need not necessarily, represent the space coordinates (x_1, x_2, x_3) occupied by the particle at a conventional epoch, usually denoted by $t = 0$. Their general définition is that of an independent set of integration constants of the differential equations of trajectories

$$\frac{dx_1}{u_1(x,t)} = \frac{dx_2}{u_2(x,t)} = \frac{dx_3}{u_3(x,t)} = dt \tag{1}$$

In this differential system, $u_i(x,t)$ is the velocity field of the particles expressed as a function of space location and time. Any set of three independent first integrals of (1)

$$a_i = a_i(x,t) \tag{2}$$

provides an implicit description at a regular point of the parametric equations of the trajectories of particles

$$x_j = X_j(a,t) \tag{3}$$

At a regular point in the field, the Jacobian determinant

$$J = \frac{D(x_1, x_2, x_3)}{D(a_1, a_2, a_3)}$$

is different from zero and the volume element containing the particles
with material coordinates $\hat{a}_i$ in the set $a_i \leq \hat{a}_i \leq a_i + da_i$ is given
by

$$d\Omega = J \, da_1 da_2 da_3 \tag{4}$$

The symbol D_t is used to denote the material time derivative ; D_i to denote
the partial derivative with respect to the material coordinate a_i under fixed
time. Both are partial derivatives of any intensive variable of the field
when expressed in the form of $f(a,t)$ and we have the commutative property

$$D_t D_i f = D_i D_t f.$$

Similarly ∂_t will be the symbol of local time derivative; ∂_j that of partial
derivative with respect to x_j. Both are partial derivatives of intensive
variable f when it is expressed in Eulerian form $f(x,t)$ and we have the
commutative property

$$\partial_t \, \partial_j f = \partial_j \, \partial_t f.$$

In general, all transformations will be performed on the Eulerian description
of variables and, for intensive variables, there follows from the definitions

$$D_t f = \partial_t f + \partial_j f \, D_t x_j$$

$$D_t x_j = \frac{\partial}{\partial t} x_j(a,t) = U_j(a,t) = u_j(x,t) \tag{5}$$

so that

$$D_t f = \partial_t f + u_j \, \partial_j f \tag{6}$$

In particular, since a material coordinate is attached to a particle throughout its motion

$$D_t a_i = \partial_t a_i + u_j \partial_j a_i = 0 \tag{7}$$

and (2) may also be considered to be a solution of the partial differential equation (7).

To compute the material time derivative of the volume element, consider the Laplace expansion of the Jacobian determinant

$$J = e_{mnp} \, D_m X_1 \, D_n X_2 \, D_p X_3$$

where use is made of the permutation symbol e_{mnp}.

$$D_t J = e_{mnp} \, (D_t D_m X_1 \, D_n X_2 \, D_p X_3 + D_m X_1 \, D_t D_n X_2 \, D_p X_3 + D_m X_1 \, D_n X_2 \, D_t D_p X_3)$$

However,

$$D_t D_m X_1 = D_m D_t X_1 = D_m U_1 = \partial_i u_1 D_m X_i$$

hence

$$e_{mnp} D_t D_m X_1 \, D_n X_2 \, D_p X_3 = \partial_i u_1 e_{mnp} \, D_m X_i \, D_n X_2 \, D_p X_3$$

$$= \partial_i u_1 \, e_{i23} J = (\partial_1 u_1) \, J$$

With a similar treatment of the other terms, ther finally comes

$$D_t J = (\partial_j u_j) \, J$$

or, in view of (4)

$$D_t \, d\Omega = (\partial_j u_j) \, d\Omega \qquad\qquad (8)$$

Formulas (6) and (8) funish a justification of the general statement about the material time derivative of any extensive quantity

$$D_t \int_\Omega f \, d\Omega = \int_\Omega D_t (f \, d\Omega) = \int_\Omega (\partial_t f + u_j \partial_j f + f \, \partial_j u_j) \, d\Omega$$

$$= \int_{\Omega} \{ \partial_t f + \partial_j (u_j f) \} \, d\Omega$$

or, after application of the divergence theorem

$$D_t \int_{\Omega} f d\Omega = \int_{\Omega} \partial_t f \, d\Omega + \int_{\partial\Omega} f(n_j u_j) \, dS \tag{9}$$

where n_j denote the direction cosines of the outward normal to the surface $\partial\Omega$ bounding the set of particles.

2. LAGRANGIAN AND EULERIAN VARIATIONS

The preceding reminder of well known results is useful in order to establish a complete analogy with similar conclusions relating material and local variations. In Hamiltonian mechanics the real motion of each particle is compared to perturbed or "varied" motions. We may, for instance, consider a family of virtual motions

$$x_j = X_j(a,t;\varepsilon) \tag{10}$$

of which the real motion (3) would correspond to the value zero for the parameter . The material or Lagrangian variation of position of a particle can then be defined as

$$\Delta X_j = \frac{\partial}{\partial\varepsilon} X_j \bigg|_{\varepsilon=0} d\varepsilon \tag{11}$$

Inversion of (10)

$$a_i = a_i(x,t;\varepsilon) \tag{12}$$

leads naturally, for the material coordinates, to a concept of local or eulerian variation, in which the space coordinates are kept fixed

$$\delta a_i = \frac{\partial}{\partial\varepsilon} a_i \bigg|_{\varepsilon=0} d\varepsilon \tag{13}$$

Since by definition $\delta x_j = 0$, we obtain from (10), considering a back-substitution of the material coordinates through (12),

$$\Delta X_j + \delta a_i \, D_i X_j = 0 \tag{14}$$

Conversely, becouse $\Delta a_i = 0$ by definition,

$$\delta a_i + \Delta X_j \partial_j a_i = 0 \tag{15}$$

which is the analogue to (7). Thus there is an equivalence by one to one correspondence between ΔX_j and δa_i; (14) and (15) are indeed the inverse relationships to one another, since for $\varepsilon = 0$

$$D_i X_j \partial_j a_m = \delta_{im} \qquad \partial_j a_i D_i X_m = \delta_{jm} \tag{16}$$

More generally, the variations of an intensive variable $f(x,t;\varepsilon)$ are related by

$$\Delta f = \delta f + \Delta X_j \, \partial_j f \tag{17}$$

which is the analogue of (6), or by

$$\delta f = \Delta f + \delta a_m \, D_m f$$

that follows immediately from its alternative representation as a function of $(a,t;\varepsilon)$.

Obviously, Δ commutes with D_i and D_t, while δ commutes with ∂_j and ∂_t. From a computation similar to that of $D_t J$, we obtain the analogue to (8)

$$\Delta J = (\partial_j \Delta X_j) \, J \qquad \text{or} \qquad \Delta d\Omega = (\partial_j \Delta X_j) \, d\Omega \tag{18}$$

and finally, the analogue to (9)

$$\Delta \int_\Omega f d\Omega = \int_\Omega \delta f d\Omega + \int_{\partial\Omega} f(n_j \Delta x_j) \, dS \tag{19}$$

The operators ∂_t and δ keeping the space coordinates fixed, we have also

$$\partial_t d\,\Omega = 0 \qquad\qquad \delta d\Omega = 0 \qquad\qquad\qquad (20)$$

3. THE HAMILTON PRINCIPLE FOR AN INVISCID FLUID

In Hamilton's principle the set of particles is kept fixed and must be followed throughout its motions, a fact that will be stressed in the formulas by writing $\Omega(t)$ and $\partial\Omega(t)$ for its volume and bounding surface. The Lagrangian per unit mass will be

$$L = \frac{1}{2} D_t X_i \, D_t X_i - U - G \qquad\qquad\qquad (21)$$

where G is a gravitational potential, assumed to be a function of the space coordinates only $G(x)$, so that

$$g_j = -\partial_j G \qquad\qquad\qquad (22)$$

is the local gravitational acceleration acting on a particle, and

$$\Delta G = -\partial_j G \Delta X_j \qquad\qquad \delta G = 0 \qquad\qquad\qquad (23)$$

The specific internal energy U is in general a function of ρ, the mass per unit volume, and of the specific entropy S. To obtain a true variational principle, it will be necessary to make the assumption that there are no heat exchanges between the particles, nor momentum exchanges. Thus, neglecting conductivity and viscosity the entropy of each particle remains the same at any time. In addition we assume, for simplicity, that the entropy of each particle is the same (homentropic flow).
It is then possible to ignore entirely the dependence of U on S; the thermodynamical pressure becomes defined by

$$p(\rho) = \rho^2 \frac{dU}{d\rho} \qquad\qquad\qquad (24)$$

and the fluid is barotropic.

Hamilton's principle asserts that

$$\Delta \int_{t_1}^{t_2} \int_{\Omega(t)} \rho L d\Omega\, dt \quad - \int_{t_1}^{t_2} \int_{\partial_2 \Omega(t)} \overline{p}\, n_j\, \Delta X_j\, dS\, dt = 0 \tag{25}$$

provided $\Delta X_j = 0$ at $t = t_1$ and $t = t_2$. The second term represents the virtual work of a prescribed $\overline{p}$ in part $\partial_2 \Omega$ of the boungary.
On the complementary part $\partial_1 \Omega$ the fluid may be assumed to glide along a fixed or moving wall so that the constraint

$$n_j\, \Delta x_j = 0 \qquad \text{on} \qquad \partial_1 \Omega \tag{26}$$

must be applied independently.

As it is stated the principle does not take care of conservation of mass that is also to be entered as a side constraint

$$\Delta(\rho d\Omega) = 0 \qquad \text{or} \qquad \Delta\rho + (\partial_j\, \Delta x_j)\rho = 0 \tag{27}$$

This gives

$$\int_{t_1}^{t_2} \int_{\Omega(t)} \rho\, \Delta L\, d\Omega\, dt - \int_{t_1}^{t_2} \int_{\partial_2 \Omega(t)} \overline{p}\, n_j \Delta x_j\, dS\, dt = 0$$

Now, in view of (23) and (24)

$$\Delta L = D_t X_j\, D_t \Delta X_j - \frac{p}{\rho^2}\, \Delta\rho - \partial_j\, G\, \Delta X_j$$

or writing u_j for $D_t X_j$ and using (27) again

$$\rho \ \Delta L \ d\Omega = \rho \ u_j \ d\Omega \ D_t \ \Delta X_j + p(\partial_j \ \Delta X_j) \ d\Omega - \rho \ \partial_j \ G \ \Delta X_j \ d\Omega$$

Because of the motion of the region occupied by the particles it is necessary to apply the operator D_t on $d\Omega$ included for any integration by parts with respect to time. This, and the application of the divergence theorem, are prepared by transforming as follows :

$$\rho \ \Delta L \ d\Omega = D_t(\rho \ u_j \ \Delta X_j \ d\Omega) - \Delta X_j \ D_t \ (\rho \ u_j \ d\Omega)$$

$$+ \ \partial_j(p \ \Delta X_j) \ d\Omega - (\partial_j \ p + \partial_j \ G) \ \Delta X_j \ d\Omega$$

We can now carry out the required integrations

$$\int_{\Omega(t_2)} \rho \ u_j \ \Delta X_j \ d\Omega - \int_{\Omega(t_1)} \rho \ u_j \ \Delta X_j \ d\Omega + \int_{t_1}^{t_2} \int_{\partial\Omega(t)} p \ n_j \ \Delta X_j \ dS \ dt$$

$$- \int_{t_1}^{t_2} \int_{\partial_2\Omega(t)} \bar{p} \ n_j \ \Delta X_j \ dS \ dt -$$

$$- \int_{t_1}^{t_2} \int_{\Omega(t)} \Delta X_j \ (D_t(\rho u_j d\Omega) + (\partial_j p + \rho \partial_j G) d\Omega) \ dt = 0$$

The two first terms vanish on account of the vanishing of ΔX_j at time limits. The Euler equation provided by the arbitrariness of ΔX_j in the last term turns out to be the correct Newtonian of motion

$$\rho \ D_t \ u_j + \partial_j p + \rho \partial_j \ G = 0 \tag{28}$$

after, again, a separate consideration of time conservation of mass

$$D_t (\rho \, d\Omega) = 0 \qquad \text{or} \qquad D_t \rho + \rho \partial_i u_i = 0 \qquad (29)$$

If due account is taken of (26), the arbitrariness of ΔX_j on $\partial_2 \Omega(t)$, yields the natural boundary condition

$$p = \bar{p} \qquad \text{on} \qquad \partial_2 \Omega(t) \qquad (30)$$

4. DERIVATION OF AN EULIRIAN PRINCIPLE

A step towards the derivation of a corresponding Eulerian principle consists in an application of formula (19) to the Hamiltonian principle. Thus we must have

$$\int_{t_1}^{t_2} \int_{\Omega(t)} \delta(\rho L) \, d\Omega \, dt = \Delta \int_{t_1}^{t_2} \int_{\Omega(t)} \rho L \, d\Omega \, dt \; - \int_{t_1}^{t_2} \int_{\partial \Omega(t)} \rho L (n_j \Delta x_j) \, dS \, dt$$

and in view of (25)

$$\int_{t_1}^{t_2} \int_{\Omega(t)} \delta(\rho L) \, d\Omega \, dt + \int_{t_1}^{t_2} \int_{\partial \Omega(t)} \rho L (n_j \Delta x_j) \, dS \, dt$$

$$- \int_{t_1}^{t_2} \int_{\partial_2 \Omega(t)} \bar{p} \, (n_j \Delta x_j) \, dS \, dt = 0$$

The correctness of this can be verified. However, the principle is not exactly Eulerian in the sense that the time integrals still imply that the motion of the volume occupied by the particles by the particles be accounted for.

The difference between the true Eulerian variation

$$\delta \int_{t_1}^{t_2} \int_{\Omega} \rho L \, d\Omega \, dt \qquad \text{and} \qquad \int_{t_1}^{t_2} \int_{\Omega(t)} \delta(\rho L) \, d\Omega \, dt$$

is in the time integrations by parts which require the ∂_t operator in the first case, the D_t operator in the second. The terms requiring integration by parts in time in $\delta(\rho L)$ are

$$\rho\, u_i\, \delta\, (D_t X_i)\ \ d\Omega$$

We have from (17)

$$\delta\, D_t X_i = \Delta D_t X_i - \Delta X_j\, \partial_j u_i = D_t \Delta X_i - \Delta X_j\, \partial_j u_i$$

whence

$$\rho\, u_i\, \delta\, (D_t X_i)\ \ d\Omega = D_t (f\ d\Omega) - E_j\, \Delta X_j\ d\Omega \tag{31}$$

where $\qquad f = \rho u_i\, \Delta X_i$

and

$$E_j = \rho\, D_t u_j + \rho u_i \partial_j u_i \tag{32}$$

Equation (31) is in the form required by the second case. It will generate two terms that vanish at the time limits and a contribution $-E_j$ to the Euler equation. If we now transform in (31)

$$D_t\, (f\ d\Omega) = (\partial_t\, f)\ d\Omega + \partial_j\, (u_j\, f)\ d\Omega$$

it follows that for the first case we generate again terms that vanish at the time limits, a surface term

$$\int_{t_1}^{t_2} \int_{\partial\Omega}\ \ (n_j u_j)\rho\, u_i \Delta X_i\ dSdt$$

and the same contribution to the Euler equation. As a consequence we do not alter the Euler equation by switching from the quasi-Eulerian functional to the really Eulerian one, that involves a fixed region Ω of space; but in so doing, we introduce a complicated combination of Lagrangian and Eulerian surfaceintegrals. This is not surprising as the boundary conditions represented by free surfaces or moving walls are essentially Lagrangian by nature. In the sequel we must be satisfied with taking Eulerian variations of the Eulerian functional

$$\int_{t_1}^{t_2} \int_{\Omega} \rho L d\Omega dt \qquad \text{and fit whatever boundary conditions that are}$$

"natural" to the situation.

5. A SELF-SUPPORTING EULERIAN VARIATIONAL PRINCIPLE

To avoid the necessity of dealing separately with conservation of mass, the functional can be augmented by means of a Lagrangian multiplier to include the satisfaction of eq. (27) . This step was taken originally by Herivel (Ref. 4) but is known to be insufficient when the variations on particle displacement are transferred to material variations of the velocity field; it restricts the flow to the irrotational case.
A logical step to remove this restriction was taken by the author (Ref. 7) in 1965, it consists in augmenting the functional further by incorporating the constraints :

$$D_t X_i - u_i = 0 \tag{33}$$

by means of a vector Lagrangian multiplier ψ_i. We thus examine the Eulerian variations of the new functional

$$\int_{t_1}^{t_2} \int_{\Omega} \left[\rho L + \theta(\partial_t \rho + \partial_i(\rho u_i)) + \rho \psi_i (D_t X_i - u_i) \right] \, d\Omega dt \tag{34}$$

where L may now be written as

$$L = \frac{1}{2} u_i u_i - U(\rho) - G(x) \tag{35}$$

and $\delta\rho$, $\delta\theta$, $\delta\psi_i$, δu_i, ΔX_i are independent variations.

The variations on the Lagrangian multipliers raise (27) and (33) to the status of Euler equations. If we collect the terms due to the variation of

$$(\frac{1}{2} u_i u_i - \frac{d}{d\rho} (\rho U) - G) \, \delta\rho + \theta\partial_t \delta\rho + \theta\partial_i(u_i \, \delta\rho) + \delta\rho\psi_i(D_t X_i - u_i),$$

take into consideration that (33) is satisfied and prepare the required integrations by parts

$$\partial_t(\theta\delta\rho) + \partial_i(\theta \, u_i\delta\rho) - \delta\rho \, (\frac{d}{d\rho} (\rho U) + \partial_t \theta + u_i\partial_i\theta + G - \frac{1}{2} u_i u_i)$$

we obtain as Euler equation

$$\delta\rho \rightarrow \qquad D_t \, \theta = \frac{1}{2} \, u_i u_i - G - \frac{d}{d\rho} (\rho U) \qquad\qquad (36)$$

The variation on u_i alone generate the terms

$$\rho u_i \, \delta u_i + \theta\partial_i(\rho\delta u_i) - \rho\psi_i \, \delta u_i = \partial_i(\theta\rho\delta \, u_i) + \rho(u_i - \partial_i\theta - \psi_i) \, \delta u_i$$

and the corresponding Euler equations

$$\delta u_i \rightarrow \qquad u_i = \partial_i\theta + \psi_i \qquad\qquad (37)$$

For ΔX_i we must manipulate

$$\delta D_t X_i = \Delta D_t X_i - (\partial_j u_i) \, \Delta X_j = \partial_t \Delta X_i + u_j\partial_j\Delta \, X_i - (\partial_j u_i)\Delta X_j$$

generating the terms

$$\rho \psi_i \delta D_t X_i = \partial_t (\rho \psi_i \Delta X_i) - \partial_t (\rho\psi_i)\Delta X_i + \partial_j (\rho\psi_i u_j \Delta X_i) - \partial_j (\rho\psi_i u_j)\Delta X_i$$

$$-\rho\psi_j\partial_i u_j \Delta X_i$$

and obtain the Euler equation

$$\Delta X_i \rightarrow \qquad \partial_t (\rho\psi_i) + \partial_j (\rho\psi_i u_j) + \rho\psi_j\partial_i u_j = 0$$

In view of (29) this simplifies to

$$\Delta X_i \quad \rightarrow \quad D_t \psi_i + \psi_j \partial_i u_j = 0 \tag{38}$$

From the various contributions we can also derive the following integral at the time limits

$$\int_\Omega (\theta \delta\rho + \rho\psi_i\Delta X_i) d\Omega \Bigg|_{t_1}^{t_2}$$

and the surface terms related to possible boundary conditions

$$\int_{t_1}^{t_2} \int_{\partial\Omega} (\theta(u_i n_i)\delta\rho + \rho\theta(n_i\delta u_i) + \rho(n_j u_j)\psi_i\Delta X_i) \, dS \, dt$$

They will be discussed later, after elimination of the only remaining Lagrangian variation ΔX_i.

It should be observed that none of the Euler equations correspond directly to the Newtonian equations of motion. They are, however, contained as combinations. First of all, we find from (24)

$$\frac{d}{d\rho} (\rho U) = U + \rho \frac{dU}{d\rho} = U + \frac{P}{\rho} = I(\rho) \tag{39}$$

the specific enthalpy considered as a function of ρ. If we then take the material derivative of (37) and eliminate ψ_i by using (38), there comes

$$D_t(u_i - \partial_i\theta) + (u_j - \partial_j\theta)\partial_i u_j = D_t u_i - (\partial_t + u_j\partial_j)\partial_i\theta - \partial_j\theta\partial_i u_j$$

$$+ \partial_i \frac{u_j u_j}{2} = D_t u_i - \partial_i D_t\theta + \partial_i \frac{u_j u_j}{2} = 0$$

or, finally, using (36)

$$D_t u_i + \partial_i(I+G) = 0 \tag{40}$$

which is a classical form of the equations of motion.

6. ELIMINATION OF THE LAGRANGIAN VARIATION OF POSITION

To eliminate the use of ΔX_i we must solve the corresponding Euler equations (38) and substitute solution into the functional. Equation (38) states that the ψ_i field is a constant circulation one; on any small segment dx_i carried by the particles the circulation of ψ_i does depend on time :

$$D_t \ (\psi_i dx_i) = 0 \tag{41}$$

Indead, this is equivalent to

$$D_t \psi_i dx_i + \psi_j D_t dx_j = D_t \psi_i dx_i + \psi_j dD_t X_j = D_t \psi_i dx_i + \psi_j du_j$$

$$= dx_i (D_t \psi_i + \psi_j \partial_i u_j) = 0$$

Equation (41) indicates that the Pfaffian form $\psi_i dx_i$ depends only on material coordinates. In any of its canonical representations

$$\psi_i dx_i = d\gamma + \alpha d\beta \tag{42}$$

the variables α, β, γ are material variables :

$$D_t \gamma = 0 \qquad D_t \alpha = 0 \qquad D_t \beta = 0 \tag{43}$$

From (42) follows then the general solution of (38)

$$\psi_i = \partial_i \gamma + \alpha \partial_i \beta \tag{44}$$

and this, substitued into (37) yields the general Clebsch representation (Ref. 1) of the rotational flow of an inviscid fluid :

$$u_i = \partial_i \phi + \alpha \partial_i \beta \tag{45}$$

where $\phi = \theta+\gamma$ is a velocity potential and α and β are material variables. If the potential ϕ is single-valued the velocity field retains the constant circulation property for closed contours carried by the flow. Moreover

$$\text{rot } \vec{u} = \text{grad}\alpha \ \times \ \text{grad}\beta \qquad\qquad (46)$$

and we have the statement that the vortex lines which are the intersections of the surface families α = constant and β= constant, are carried by the flow. The constant circulation property indicates the existence of an acceleration potential, indeed,

$$D_t u_i = \partial_t(\partial_i\phi+\alpha\partial_i\beta) + u_j\partial_j(\partial_i\phi+\alpha\partial_i\beta)$$

$$= \partial_i(\partial_t\phi+\alpha\partial_t\beta) + \partial_t\alpha\partial_i\beta - \partial_i\alpha\partial_t\beta + u_j\partial_i(\partial_j\phi+\alpha\partial_j\beta)$$

$$+ u_j\partial_j\alpha\partial_i\beta - u_j\partial_i\alpha\partial_j\beta$$

$$= \partial_i(\partial_t\phi+\alpha\partial_t\beta + \frac{u_ju_j}{2}) + \partial_i\beta D_t\alpha - \partial_i\alpha D_t\beta$$

and the two last terms vanish by (43). With this result and (40), we obtain the generalized Bernoulli integral of the equations of motion

$$\partial_t\phi+\alpha\partial_t\beta + \frac{u_ju_j}{2} + I + G = h(t) \qquad\qquad (47)$$

Since the potential ϕ contains an arbitrary additive function of time only, there is no restriction in making $h(t) = 0$.

To eliminate ΔX_i replace in (34)

$$\psi_i D_t X_i = (\partial_i\gamma+\alpha\partial_i\beta)D_t X_i = D_t\gamma - \partial_t\gamma +\alpha(D_t\beta -\partial_t\beta) = -(\partial_t\gamma+\alpha\partial_t\beta)$$

Then

$$\rho\psi_i(D_t X_i-u_i) = -\rho(\partial_t\gamma+\alpha\partial_t\beta) - \rho u_i(\partial_i\gamma+\alpha\partial_i\beta)$$

If in addition we transform

$$\theta\{ \partial_t\rho + \partial_i(\rho u_i)\} = \partial_t(\rho\theta) + \partial_i(\theta\rho u_i) - \rho\partial_t\theta - \rho u_i\partial_i\theta$$

the two contributions combine into

$$\theta\{\partial_t\rho + \partial_i(\rho u_i)\} + \rho\psi_i(D_t X_i - u_i) = \partial_t(\rho\theta) + \partial_i(\theta\rho\, u_i) - \rho D_t\phi - \rho\alpha\, D_t\beta$$

Thus discarding the term that goes to time limits and the divergence term, we obtain a functional

$$\int_{t_1}^{t_2} \int_\Omega \rho K d\Omega dt$$

$$K = \frac{1}{2}\, u_i u_i - U - G - D_t\phi - \alpha D_t\beta \qquad (D_t = \partial_t + u_i\partial_i) \tag{48}$$

The Eulerian variation of this functional yields the following Euler equations :

$$\delta\rho \quad \rightarrow \quad \frac{u_i u_i}{2} - I - G - D_t\phi - \alpha D_t\beta = 0 \tag{49}$$

$$\delta u_i \quad \rightarrow \quad \rho(u_i - \partial_i\phi - \alpha\partial_i\beta) = 0 \tag{50}$$

$$\delta\phi \quad \rightarrow \quad \partial_t\rho + \partial_i(\rho u_i) = 0 \tag{51}$$

$$\delta\alpha \quad \rightarrow \quad D_t\beta = 0 \tag{52}$$

$$\delta\beta \quad \rightarrow \quad \partial_t(\rho\alpha) + \partial_i(\rho\alpha\, u_i) = 0 \tag{53}$$

The last being equivalent to $D_t\alpha = 0$ in view of (51).
The time limits term is

$$-\int_\Omega \rho(\delta\phi + \alpha\delta\beta)d\Omega \quad \Big|_{t_1}^{t_2}$$

and can be deleted under the vonvention that the variations on α and β

vanish for $t = t_1$ and $t = t_2$.

The surface terms are

$$- \int_{t_1}^{t_2} \int_{\partial\Omega} \rho(\delta\phi + \alpha\delta\beta)(n_i u_i) \, dS \, dt$$

The simpler boundary condition is that of a fixed wall. Freedom in the variations of either ϕ or β at such a boundary wall will give the corresponding requirement $n_i u_i = 0$ as a natural boundary condition. If a normal velocity $(n_i \overline{u_i})$ is imposed at a part $\partial_1\Omega$ of the boundary we add to the functional the term

$$\int_{t_1}^{t_2} \int_{\partial_1\Omega} \rho\phi(n_i u_i) \, dS \, dt$$

and the free variation of ϕ will impose $n_i u_i = n_i \overline{u_i}$ on $\partial_1\Omega$. The free variation on ρ in this additional term will cause $\phi = 0$ on the same boundary.

7. THE PRESSURE INTEGRAL

Consider the energy per unit volume

$$f(\rho) = \rho U(\rho)$$

The variable conjugate to ρ is by definition

$$\frac{df}{d\rho} = U + \frac{p}{\rho} = I$$

A co-energy is then defined as in elasticity theory by the Legendre transformation

$$\rho I - f = \rho(I-U) = p(I) \tag{54}$$

and turns out to be the pressure to be considered as a function of the enthalpy. Differentiation of (54) produces then the involutory of conjugate variables

$$\rho dI = dp \tag{55}$$

Precisely, when (49) is used to eliminate the consideration of   as a variable, the Kernel of the functional reduces to

$$\rho K = \rho(I-U) = p$$

We obtain one of Bateman's Eulerian variational principles (Ref. 3), the so-called pressure integral

$$\int_{t_1}^{t_2} \int_\Omega p(I)d\Omega\, dt \tag{56}$$

in which the enthalpy must be considered to be expressed through (49) as

$$I = \frac{u_i u_i}{2} - G - D_t \phi - \alpha D_t \beta \tag{57}$$

From (55) and (57) there comes

$$\delta p = \frac{dp}{dI}\, \delta I = \rho(u_i \delta u_i - \delta D_t \phi - \delta \alpha D_t \beta - \alpha \delta D_t \beta)$$

and the Eulerian equations are still given by (50), (51), (52) and (53); nothing is changed concerning the time limit and surface terms.

The pressure integral can be further simplified by accepting a priori the Euler equation (50). Hence substituting the Clebsch representation

$$u_i = \partial_i \phi + \alpha \partial_i \beta$$

we obtain the pressure integral (56) with the enthalpy given this time by

$$I = -(\frac{1}{2}(\partial_i \phi + \alpha \partial_i \beta)(\partial_i \phi + \alpha \partial_i \beta) + G + \partial_t \phi + \alpha \partial_t \beta) \tag{58}$$

as would result from the generalized Bernoulli integral (47).

This principle depends only on the potential ϕ and the Lagrangian variables of the Clebsch representation. Their variations produce the Euler equations (51), (52) and (53); again the time limits and surface terms are unaltered.

The principle has received a good deal of attention in the problem of the perturbation of a uniform flow of a compressible fluid by aerodynamic bodies. Using a pressure coefficient

$$P = \frac{p - p_\infty}{\frac{1}{2} \rho_\infty U^2}$$

in place of the pressure itself; introducing the Mach number

$$M = \frac{U}{a_\infty} \qquad \text{with } a_\infty^2 = \gamma \frac{p_\infty}{\rho_\infty}$$

and the variable

$$Z = \frac{I_\infty - I}{U^2}$$

instead of the enthalpy itself, the relationships

$$I - I_\infty = c_p (T - T_\infty) \qquad \text{and} \qquad \frac{p}{p_\infty} = (\frac{T}{T_\infty})^{\frac{\gamma}{\gamma - 1}}$$

of gas dynamics yield the explicit law

$$P = \frac{2}{\gamma M^2} \left[(1 - (\gamma - 1)M^2 Z)^{\frac{\gamma}{\gamma - 1}} - 1 \right] \tag{59}$$

While, if the flow depends only on the potential ϕ, and a perturbation potential η of the uniform flow along x_1 be introduced by

$$\phi = U(x_1 + \eta)$$

$$Z = \partial_1 \eta + \frac{1}{U} \partial_t \eta + \frac{1}{2} \partial_i \eta \partial_i \eta \tag{60}$$

An approximate determination of the perturbation potential is possible
by application of the Rayleigh-Ritz method via the variational principle

$$\delta \int_{t_1}^{t_2} \int_{\Omega} \left[(1-(\gamma-1)M^2 Z)^{\frac{\gamma}{\gamma-1}} -1 \right] d\Omega \, dt = 0$$

with Z given by (60). The principle is also useful as a theoretical tool
for delivering coherent approximations to the field equation governing the
perturbation potential and its boundary conditions, by assuming small pressure and velocity perturbations and expanding the Kernel by the binomial
theorem.

8. THE PARTICULAR CASE OF INCOMPRESSIBLE FLOW

Incompressible flow is an idealized case where the pressure is no more
of thermodynamical origin but constitutes a purely mechanical reaction
against changes of volume. One can, however, consider it as a limiting
case of the enthalpy formula through (55) since then

$$I = \int \frac{dp}{\rho} = \frac{p}{\rho}$$

and $\frac{p}{\rho}$ is sometimes called the specific pressure energy.
Thus the general pressure formulation applies here in the form (56) with

$$p = \rho \left(\frac{u_i u_i}{2} - G - D_t \phi - \alpha D_t \beta \right) \tag{61}$$

There is no loss in generality in dropping the constant factor ρ in the
Kernel of the principle. It will also be observed that G plays no role in
the variational equations and may be dropped in the Kernel that becomes :

$$\delta \int_{t_1}^{t_2} \int_{\Omega} \left(\frac{u_i u_i}{2} - D_t \phi - \alpha D_t \beta \right) \, d\Omega \, dt = 0 \tag{62}$$

However, G retains its role as additional hydrostatic pressure when the
pressure is computed from (61) after the potential and Lagrangian functions
have been determined.

Consider now the very special case of stationary potential flow of an incompressible inviscid fluid. The assumptions can be summarized in

$$\partial_t \vec{u} = 0 \qquad \text{rot } \vec{u} = 0 \qquad \text{div } \vec{u} = 0$$

and the problem is almost purely geometrical in nature.

Because of the stationarity assumption, the time integral may be dropped in the variational principle. The second assumption allows to retain only ϕ, and (62) degenerates into

$$\int_\Omega (u_i \partial_i \phi - \frac{u_i u_i}{2}) d\Omega - \int_{\partial_2 \Omega} \phi \overline{u_\nu}\ dS - \int_{\partial_1 \Omega} (n_i u_i)(\phi - \overline{\phi}) dS$$

$$\left. \begin{array}{c} \min \\ \phi \end{array} \right| \left. \begin{array}{c} \max \\ u_i \end{array} \right| \qquad (63)$$

The sign of the functional has been changed and surface terms added to provide for natural boundary conditions throughout.

Those and the Euler equations are in fact

$$\delta u_i \qquad \text{in} \quad \Omega \quad \rightarrow \quad u_i = \partial_i \phi$$

$$\delta u_i \qquad \text{on} \ \partial_1 \Omega \quad \rightarrow \quad \phi = \overline{\phi}$$

which is equivalent to the imposition of the velocity components tangent to this boundary,

$$\delta \phi \qquad \text{in} \quad \Omega \quad \rightarrow \quad \partial_i u_i = 0$$

$$\delta \phi \qquad \text{on} \ \partial_2 \Omega \quad \rightarrow \quad n_i u_i = \overline{u}_\nu$$

the imposition of the velocity component normal to this boundary.

Principle (63) is in the so-called canonical (in the sense of Hamilton) or involutory form advocated by Friedrichs and whose analogue in elasticity theory is better known under the name of Reissner. It is a saddle point principle in which, after looking for a maximizing choice of the velocity field under a given potential, one looks after the minimum of all those maxima for the choice of the potential.

From it, two simpler single-field principles may be derived, whose dual character enables the kinetic energy estimates of the flow, obtained from Rayleigh-Ritz approximations, to be bounded from below and from above respectively.

The first is obtained by accepting a priori the potential character of the flow. If we add to this the a priori satisfaction of $\phi = \tilde{\phi}$ on $\partial_1 \Omega$, we obtain

$$\int_\Omega \frac{1}{2} \, \partial_i \phi \partial_i \phi \ d\Omega \ - \int_{\partial_2 \Omega} \phi \, \bar{u}_\nu \ dS \qquad \min_\phi \qquad (64)$$

and the principle accounts simply for the incompressibility condition

$$\partial_i u_i = \partial_i \partial_i \phi = 0$$

and for the boundary conditions on $\partial_2 \Omega$.

If, on the contraty, we want to simplify (63) by a priori satisfaction of the incompressibility condition plus the boundary condition in $\partial_2 \Omega$, it becomes necessary to transform the functional by an integration by parts :

$$- \int_\Omega (\phi \partial_i u_i + \frac{u_i u_i}{2}) \, d\Omega + \int_{\partial_2 \Omega} \phi(n_i u_i - \bar{u}_\nu) \ dS + \int_{\partial_1 \Omega} n_i u_i \phi dS$$

Accepting now a priori the constraints

$$\partial_i u_i = 0 \qquad \text{and} \quad n_i u_i = \bar{u}_\nu \qquad \text{on} \quad \partial_2 \Omega$$

we obtain the dual single-field principle (the sign has again be changed)

$$\int_\Omega \frac{1}{2} \, u_i u_i \ d\Omega \ - \int_{\partial_1 \Omega} n_i u_i \, \tilde{\phi} dS \qquad \begin{array}{c} \min \\ u_i \text{ constrained} \end{array} \qquad (65)$$

9. THE VECTOR POTENTIAL

The implementation of the incompressibility constraint on the u_i field calls naturally for the use of a vector potential $\vec{A}$:

$$\vec{u} = \text{rot } \vec{A} \; \to \; \text{div } \vec{u} = 0 \tag{66}$$

but introduces interpretation difficulties for the boundary terms.
For this reason we carry out the required transformations on (63) instead of (65). Since the use of a vector potential

$$u_i = e_{ipq} \, \partial_p \, A_q$$

automatically entails $\partial_i u_i = 0$, the functional in (63) may already be transformed to

$$-\int_\Omega \frac{u_i u_i}{2} \; d\Omega \; + \int_{\partial_2 \Omega} \phi(n_i u_i - \bar{u}_\nu) \; dS + \int_{\partial_1 \Omega} \bar{\phi} n_i u_i dS$$

Or, with the understanding that $\phi = \bar{\phi}$ on $\partial_1 \Omega$,

$$-\int_\Omega \frac{u_i u_i}{2} \; d\Omega \; + \int_{\partial\Omega} \phi n_i u_i \; dS \; - \int_{\partial_2 \Omega} \phi \bar{u}_\nu \; dS$$

The free variation $\delta\phi$ on $\partial_2\Omega$ produces the natural boundary condition

$$\vec{n}.\vec{u} = \bar{u}_\nu \qquad \text{on } \partial_2\Omega,$$

The variation $\delta u_i = e_{ipq} \, \partial_p \delta A_q$ gives

$$-\int_\Omega \delta\vec{A}.\text{rot } \vec{u} \, d\Omega \; - \int_{\partial\Omega} n u_i e_{ipq} \delta A_q dS \; + \int_{\partial\Omega} \phi n_i e_{ipq} \, \partial_p \delta A_q dS = 0$$

The Euler equation is obviously rot $\vec{u} = 0$, as was to be expected. The last term, that contains derivatives of the variations of the vector potential, is transformed as follows :

$$\int_{\partial\Omega} \phi\vec{n}.\text{rot } \delta\vec{A} \; dS \; = \int_{\partial\Omega} \vec{n}.\text{rot}(\phi\delta \, \vec{A}) \; dS \; - \int_{\partial\Omega} \vec{n}.(\text{grad}_\phi \times \delta A) \; dS$$

where, on account of

$$\vec{n} \cdot (\text{grad}\phi \times \delta\vec{A}) = \delta\vec{A} \cdot (\vec{n} \times \text{grad}\phi)$$

the scalar potential has only to be defined on the surface.

As $\int_{\partial\Omega} \vec{n} \cdot \text{rot} (\phi\delta\vec{A}) \, dS = 0$, we finally obtain for the surface terms

$$\int_{\partial\Omega} \delta\vec{A} \cdot \left[\vec{n} \times (\vec{u} - \text{grad}\phi) \right] \, dS = 0 \qquad \text{with } \phi = \bar{\phi} \text{ on } \partial_1\Omega$$

The boundary conditions are thus finally obtained in the form

$$\vec{n} \times \vec{u} = \vec{n} \times \text{grad}\bar{\phi} \qquad \text{on} \qquad \partial_1\Omega$$

$$\vec{n} \cdot \vec{u} = \bar{u}_\nu \qquad \text{on} \qquad \partial_2\Omega$$

10. ORTHOGONALITY ASPECTS OF ISOCHORIC AND IRROTATIONAL FLOW

Consider on the one hand an irrotational flow described by a scalar potential :

$$u_i = \partial_i\phi \qquad \rightarrow \qquad \text{rot } \vec{u} = 0$$

on the other hand, and isochoric flow, described by a vector potential :

$$v_i = e_{ipq} \partial_p A_q \qquad \rightarrow \qquad \text{div } \vec{v} = \partial_i v_i = 0$$

and define a scalar product between the two as

$$(u,v) = \int_\Omega u_i v_i \, d\Omega$$

From a first type of integration by parts

$$(u,v) = \int_\Omega v_i \partial_i\phi \, d\Omega = \int_{\partial\Omega} \phi n_i v_i \, dS - \int_\Omega \phi \partial_i v_i \, d\Omega = \int_{\partial\Omega} \phi n_i v_i \, dS \qquad (67)$$

It is apparent that this scalar product vanishes if the bounding surface is subdivided in parts :

$$\partial_1\Omega \qquad \text{over which} \qquad \phi = 0$$

$\partial_2\Omega$ over which $n_i v_i = \vec{n}.\text{rot }\vec{A} = 0$

The second type of integration by parts

$$(u,v) = \int_\Omega u_i e_{ipq} \partial_p A_q \, d\Omega = \int_{\partial\Omega} n_p u_i e_{ipq} A_q \, dS + \int_\Omega \vec{A}.\text{rot }\vec{u} \, d\Omega \, , \quad \text{or}$$

$$(u,v) = -\int_{\partial\Omega} \vec{A} \ (\vec{n}x\vec{u}) \, dS = + \int_{\partial\Omega} \vec{u}.(\vec{n}x\vec{A}) \, dS \qquad (68)$$

leads to the same conclusions. On $\partial_1\Omega$ the imposition of $\phi = 0$ is equivalent to the requirement $\vec{n}x\vec{u} = 0$. On $\partial_2\Omega$ the requirement $\vec{n}.\text{rot }\vec{A} \neq 0$ is satisfied by the somewhat stronger one $\vec{n}x\vec{A} = 0$.

The equivalence between the two surface integrals to which the scalar product reduces follows also from the general statement

$$\int_\Omega \text{div}\ \text{rot }\vec{B} \, d\Omega = \int_{\partial\Omega} \vec{n}.\text{rot }\vec{B} \, dS = 0$$

with $\vec{B} = \phi A$

This gives indeed, from

$$\text{rot}(\phi\vec{A}) = \phi \ \text{rot }\vec{A} + \text{grad}\phi x\vec{A}$$

the result

$$\int_\Omega \phi\vec{n}.\text{rot }\vec{A} \, dS = \int_{\partial\Omega} \text{grad } \phi.(\vec{n}x\vec{A}) \, dS \qquad (69)$$

The orthogonality property is thus found to hold between an irrotational flow, whose tangential velocity component vanishes on $\partial_1\Omega$ and isochoric flow whose normal velocity component vanishes on the complementary part $\partial_2\Omega$.

11. BOUNDING OF THE KINETIC ENERGY

Consider a flow that is both irrotational and isochoric and satisfies non homogeneous boundary conditions on $\partial_1\Omega$, where the tangential velocity component is specified, and $\partial_2\Omega$ where the normal component is specified.

In keeping with the preceding section, denote by u an irrotational flow that satisfies the homogeneous condition on $\partial_1\Omega$ (no tangential velocity) but is left unspecified on $\partial_2\Omega$, while u_o denotes any particular irrotational flow complying with the non homogeneous data on $\partial_1\Omega$.

Similarly, v will denote any isochoric flow satisfying the homogeneous condition on $\partial_2\Omega$ (no normal velocity component) and without specification on $\partial_1\Omega$, while v_o will denote any particular isochoric flow satisfying the non homogeneous data on $\partial_2\Omega$. We then find

$$(u_o,v) = \int_{\partial\Omega} \overline{\phi} \; (\vec{n}.\mathrm{rot}\; \vec{A}) \; dS = +\int_{\partial_1\Omega} \overline{\vec{A}.(\vec{n}\times\mathrm{grad}\phi)} \; dS \qquad (70)$$

$$(u,v) = \int_{\partial_2\Omega} \phi \; \overline{\dfrac{u}{v}} \; dS = -\int_{\partial_2\Omega} \overline{\mathrm{grad}\phi.(\vec{n}\times\vec{A})} \; dS \qquad (71)$$

In approximating the flow by a numerical analysis of Rayleigh-Ritz type, we may either consider the irrotational flow to contain adjustable parameters in u to satisfy in some best sense the incompressibility condition and the boundary data on $\partial_2\Omega$, or the isochoric flow to contain adjustable parameters in v to satisfy in some best sense the irrotationality condition and the boundary data on $\partial_1\Omega$.

Both viewpoints are combined in the requirement that the squared "distance" between the two adjustable fields by minimized :

$$(u+u_o-v-v_o, \; u+u_o-v-v_o) = (u+u_o,u+u_o)+(v+v_o,v+v_o)-2(u+u_o,v+v_o)$$
$$\mathrm{minimum}$$

Because of the orthogonality property $(u,v) = 0$, this condition naturally splits into the two independant requirements

$$(u+u_o,u+u_o) - 2(u,v_o) \qquad\qquad \underset{u}{\mathrm{minimum}} \qquad\qquad\qquad (72)$$

$$(v+v_o,v+v_o) -2(v,u_o) \qquad\qquad \underset{v}{\mathrm{minimum}} \qquad\qquad\qquad (73)$$

The term $-2(u_o,v_o)$ has been dropped as constant.

Now $u + u_o$ is the potential flow $u_i = \partial_i \phi$ where $\phi = \tilde{\phi}$ as specified on $\partial_1 \Omega$, and the first requirement is identical to our previous variational principle (64) :

$$\frac{1}{2} \int_\Omega \partial_i \phi \partial_i \phi d\Omega - \int_{\partial_2 \Omega} \phi \, \overline{u_\nu} \, dS \qquad \text{minimum}$$

Similarly, $v + v_o$ is the isochoric flow $\vec{v} = \text{rot } \vec{A}$ where $\vec{n}.\text{rot } \vec{A} = \overline{u_\nu}$ is specified on $\partial_2 \Omega$, and the second requirement is identical to the irrotational principle (65) implemented by a vector potential

$$\frac{1}{2} \int_\Omega \text{rot } \vec{A} \text{ rot } \vec{A} \, d\Omega - \int_{\partial_1 \Omega} \tilde{\phi} \, (\vec{n}.\text{rot } \vec{A}) \, dS \qquad \text{minimum}$$

We obtain a bounding of the kinetic energy of the flow by considering the separate problems :

Problem 1 : The data specified on $\partial_1 \Omega$ are non homogeneous ($u_o \neq 0$) but homogeneous on $\partial_2 \Omega$ ($v_o = 0$)

Problem 2 : The complementary problem : $u_o = 0$, $v_o \neq 0$.

We may note that the general problem can always be handled by linear superposition of problems 1 and 2.

In problem 1 we must find the best approximations to

$$(u + u_o, u + u_o) \qquad \text{minimum} \tag{74}$$

$$(v,v) - 2(v,u_o) \qquad \text{minimum} \tag{75}$$

Set $\quad u = \sum_1^n \alpha_j u_j$

where each u_j field is generated by an assumed potential that is zero on $\partial_1 \Omega$. The best coefficients $\hat{\alpha}_j$ are given by equating to zero the partial derivatives of the quadratic form

$$(u,u) + 2(u,u_o) = \Sigma\Sigma \alpha_j \alpha_k (u_j,u_k) + 2\Sigma \, \alpha_j (u_j,u_o)$$

Thus

$$\Sigma \ \hat{a}_k(u_j, u_k) + (u_j, u_o) = 0 \qquad\qquad j = 1, 2, \ldots n$$

Denoting by

$$\hat{u} = u_o + \Sigma \ \hat{a}_k \ u_k$$

The best approximation, those equations are equivalent to

$$(\hat{u}_j, \hat{u}) = 0 \qquad\qquad j = 1, 2, \ldots, n$$

Multiplying each by its coefficient $\hat{a}_j$ and summing

$$(\hat{u} - u_o, \hat{u}) = 0 \qquad\qquad \text{or} \ (\hat{u}, \hat{u}) = (\hat{u}, u_o) \qquad\qquad (76)$$

The exact solution s, which is both irrotational and isochoric satisfies the similar equation

$$(s, s) = (s, u_o) \qquad\qquad (77)$$

Indeed, s is both a v-type field ($v_o = 0$) and simultaneously $s - u_o$ is a u-type field, so that $(s - u_o, s) = 0$ by orthogonality.
Furthermore, since the minimum in (74) is not necessarily reached,

$$(\hat{u}, \hat{u}) \geqslant (s, s) \qquad\qquad (78)$$

We can give a similar treatment to

$$v = \sum_1^n \ \beta_j \ v_j$$

each v_j field being generated by a vector potential such that $\vec{n}.\text{rot}\vec{A} = 0$ on $\partial_2 \Omega$.

$$(v, v) - 2(v, u_o) = \Sigma\Sigma\beta_j\beta_k(v_j, v_k) - 2\Sigma \ \beta_j(v_j, u_o) \ \text{minimum}$$

furnishes the linear system

$$\Sigma \ \hat{\beta}_k(v_j, v_k) - (v_j, u_o) = 0$$

or with $\hat{v} = \Sigma \hat{\beta}_k v_k$, the best approximation,

$$(v_j, \hat{v}) - (v_j, u_o) = 0 \qquad\qquad j = 1, 2, \ldots, n$$

Multiplying by each $\hat{\beta}_j$ and adding

$$(\hat{v}, \hat{u}) - (\hat{v}, u_o) = 0 \qquad\qquad\qquad\qquad (79)$$

Since the minimum of (75) is not necessarily reached

$$(\hat{v}, \hat{v}) - 2(\hat{v}, u_o) \geqslant (s, s) - 2(s, u_o)$$

This inequality is transformed by (79) and (77) into

$$- (\hat{v}, \hat{v}) \geqslant - (s, s)$$

and this result combined with (78) gives the kinetic energy bounding

$$(\hat{v}, \hat{v}) \leqslant (s, s) \leqslant (\hat{u}, \hat{u}) \qquad\qquad\qquad\qquad (80)$$

A similar treatment of Problem 2 yields the reverse bounding

$$(\hat{u}, \hat{u}) \leqslant (s, s) \leqslant (\hat{v}, \hat{v}) \qquad\qquad\qquad\qquad (81)$$

where $\hat{u} = \Sigma \hat{\alpha}_j u_j$ and $\hat{v} = v_o + \Sigma \hat{\beta}_j v_j$

12. FINITE ELEMENT IMPLEMENTATION OF THE RAYLEIGH-RITZ PROCESSES

The simply connected region Ω is divided into adjacent subdomains Ω_α, the so-called finite elements. Any integral extended over the whole region is understood to be the sum of integrals over the Ω_α. Whenever an integration by parts is applied, the boundary terms involve the whole set of boundaries $\partial\Omega_\alpha$ of each subdomain and can be regrouped as a sum of integrals covering the external boundary $\partial\Omega$ of Ω and a sum of integrals involving the two faces of interfaces I_β of the subdomains

$$\underset{\alpha}{\Sigma} \int_{\partial\Omega_\alpha} n_i f_i dS = \int_{\partial\Omega} n_i f_i dS + \underset{\beta}{\Sigma} \int_{I_p} n_i (f_i^+ - f_i^-) dS$$

In the last terms the convention is adopted that the normal n_i to the interface is that of one of its faces, denoted by the superscript +. Since for outward normals

$$n_i^- = -n_i^+ = -n_i$$

the minus sign of the contribution of the other face is understood.

Consider a first subdivision into finite elements, in each of which a scaler potential ϕ_α is defined, usually in the form of a complete polynomial of chosen degree with unknown coefficients. The transition conditions to be satisfied at the interfaces can be found by examination of the orthogonality condition, generalizing (67)

$$(u,v) = \Sigma \int_{\Omega_\alpha} v_i \partial_i \phi_\alpha \, d\Omega = - \Sigma \int_{\Omega_\alpha} \phi_\alpha \partial_i v_i d\Omega + \int_{\partial\Omega} \phi_\alpha n_i v_i dS +$$

$$+ \Sigma \int_{I_\beta} n_i (\phi_\alpha^+ v_i^+ - \phi_\alpha^- v_i^-) dS$$

To obtain orthogonality with an isochoric flow ($\partial_i v_i = 0$) satisfying $n_i v_i = 0$ on the part $\partial_2\Omega$ of the outer boundary, while $\phi_\alpha = 0$ on $\partial_1\Omega$, we must still ensure that at the interfaces

$$\Sigma \int_{I_\beta} n_i (\phi_\alpha^+ v_i^+ - \phi_\alpha^- v_i^-) dS = 0$$

This must hold in particular for a continuous isochoric flow, hence $v_i^+ = v_i^-$ at the interfaces and

$$\Sigma \int_{I_\beta} n_i v_i (\phi_\alpha^+ - \phi_\alpha^-) dS = 0$$

It is thus sufficient, although admittedly not necessary, that there be continuity of the scalar potentials at the interfaces. It turns out that in the case of complete polynomials, it is extremely easy to enforce the interface continuity conditions, both in 2 and 3 dimensions. For the two-dimensional case, taking triangular finite elements, the coefficients

of the polynomial defining the scalar potential inside can be determined
in terms of local values of the potential at the vertices and along the
sides, plus, as turns out ot be the case for degrees higher or equal to 3,
at some interior points. If, at an interface, the local values of the potentials ϕ^+ and ϕ^- coincide, the potentials coincide along the whole
interface.

Similarly, for the same or another subdivision into finite elements,
for each of which a vector potential is assigned containing unknown
coefficients, the transition conditions follow from (68)

$$(u,v) = \Sigma \int_{\Omega_\alpha} \vec{u}.\text{rot } \vec{A}_\alpha \; d\Omega = \Sigma \int_{\Omega_\alpha} \vec{A}_\alpha.\text{rot } \vec{u} \; d\Omega - \int_{\partial\Omega} \vec{A}.(\vec{n}x\vec{u}) \; dS$$
$$+\Sigma \int_{I_\beta} (\vec{A}.(\vec{n}x\vec{u})^+ + \vec{A}.(\vec{n}x\vec{u})^-) dS$$

To obtain orthogonality with an irrotational flow (rot $\vec{u} = 0$) satisfying
$\vec{n}x\vec{u} = 0$ on part $\partial_1\Omega$ of the outer boundary, while $\vec{A}_\alpha = 0$ on part $\partial_2\Omega$,
it is sufficient to have continuity of the vector potential at the inter-
faces.

Again, this is quite easily implemented for polynomial approximations
of the vector potential. Here in the two-dimensional cases of plane flow
or axisymmetric flow, a scalar stream function replaces the vector potential.

13. VARIATIONAL PRINCIPLES FOR VISCOUS FLOW

There are no true variational principles yielding as Euler equations
the Navier-Stokes general equations. There is, however, a principle that
governs the dissipation in steady state flow for cases of such low Reynolds
numbers that the acceleration terms are negligible. Limiting ourselves to
incompressible fluids, the dissipation

$$F = \mu\theta_{ij}\theta_{ij}$$

$$\theta_{ij} = \frac{1}{2}(\partial_i u_j + \partial_j u_i)$$

and the functional yielding the equations of motion with negligible inertia
force terms can be taken as

$$J = \int_{\Omega} (\rho g_i u_i + p(\partial_i u_i) - F) \, d\Omega + \oint_{\partial_2 \Omega} \overline{t}_i u_i \, dS$$

with $u_i = \overline{u_i}$ on $\partial_1 \Omega$.

The pressure p appears here as a Lagrangian multiplier, whose variations enforce the incompressibility condition. With the viscous stresses

$$\sigma_{ij} = \frac{\partial F}{\partial \theta_{ij}}$$

the Euler equations stemming from variations on the velocity field are

$$\rho \, g_i - \partial_i p + \partial_j \sigma_{ij} = 0$$

and the natural boundary conditions on $\partial_2 \Omega$ are

$$n_j (\sigma_{ij} - p\delta_{ij}) = \overline{t}_i$$

As shown by Debongnie (Ref. 9), this principle can be extended by the Friedrichs technique to a canonical form involving simultaneously variations on the viscous stresses themselves.

Applications have been made in several directions. To biomechanics by P. Tong and Y.C. Fung (Ref. 8). To oil bearing problems and flow over deep wells by Debongnie (Ref. 9).
In applying the finite element methods to the two-dimensional cases, advantage may be gained from the remarkable analogy with Kirchhoff plate bending problems. The analogy comprises that between the stream function of the flow and the transverse plate flexure, the viscosity stresses and the bending moments tensor.

14. REFERENCES

1. CLEBSCH, A. : J. Reine und Angew.Math. 54, 293, 1857.
                                        56, 1,   1959.

2. LICHTENSTEIN, L. : Grundlagen der Hydrodynamik. Chap. 9.
                      Berlin, Springer, 1929.

3. BATEMAN, H. : Proc. Royal Soc. London, Ser. A, 125, 598, 1929.

4. HERIVEL, J.W. : Proc. Cambridge Phil. Soc., 51, 344, 1955.

5. SERRIN, J. : Handbuch der Physik, Band VIII/1, 125.
               Berlin, Springer, 1959.

6. ECKART, C. : The Physics of Fluids, 3, 421, 1960.

7. FRAEIJS de VEUBEKE, B. : Variational principles in fluid mechanics. in
                            Fluid Dynamics Transactions, vol. 3,
                            (Sumposium-Jurata, 1965), p 111, Polish
                            Academy of Sciences.

8. TONG, P. and FUNG, Y.C. : Slow particulate viscous flow in channels
                             and tubes. Applications to biomechanics.
                             J. of Applied Mechanics, 1971, p. 721.

9. DEBONGNIE, J.F. : Application de la méthode des éléments finis en
                     mécanique des fluides.
                     Université de Liège, Faculté des Sciences Appliquées,
                     1973.

RECENT DEVELOPMENTS OF FINITE-DIFFERENCE APPROXIMATIONS FOR BOUNDARY-LAYER EQUATIONS

E. KRAUSE

AERODYNAMISCHES INSTITUT

RHEINISCH-WESTFÄLISCH TECHNISCHE HOCHSCHULE AACHEN, GERMANY

ABSTRACT

Finite-difference solutions for Prandtl's boundary-layer equations are described for steady, two- and three-dimensional laminar and turbulent flows. For three-dimensional flows only boundary sheets are considered and curvature effects in the direction normal to the wall are being neglected. The governing equations are presented in form of a matrix-vector equation. Its numerical stability is discussed for elementary finite-difference molecules. Non-orthogonal coordinates are shown to affect the stability limits for the convective terms. If the momentum equations and/or the energy equation are decoupled by splitting the main part of the differential equations additional conditions must be observed for stable solutions. Finite-difference approximations with truncation error of fourth order are introduced to enable either increased accuracy or shortened calculation times, in particular, for three-dimensional problems. Studies of the behaviour of the overall error of the solution and several applications to real flow situations supplement the general considerations. Finally, a brief discussion is given for second-order closure problems.

1. INTRODUCTION

Although considerable effort has been spent on the analysis of turbulent boundary layers, our knowledge about such flows and our ability to predict them is rather limited. The major - so far unsurmountable - difficulties nest in the closure problem. Solutions cannot be obtained unless some empiricism is introduced at some point in the development of the solution. Aside from the physical problems there are computa-

---

This lecture is based on a paper [1] presented the AIAA-Computational Fluid Dynamics Conference, Palm Springs, Cal., July 19 - 20, 1973 and extensions thereof.

tional problems, if a solution is sought by means of finite-difference techniques. The computational problems are caused by the very nature of turbulent flow: A steep increase of the time averaged tangential velocity components in the immediate vicinity of the wall and the presence of large Reynolds stresses almost throughout the entire boundary layer. Their maximum value is much larger than that of the Newtonian stresses and numerical solutions which do not account for this peculiar behaviour of the dependent variables and of the coefficients of the differential equations are bound to be ineffective, in particular for three-dimensional flow problems.

This paper deals mainly with problems concerning the numerical integration of the boundary-layer equations for turbulent flows. Closure assumptions are only briefly discussed as the analysis presented herein is centered on an attempt to improve the numerical tools and techniques in order to make existing solution procedures more efficient. A survey of finite-difference methods was recently given in Ref. [2]. The finite-difference solutions presently being used can be divided into two groups: The first group is characterized by the common assumption of hyperbolic differential equations for the description of the time averaged velocity components and of the Reynolds stresses. These methods are described in [3], [4] and [5]. The change of the type of the differential equations is introduced by a particular approximation of the equation for the turbulent kinetic energy. As the slope of the characteristics at the wall is infinite the numerical solution fails there and a law of the wall must be matched to the numerical integration procedure. The method can only be applied to fully turbulent flows as the Newtonian part of the stress tensor has been completely neglected. An attempt to improve the law of the wall for three-dimensional flows was recently made in [6].

The second group of solutions uses parabolic equations in conjunction with a gradient-type representation for the cross-correlations. The approximations are, in general, extensions of first-order closures used in the analysis of two-dimensional flows. The validity of these concepts must be considered very critically when they are applied to three-dimensional flows. There is experimental evidence [7] that first-order closure can only be employed in mildy three-dimensional flows. Description of these solutions may be found in [2], [8], [9] and [10]. Numerical improvements of the accuracy of the solutions for parabolic equations were presented in [11] and [12]. For laminar flows [13] considerable savings of computation time were obtained for constant accuracy. In this lecture two fourth-order discretization procedures will be derived for parabolic equations which are written in matrix-vector form for non-orthogonal curvilinear coordinates. Both developments can be casted into a form identical with that of the difference equations for the second-order solution. Only the coefficients have to be redefined. If the differential equations are locally linearized, the coefficient matrix is tridiagonal for implicit formulation and the commonly used recursion relations can be employed in the solution.

2. GOVERNING EQUATIONS FOR BOUNDARY SHEETS

We begin the development by listing the momentum equations for those threedimensional boundary-layers which are categorized as boundary sheets. Curvature effects in the direction normal to the wall will not be included. Let V_1, V_2, V_3 be the three time averaged velocity components, V_3 being the normal component and V_1 and V_2 the two tangential components, and q_1, q_2, q_3 corresponding non-orthogonal curvilinear coordinates. With the boundary-layer assumption of constant pressure normal to the wall, the two momentum equations for the directions of q_1 and q_2 may be written as a matrix-vector equation of the following form (the flow is assumed to be steady and incompressible):

$$A_0 F + A_1 \frac{\partial F}{\partial q_1} + A_2 \frac{\partial F}{\partial q_2} + A_3 \frac{\partial F}{\partial q_3} + A_4 \frac{\partial^2 F}{\partial q_3^2} + B = 0 \tag{2.1}$$

All quantities are properly normalized and F is a column vector with components V_1 and V_2; A_0, A_1, A_2, A_3 and A_4 are matrices and B a vector defined below:

$$A_0 = \begin{vmatrix} k_1 v_1 + k_2 v_2 & k_3 v_2 \\ l_1 v_1 + l_2 v_2 & l_3 v_2 \end{vmatrix} \tag{2.2}$$

$$A_1 = \frac{v_1}{h_1} I \quad ; \quad A_2 = \frac{v_2}{h_2} I \tag{2.3}$$

where I is the identity matrix of order two. The matrices A_3 and A_4 are of the following form

$$A_3 = \left(v_3 - \frac{\partial \varepsilon_1}{\partial q_3} \right) \begin{pmatrix} 1 & 0 \\ 0 & 1+e_1 \end{pmatrix} \tag{2.4}$$

$$A_4 = -\left(1 + \varepsilon_1 \right) \begin{pmatrix} 1 & 0 \\ 0 & 1+e_2 \end{pmatrix} \tag{2.5}$$

In the last equation, ε_1 is the component of the eddy viscosity in the direction of q_1. The excentricities e_1 and e_2, being defined by

$$e_1 = \left[\frac{\partial}{\partial q_3} (\varepsilon_1 - \varepsilon_2) \right] / \left(v_3 - \frac{\partial \varepsilon_1}{\partial q_3} \right) \tag{2.6}$$

$$e_2 = (\varepsilon_1 - \varepsilon_2) / (1 + \varepsilon_1) \tag{2.7}$$

vanish, if the eddy viscosity is assumed to be a scalar (i.e. $\varepsilon_1 = \varepsilon_2 = \varepsilon$). Then the matrices A_3 and A_4 can again be defined in terms of the identity matrix. Because of the introduction of non-orthogonal curvilinear coordinates both components of the vector B contain the pressure gradients for the directions q_1 and q_2:

$$B = - \frac{h_1 h_2}{g m} \begin{vmatrix} \frac{h_2}{h_1} & -\frac{g}{h_2} \\ -\frac{g_1}{h_1} & \frac{h_1}{h_2} \end{vmatrix} \begin{bmatrix} \frac{\partial p}{\partial q_1} \\ \frac{\partial p}{\partial q_2} \end{bmatrix} \tag{2.8}$$

The metric coefficients $k_1 - k_3, l_1 - l_3, h_1, h_2, g$ and m are defined in [1] They will not be repeated here. Eq. (2.1) can easily be adapted for twodimensional flows by identifying F with V, and setting A_2 equal to zero. If compressible flows are considered, the energy equation can also be written in a form identical to that of Eq. (2.1), and the components of F are V_1 and either h, h_s or T depending on what variable is used. For threedimensional compressible flows, F contains the three components V_1, V_2 and T, or h_s, the stagnation enthalpy.

The normal velocity component is determined from the continuity equation which for three-dimensional flows reads

$$\frac{\partial}{\partial q_1} \left(\frac{m}{h_1} v_1 \right) + \frac{\partial}{\partial q_2} \left(\frac{m}{h_2} v_2 \right) + m \frac{\partial v_3}{\partial q_3} = 0 \tag{2.9}$$

The boundary conditions to be imposed on Eqs. (2.1) and (2.9) are

$$F(q_1, q_2, 0) = 0 \quad ; \quad \lim_{q_3 \to \infty} F = Fe \tag{2.10}$$

$$v_3 (q_1, q_2, 0) = f_1 (q_1, q_2) \tag{2.11}$$

where $f_1(q_1, q_2)$ must be prescribed in terms of the surface coordinates q_1 and q_2; $f_1(q_1, q_2)$ vanishes identically for an impermeable wall, is negative for the case of suction and positive for normal injection. As Eqs. (2.1) and (2.9) are given in a nondimensionless form in which the normal coordinate and the normal velocity component are stretched by the square root of the characteristic Reynolds number, the function f_1 must be of order unity so that the boundary-layer assumptions are satisfied. Initial conditions are to be prescribed for F along some initial surface such that

$$F(q_{1i}, q_{2i}, q_3) = F_i(q_{1i}, q_{2i}, q_3) \tag{2.12}$$

where F_i is a known vector function specifying its components on a surface normal to the body erected over the line specified by q_{1i} and q_{2i}. For twodimensional problems the surface on which the initial conditions are specified collapses into the surface normal of the body.

For closure of the problem, the eddy viscosity introduced earlier may be approximated to first order by a scalar function of the form

$$\varepsilon = l^2 \left[\left(\frac{\partial v_1}{\partial q_3} \right)^2 + \left(\frac{\partial v_2}{\partial q_3} \right)^2 \right]^{1/2} \tag{2.13}$$

where l is the mixing length, which may be taken proportional to q_3 in the inner part of the boundary layer and equal to a constant in the outer. Several models are compared in [2]. The validity of Eq. (2.13) must be considered very critically as there is experimental evidence that this assumption does not hold true in general [7]. As the closure assumption is not relevant for the following discussion, no justification for the adoption of the mixing length hypothesis and the eddy viscosity concept is given; it is only mentioned that most finite-difference solutions derived for parabolic euquations rest on the validity of Eq. (2.13). If compressible flows are considered, an equation of state and laws specifiying the laminar viscosity and the thermal conductivity in terms of the temperature must be given. First-order closure for turbulent flows would require a turbulent Prandtl number.

The use of non-orthogonal coordinates for the numerical solution is advocated mainly for two reasons: If an algorithm is derived from implicit linearized difference equations the coordinates used affect only the coefficients but do not change the form of the recursion relations. The more complicated expression for the pressure gradient term is of little importance as it can be calculated before the integration is initiated and need not be determined repeatedly. The second reason is that it can only be decided, which coordinates are suitable

when the particular problem to be investigated is specified.
Eqs. (2.3) - (2.9) can always be reduced to an appropriate form and
for example, orthogonal - either external flow or surface orientated-
coordinates are obtained by setting g in Eqs. (2.8) - (2.9) equal to
zero. Although non-orthogonal curvilinear coordinates offer the possibility of applying the solution to a wide range of problems, they also
influence the stability conditions, which will be briefly discussed
in the next section.

3. REMARKS ON THE NUMERICAL STABILITY

The stability of the finite-difference approximations for Eq.(2.1) has
been discussed repeatedly, e.g. in [2] and earlier in [14] and [15].
Therein it was shown that the implicit finite-difference approximations
of Eq. (2.1) are only stable if the Courant-Friedrichs-Lewy condition
is satisfied for the convective terms; that is to say that on the surfaces parallel to the surface of the body the numerical domain of dependence must include the domain of dependence of the differential
equation. The latter is defined by the "Raetz-influence-principle" [5],
which will be repeated here to clarify matters: If initial conditions
for V_1 and V_2 are specified on a surface normal to the surface of the
body and if the line of intersection is of finite length and does not
coincide with the projection of a streamline, then the solution of
Eqs. (2.1) and (2.9) is completely defined in terms of the initial
conditions only over a region of influence of finite extent. The
region of influence is bounded by the surface on which the initial
conditons are specified and two other surfaces which are normal to the
surface of the body and intersect the initial surface in its end points.
The bounding surfaces are erected over a limiting and/or the projection
of an external streamline such that the enclosed area on the surface
of the body attains a minimum. The flow outside of the domain of dependence cannot be computed as it is advected from points for which
no initial data are available.

Within the domain of dependence the implicit solution is stable only
if the convective terms in Eq. (2.1) are substituted for by advective
finite-difference approximations [16], i.e. the integration must always follow the main direction of the flow. In [14] and [15] the von-
Neumann analysis was shown to predict the stability limits accurately
for cartesian coordinates. The solution became immediately unstable

as soon as the stability limits were exceeded. In the following we outline the von-Neumann test for curvilinear coordinates. By freezing the coefficients in Eq. (2.1) one can derive locally valid stability limits which are defined in terms of the stability parameter

$$\Gamma = \frac{v_2\, h_1\, \Delta q_1}{v_1\, h_2\, \Delta q_2} \tag{3.1}$$

In comparison to cartesian coordinates Γ in Eq. (3.1) may be more restricted in stable regions. This is caused by the appearance of the length parameter h_1 and h_2 in the expression for Γ . In addition for curvilinear non-orthogonal coordinates lower-order instabilities may arise from the curvature terms: Consider a simple difference molecule depicted in Fig. 1. In this case the derivatives in the direction of q_3 are approximated with second-order accuracy and those in the directions of q_1 and q_2 with first order.

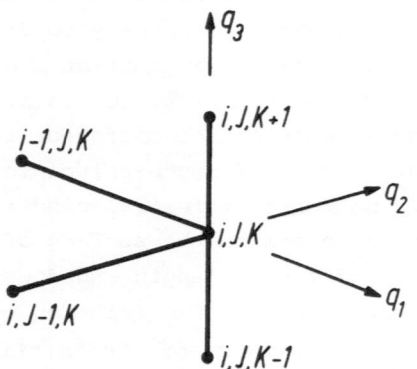

If the pressure gradient term in Eq. (2.1) is neglected and the matrices A_1 - A_4 are assumed to be constant with $e_1 = e_2 = 0$ the amplification matrix for the direction of q_1 is

Fig. 1 Simple finite difference molecule
for implicit integration

$$G\,(\underline{K}.\,\Delta q_1) = \frac{h_1\, \Delta q_1}{D} \begin{vmatrix} E + l_3\, \frac{v_2}{v_1} & -K_3\, \frac{v_2}{v_1} \\ -\left(l_1 + l_2\, \frac{v_2}{v_1}\right) & E + K_1 + K_2\, \frac{v_2}{v_1} \end{vmatrix} \tag{3.2}$$

where E is an abbreviation of the complex expression

$$E = \frac{1}{h_1\, \Delta q_1} \left[1 + \Gamma\, Z_1 - \frac{2(1+\varepsilon)}{(\Delta q_3)^2}\, \frac{h_1 \Delta q_1}{v_1}\, Z_2 + \left(v_3 - \frac{\partial \varepsilon}{\partial q_3}\right) \frac{h_1 \Delta q_1}{v_1 \Delta q_3}\, Z_3 \right] \tag{3.3}$$

with

$$Z_1 = 1 - \cos(K_2\, \Delta q_2) + i\, \sin(K_2\, \Delta q_2) \tag{3.4}$$

$$Z_2 = 1 - \cos(K_3\, \Delta q_3) \tag{3.5}$$

$$Z_3 = i \sin (K_3 \Delta q_3)$$ (3.6)

The vector $\underline{K}$ in Eq. (3.2) is defined by the two integer components K_2 and K_3. The quantity D is the determinant of the matrix in Eq. (3.2). From Eq. (3.2) and Eq. (3.3) it is clear that the absolute values of the complex eigenvalues satisfy the von-Neumann condition as long as Γ is properly restricted (as shown in [14]) and the curvature terms are of order unity. However, small step sizes Δq_1 may be necessary when the curvature terms are large. This can easily be verified by setting Z_1, Z_2 and Z_3 equal to zero. Then Eq.(2.1) reduces to an ordinary differential equation [17], which for large curvature terms requires small step sizes Δq_1, if the direction of q_1 is assumed to be the marching direction. These results imply that the length elements h_1 and h_2 should not be too different from each other; in addition, the curvature terms should be of order one so that the step size in the direction of the tangential coordinates does not have to be restricted unnecessarily.

Another difficulty may be encountered, if the main part of the differential equation (2.1) is split in order to decouple the equations. Then the second-order derivatives of all components of the solution vector may appear in each of the equations (2.1). It is, in general, possible to solve the resulting difference equations simultaneously without decoupling. Since this method of solution would require matrix inversion the equations are often decoupled -for the sake of convenience- so that they can be solved one after the other. The component of the solution vector in question is then written in implicit formulation and the rest of the main part of the differential equation is discretized explicitly. These terms are evaluated from the last station computed, and act as a forcing function, comparable to pressure-gradient terms in the momentum equations. Such a decoupling may be necessary when the eddy viscosity is not assumed to be a scalar. To demonstrate the point we consider a simplified form of Eq. (2.1)

$$\frac{\partial F}{\partial q_1} + A_4 \frac{\partial^2 F}{\partial q_3^2} = 0 ,$$ (3.7)

the matrix A_4 is assumed to have four positive elements a_{ij} instead of two as in Eq. (2.5). If the terms with coefficients a_{11} and a_{22} are written in implicit formulation and those with a_{12} and a_{21} explicitly,

12

the amplification matrix of (3.7) becomes

$$G(\underline{K}, \Delta q_1) = \begin{vmatrix} \dfrac{1}{1+g_{11}} & \dfrac{-g_{12}}{1+g_{11}} \\[2mm] \dfrac{-g_{21}}{1+g_{22}} & \dfrac{1}{1+g_{22}} \end{vmatrix} \tag{3.8}$$

The abbreviations g_{ij} are defined as

$$g_{ij} = \frac{2\,a_{ij}\,\Delta q_1}{(\Delta q_3)^2}\,(1-\cos K_3 \Delta q_3) \tag{3.9}$$

The two eigenvalues of Eq. (3.8) are of the form

$$\lambda_{1,2} = \frac{1}{2}\left(\frac{1}{1+g_{11}} + \frac{1}{1+g_{22}}\right) \pm \sqrt{\frac{1}{4}\left(\frac{1}{1+g_{11}} - \frac{1}{1+g_{22}}\right)^2 + \frac{g_{12}\,g_{21}}{(1+g_{11})(1+g_{22})}} \tag{3.10}$$

In order to satisfy the von-Neumann condition the absolute values of the eigenvalues must be less than unity. This is always true when the matrix elements a_{12} and a_{21} are equal to zero. Since g_{11} and g_{22} are non-negativ for all values of $k_3 \Delta q_3$,the eigenvalues are positive and less than unity. This is the expected result and the finite-difference solution in implicit formulation is stable. On the other hand, the terms along the Spur may vanish, i.e. $a_{11} = a_{22} = 0$. This condition is encountered if the eddy viscosity does not depend on the component of the solution vector considered but only on the forcing function. The eigenvalues become for $a_{11} = a_{22} = 0$

$$\lambda_{1,2} = 1 \pm \sqrt{g_{12}\,g_{21}} \tag{3.11}$$

The second eigenvalue yields the stability condition

$$\Delta q_1 \le \frac{1}{2}\,\frac{(\Delta q_3)^2}{\sqrt{a_{12}\,a_{21}}} \tag{3.12}$$

for explicit integration [+)], but the first eigenvalue is always larger than unity so that the solution of the difference equations is un-stable, if the elements along the Spur vanish. If neither one of the a_{ij} is zero, the first eigenvalue yields the condition that the product $a_{12}\,a_{21}$ must be less than $a_{11}\,a_{22}$ otherwise λ_1 (which is positive) would exceed unity. The second eigenvalue λ_2 leeds to a condition

$$\left[(a_{12}\,a_{21} - a_{11}\,a_{22})\frac{\Delta t}{(\Delta q_3)^2} - \frac{1}{2}(a_{11} + a_{22})\right]\frac{\Delta t}{(\Delta q_3)^2} < \frac{1}{4} \tag{3.13}$$

[+)] It was pointed out by J.J. Smolderen that with $a_{11} = a_{22} = 0$ the solution should always be unstable. This is clearly confirmed by the first eigenvalue of Eq. (3.11).

Since a_{11} and a_{22} are positive the bracketed term in Eq.(3.13) is negative if

$$a_{11} \, a_{22} \geqq a_{12} \, a_{21} \tag{3.14}$$

so that stability can be guaranteed if Eq.(3.14) is satisfied. From physical considerations there is no need to impose the restriction given by Eq. (3.14) on the elements of the matrix A_4, as they are independent of each other and can be chosen arbitrarily i.e. $a_{ij} > 0$. Although numerical tests have not been carried out the above derivations show that decoupling by explicit-implicit formulation may not always leed to stable difference equations. These considerations also apply to multidiffusion problems, if the diffusional flux of each one of the chemical species is expressed though the sum of the others and the decoupling of the equations is carried out as described above. The terms which are incorporated in the finite-difference approximation as explicit terms may therefore have a destabilizing effect as they act as forcing functions. We next turn now to a description of higher-order schemes.

4. FOURTH ORDER DIFFERENCE APPROXIMATIONS

In numerical solutions of the boundary layer equations (2.1) and (2.9) second-order finite-difference approximations have been used predominantly. Higher-order solutions are applied in order to achieve an increased accuracy or to reduce the computation time when only moderate accuracy is desired. Although both problems seem to be identical, experience has shown that this is only true when all flow variables and their derivatives are of order one. In turbulent boundary layers, not only the derivatives of the tangential velocity components are large (in the vicinity of the wall) but also the dimensionless turbu-

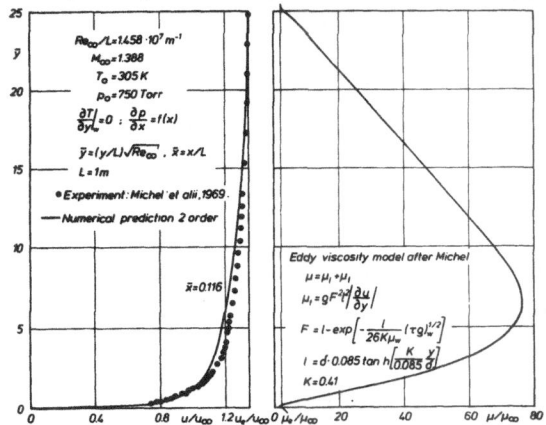

Fig. 2 Typical profile of tangential velocity component and turbulent viscosity

lent viscosity is large compared to unity and exhibits steep gradients on both sides of its maximum value. Even in the outer portion of the boundary layer where the velocity gradients are small do the gradients of the turbulent viscosity remain large. This behaviour is common to all viscosity models and can be seen in Fig. 2 where a sample calculation of [18] is shown for a twodimensional compressible boundary layer and compared to experimental data. Depicted is a profile of the tangential velocity component and of the effective turbulent viscosity, which was calculated from Michel's equation for the mixing length

$$\frac{l}{\delta} = 0.085 \; th\left(\frac{k}{0.085} \; \frac{q_3}{\delta}\right)$$

(4.1)

and Eq. (2.13) reduced to two-dimensional flows. The behaviour near the wall is determined by the van Driest damping factor

$$F = 1 - exp\left[-\frac{l}{26k\mu} \; (\tau \varrho)^{1/2}\right]$$

(4.2)

Other approximations [2] for the eddy viscosity yield comparable data. This is shown in Fig. 3 where for a typical case, Equs. (2.13), (4.1) and (4.2) are compared with Pletcher's polynomial curve fit of experimental data:

$$\left(\frac{l}{\delta}\right)_i = k\left[1 - exp\left(-q_3^*/\delta\right)\right]\left(q_3/\delta\right)$$

(4.3)

$$0 \leq q_3/\delta \leq 0.1$$

$$\left(\frac{l}{\delta}\right) = \left(\frac{l}{\delta}\right)_i + \sum_{n=2}^{n=4} a_n \left(q_3/\delta - 0.1\right)^n$$

(4.4)

$$0.1 < q_3/\delta \leq 0.6$$

$$a_2 = -1.53506; \qquad a_3 = 2.75625; \qquad a_4 = -1.88425$$

$$\frac{l}{\delta} = 0.089 \qquad\qquad 0.6 < q_3/\delta$$

(4.5)

Often Klebanov's intermittency factor is used to describe the eddy viscosity in the outer portion of the boundary layer:

$$\varepsilon_0 = k_1 \, V_e \; \delta^*/\left[1 + 5.5 \left(q_3/\delta\right)^6\right]$$

(4.6)

In the above relations k is von Kármán's constant, $k_1 = 0.0168$; q_3^* is defined in terms of the shear velocity. It is this characteristic dependence of the effective turbulent viscosity and the form of the tangential velocity profile, that complicates the integration of the

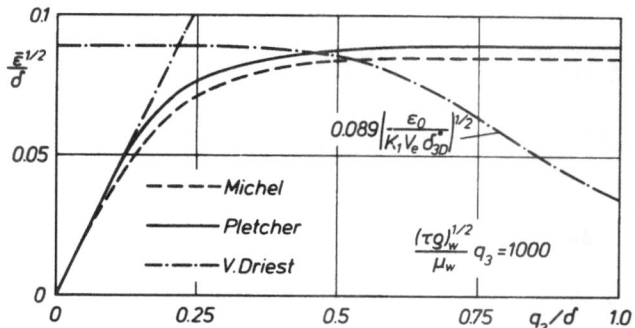

Fig.3 Comparison of several approximations
for the eddy viscosity.

boundary-layer equations for turbulent flows. For this reason it is necessary to construct accurate finite-difference approximations as otherwise numerical errors can be attributed to the failure of the closure relations chosen and may leed to wrong conclusions.

There are several ways to improve the accuracy of numerical solutions. Richardson extrapolation has been used successfully, for example, in Refs. [19] and [20]. Here an alternative approach will be discussed. The basic idea is that the first higher-order derivatives which are neglected in the second-order solution are eliminated through auxiliary relations obtained either by Taylor series development or from the differential equations. In order to show the major steps of the development we will start with the Taylor series and define the two difference expressions.

$$(\delta F)_k = F_{k+1} - F_{k-1} \tag{4.7}$$

$$(\delta^2 F)_k = F_{k+1} - 2 F_k + F_{k-1} \tag{4.8}$$

where the subscript K designates a net point and K + 1 and K - 1 neighbouring points being a constant Δq away from K. The expressions δF and $\delta^2 F$ possess a Taylor series representation of the form

$$(\delta F)_k = 2 \left[\left(\frac{\partial F}{\partial q} \right)_k \Delta q + \frac{1}{3} \left(\frac{\partial^3 F}{\partial q^3} \right)_k (\Delta q)^3 + \cdots \cdots \right] \tag{4.9}$$

$$(\delta^2 F)_k = 2 \left[\frac{1}{2!} \left(\frac{\partial^2 F}{\partial q^2} \right)_k (\Delta q)^2 + \frac{1}{4!} \left(\frac{\partial^4 F}{\partial q^4} \right)_k (\Delta q)^4 + \cdots \right] \tag{4.10}$$

In the second-order solution only the first term on the right of the last two equations is retained, while all other terms are neglected. The finite-difference representation may now be improved, if the third- and fourth-order derivatives are eliminated. First the series (4.9) and (4.10) are repeated for the first- and second-order derivatives: There result the four expressions [11]

$$\left[\delta\left(\frac{\partial F}{\partial q}\right)\right]_k \Delta q = 2\left[\left(\frac{\partial^2 F}{\partial q^2}\right)_k (\Delta q)^2 + \frac{1}{3!}\left(\frac{\partial^4 F}{\partial q^4}\right)_k (\Delta q)^4 + \ldots\ldots\right] \tag{4.11}$$

$$\left[\delta^2\left(\frac{\partial F}{\partial q}\right)\right]_k \Delta q = 2\left[\frac{1}{2!}\left(\frac{\partial^3 F}{\partial q^3}\right)_k (\Delta q)^3 + \frac{1}{4!}\left(\frac{\partial^5 F}{\partial q^5}\right)_k (\Delta q)^5 + \ldots\ldots\right] \tag{4.12}$$

$$\left[\delta\left(\frac{\partial^2 F}{\partial q^2}\right)\right]_k (\Delta q)^2 = 2\left[\left(\frac{\partial^3 F}{\partial q^3}\right)_k (\Delta q)^3 + \frac{1}{3!}\left(\frac{\partial^5 F}{\partial q^5}\right)_k (\Delta q)^5 + \ldots\right] \tag{4.13}$$

$$\left[\delta^2\left(\frac{\partial^2 F}{\partial q^2}\right)\right](\Delta q)^2 = 2\left[\frac{1}{2!}\left(\frac{\partial^4 F}{\partial q^4}\right)_k (\Delta q)^4 + \frac{1}{4!}\left(\frac{\partial^6 F}{\partial q^6}\right)_k (\Delta q)^6 + \ldots\ldots\right] \tag{4.14}$$

Eqs. (4.9) - (4.14) can now be combined to yield the following four linearly independent expressions:

$$(\delta F)_k = 2\left(\frac{\partial F}{\partial q}\right)_{k+1}\Delta q - \frac{2}{3}\left[\left(\frac{\partial^2 F}{\partial q^2}\right)_{k+1} + 2\left(\frac{\partial^2 F}{\partial q^2}\right)_k\right](\Delta q)^2 + \frac{2}{45}\left(\frac{\partial^5 F}{\partial q^5}\right)_k (\Delta q)^5 + \cdots \tag{4.15}$$

$$(\delta F)_k = 2\left(\frac{\partial F}{\partial q}\right)_{k-1}\Delta q + \frac{2}{3}\left[\left(\frac{\partial^2 F}{\partial q^2}\right)_{k-1} + 2\left(\frac{\partial^2 F}{\partial q^2}\right)_k\right](\Delta q)^2 + \frac{2}{45}\left(\frac{\partial^5 F}{\partial q^5}\right)_k (\Delta q)^5 + \cdots \tag{4.16}$$

$$(\delta F)_k = 2\left(\frac{\partial F}{\partial q}\right)_k \Delta q + \frac{1}{6}\left[\delta\left(\frac{\partial^2 F}{\partial q^2}\right)\right]_k (\Delta q)^2 + \frac{1}{180}\left(\frac{\partial^5 F}{\partial q^5}\right)_k (\Delta q)^5 + \ldots \tag{4.17}$$

$$(\delta^2 F)_k = \frac{1}{12}\left[\left(\frac{\partial^2 F}{\partial q^2}\right)_{k+1} + 10\left(\frac{\partial^2 F}{\partial q^2}\right)_k + \left(\frac{\partial^2 F}{\partial q^2}\right)_{k-1}\right](\Delta q)^2 - \frac{1}{240}\left(\frac{\partial^6 F}{\partial q^6}\right)_k (\Delta q)^6 + \ldots \tag{4.18}$$

If now all terms of fifth and higher order are neglected the last four equations contain three unknown first and three unknown second-order derivatives. The truncation error has been reduced to fourth order. The finite-difference representation just given will now be used to replace the derivatives in the direction normal to the wall.

For the elimination of the six unknown derivatives in Eqs. (4.15) - (4.18) three additional relations are necessary. They are obtained from the momentum equation (2.1). If it is assumed that appropriate difference approximations have been introduced for the derivatives in the direction of q_1 and q_2, the momentum equations can be written in the following form for the points $K + 1$, K, and $K - 1$:

$$G_o + A_{3l} \left(\frac{\partial F}{\partial q_3} \right)_l + A_{4l} \left(\frac{\partial^2 F}{\partial q_3^2} \right)_l = 0 \qquad l = K-1, K, K+1 \qquad (4.19)$$

where G is the finite-difference approximation of the first three terms and of B in Eq. (2.1)

$$G = A_o F + A_1 \frac{\partial F}{\partial q_1} + A_2 \frac{\partial F}{\partial q_2} + B \qquad (4.2o)$$

Eqs. (4.15) - (4.2o) yield now a set of difference equations which have the same form as those of the second-order implicit solution

$$M_{1K} F_{K+1} + M_{2K} F_K + M_{3K} F_{K-1} + M_{4K} = 0 \qquad (4.21)$$

and can be solved with the algorithm

$$F_K = P_j F_{K+1} + Q_j \qquad (4.22)$$

The definitions of P_j and Q_j follow from Eq. (4.21). The major difference is, however, that F is determined with an error

$$\varepsilon = 0 \left[(\Delta q_3)^4 \right] + 0 \left[(\Delta q_1)^2 \right] + 0 \left[(\Delta q_2)^2 \right] ,$$

if the matrices A_1 - A_4 are calculated to the same degree of accuracy. In order to do so five net points will be decessary.

This method derives its name ("Mehrstellen"-method) from the fact that collocation is enforced at three net points instead of one for the second-order solution . Approximations similar to the one just outlined were described earlier in [21] and [22], and the method has been applied to boundary layers in [11], [13], and [23] .

In [17] it is shown how a fourth-order solution can be obtained for the heat conduction equation in a manner different from the one just outlined. That approach can also be extended to the boundary-layer equations and a brief derivation will be given here: While in the development of the "Mehrstellen"-method auxiliary relations for the elimination of the third- and fourth-order derivatives are obtained from Taylor series expansions, they may also be obtained from the differential equation (2.1). If G is redefined as

$$G = -A_4^{-1} \left[A_o F + A_1 \frac{\partial F}{\partial q_1} + A_2 \frac{\partial F}{\partial q_2} + B \right] \qquad (4.23)$$

and

$$H = A_4^{-1} A_3 \qquad (4.24)$$

Eq. (4.19) can be written as

$$G = H \frac{\partial F}{\partial q_3} + \frac{\partial^2 F}{\partial q_3^2}$$

(4.25)

This expression can be differentiated with respect to q_3; after elimination of the second-order derivatives there results

$$\frac{\partial^3 F}{\partial q_3^3} = -HG + \frac{\partial G}{\partial q_3} + \left(H^2 - \frac{\partial H}{\partial q_3}\right) \frac{\partial F}{\partial q_3}$$

(4.26)

and a second differentiation gives

$$\frac{\partial^4 F}{\partial q_4} = \frac{\partial^2 G}{\partial q_3^2} - \frac{\partial}{\partial q_3}(HG) - \left[\frac{\partial}{\partial q_3}\left(\frac{\partial H}{\partial q_3} - H^2\right)\right] H^{-1}G$$

$$- \left\{\frac{\partial H}{\partial q_3} - H^2 - \left[\frac{\partial}{\partial q_3}\left(\frac{\partial H}{\partial q_3} - H^2\right)\right] H^{-1}\right\} \frac{\partial^2 F}{\partial q_3^2}$$

(4.27)

With the following abbreviations introduced

$$B_0 = -HG + \frac{\partial G}{\partial q_3}$$

(4.28)

$$B_1 = H^2 - \frac{\partial H}{\partial q_3}$$

(4.29)

$$C_0 = \frac{\partial^2 G}{\partial q_3^2} - \frac{\partial}{\partial q_3}(HG) - \left[\frac{\partial}{\partial q_3}\left(\frac{\partial H}{\partial q_3} - H^2\right)\right] H^{-1}G$$

(4.30)

$$C_1 = -\frac{\partial H}{\partial q_3} + H^2 + \left[\frac{\partial}{\partial q_3}\left(\frac{\partial H}{\partial q_3} - H^2\right)\right] H^{-1}$$

(4.31)

in Eqs. (4.9) and (4.1o) one obtains for the derivatives of first and second order

$$\left(\frac{\partial F}{\partial q_3}\right)_k = \left[I + \frac{(\Delta q_3)^2}{6} B_{1k}\right]^{-1} \left[\frac{(\delta F)_k}{2\Delta q_3} - \frac{(\Delta q_3)^2}{6} B_{0k}\right] + 0\left[(\Delta q_3)^4\right]$$

(4.32)

$$\left(\frac{\partial^2 F}{\partial q_3^2}\right)_k = \left[I + \frac{(\Delta q_3)^2}{12} C_{1k}\right]^{-1} \left[\frac{(\delta^2 F)_k}{(\Delta q_3)^2} - \frac{(\Delta q_3)^2}{12} C_{0k}\right] + 0\left[(\Delta q_3)^4\right]$$

(4.33)

After substitution of the last two expressions into Eq. (4.25) there results again a finite-difference approximation with a truncation error

$$\varepsilon = 0\left[(\Delta q_3)^4\right] + 0\left[(\Delta q_1)^2\right] + 0\left[(\Delta q_2)^2\right] \quad .$$

Just as before, only three net points are necessary, if the quantities

B_0, B_1, C_0, C_1 contain no higher than second-order derivatives. By comparing Eq. (2.4) with Eσ. (4.3o) it is seen that third-and fourth-order derivatives appear in Eqs. (4.29) - (4.31). In Ref. [17] a transformation is suggested to overcome this difficulty for the case that the excentricities in Eqs. (2.4) and (2.5) are zero, i.e. $\varepsilon_1 = \varepsilon_2 = \varepsilon$

$$\eta = \int \frac{1}{(1+\varepsilon)} \, dq_3 \qquad (4.34)$$

Then Eq. (2.1) can be written as

$$(1+\varepsilon)\left(A_0 F + A_1 \frac{\partial F}{\partial q_1} + A_2 \frac{\partial F}{\partial q_2}\right) + A_3 \frac{\partial F}{\partial \eta} - \frac{\partial^2 F}{\partial \eta^2} + (1+\varepsilon) B = 0 \qquad (4.35)$$

$$A_3 = (1+\varepsilon)\left(\frac{\partial \eta}{\partial q_1} A_1 + \frac{\partial \eta}{\partial q_2} A_2\right) + v_3 I \qquad (4.36)$$

For flow fields for which (4.34) is applicable, the quantities B_0, B_1, C_0, and C_1 need only be evaluated with second-order accuracy as they are multiplied by $(\Delta q_3)^2$. Recently, W. Kordulla investigated the transformation (4.34) in his doctoral dissertation 24 . He showed that the transformation can only be employed when ε is of order unity. This result was obtained from the following test case: A known tangential velocity profile was used to evaluate the turbulent effective viscosity from Michel's approximation. Eq. (4.34) was then integrated numerically yielding η as a function of y. In Fig. 4 the inverse transformation is shown as a function of η . Because of the steep increase of the effective turbulent viscosity there is a very large contraction in η ; while y increases from zero to six, η varies only from zero to 0.5. For larger values of y the contraction is zero and y

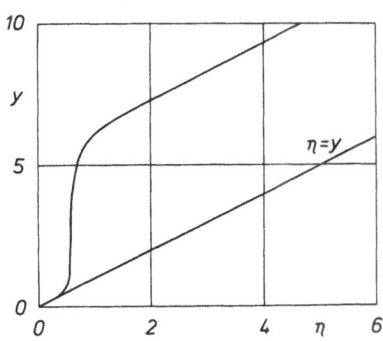

Fig. 4   The inverse transformation Eq. (4.34)

increases linearly with η . The effect of the contraction on the velocity profile is demonstrated in Fig. 5, where in customary representation the profile of the dimensionless tangential velocity component is shown as a function of the normal coordinate y and the transformed coordinate η . The contraction changes the velocity profile in two ways: first, the external flow is reached for $\eta = 0,7$ while for the same value of y, the value of the velocity in untransformed coordinates is only about 0,25. Secondly the contraction distortes the velocity profile completely, causing an unusual S-shaped form.

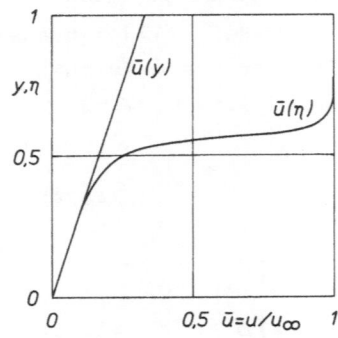

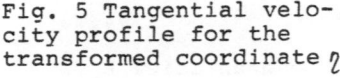

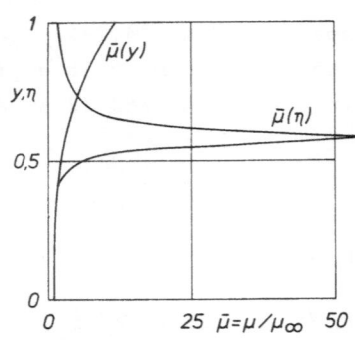

The integration of the momentum equation in the form of Eq. (4.35) could not be facilitated. The failure is due to the shape of the viscosity profile, which is depicted in Fig. 6.

Fig. 5 Tangential velocity profile for the transformed coordinate η

Fig. 6 Shape of the profile of the effective turbulent viscosity for the transformed coordinate η.

It is seen that $\mu(\eta)$ shows a sharp peak near $\eta = 0,6$ with very large gradients on both sides of the maximum. It is clear that this function is not suitable for numerical integration, unless particularly fine resultion is used. Thus the transformation suggested by Richtmyer and Morton is not applicable to turbulent flows, but may be in solutions for laminar flows as long as the dimensionless viscosity is of order unity. This result does by no means invaliditate the use of Eq. (4.25) together with the fourth-order approximations for the first-and second-order derivatives normal to the wall, Eqs. (4.32) and (4.33). It is only necessary to evaluate the third-and fourth-order derivates in the coefficients Eq. (4.28) - (4.31) with second-order accuracy, which requires five net-points. In the field the usual difference approximations are

$$\left(\frac{\partial^3 F}{\partial q^3}\right)_k = \frac{1}{2}\left[F_{k+2} - 2\left(F_{k+1} - F_{k-1}\right) - F_{k-2}\right]\Big/(\Delta q)^3 + 0\left[(\Delta q)^2\right] \qquad (4.37)$$

$$\left(\frac{\partial^4 F}{\partial q^4}\right)_k = \left[F_{k+2} - 4F_{k+1} + 6F_k - F_{k-1} + F_{k-2}\right]\Big/(\Delta q)^4 + 0\left[(\Delta q)^2\right] \qquad (4.38)$$

Eqs. (4.37) and (4.38) cannot be used for the first net point close to the wall, because of the appearance of F_{k-2}. Here the coefficients are determined from the law of the wall. This assumption is, as far as the order of the truncation error is concerned, at the first glance in-

consistent with the above derivations, but can be made compatible with
the fourth-order accuracy scheme: The law of the wall results from the
exact momentum equations Eq. (4.25) by setting the first three terms
in Eq. (4.23) equal to zero. This simplification is justified as all
velocity components and the effective turbulent viscosity vanish for
$q_3 \rightarrow 0$. Thus in the frame of the finite-difference approximation the
law of the wall is consistant with the fourth-order scheme as long as

$$\left| H \frac{\partial F}{\partial q_3} - A_4^{-1} \left[A_0 F + A_1 \frac{\partial F}{\partial q_1} + A_2 \frac{\partial F}{\partial q_2} \right] \right| \leq c_1 (\Delta q_3)^4 \qquad (4.39)$$

The left-hand side of Eq. (4.39) can always be evaluated from the initial profile. Then the location of the point nearest to the wall must
be chosen so that the above condition is satisfied. This approximation
should not be confused with the use of the law of the wall in [3].It
was pointed out earlier that in the solution of Ref. [3] the slope of
the characteristics at the wall is infinite and the numerical solution
fails there. For that reason the law of the wall must be employed over
a distance of finite extent in the direction normal to the wall until
the order of magnitude of the slopes of the characteristics approaches
unity, so that the truncation errors become of order $0\left[(\Delta q_3)^2 \right]$. In the
present approximation the distance between the first net point and the
wall can be made arbitary small, i.e., the smaller it is the better
the approximation.

The integration of the continuity equation (2.9) can be carried out
with fourth-order accuracy with the iteration method developed in [25]
for two-dimensional flows. For second- and fourth order accuracy the
integration procedure has been tested successfully. According to this
method, the integration starts with an initial guess for the profile
of the normal velocity component, say $v_3=0$, and all velocity components
are then calculated iteratively from the continuity equation and momentum equations until the solution converges within prescribed error
bounds.They must again be of order $0\left[(\Delta q_3)^4 \right]$.

The difference form of Eq. (2.9) is

$$a_{k+1} \, v_{3k+1} + a_k \, v_{3k} + a_{k-1} \, v_{3k-1} = b_{k+1} \, c_{k+1} + b_k \, c_k + b_{k-1} \, c_{k-1} \qquad (4.40)$$

$$c_k = - \frac{1}{m} \left[\frac{\partial}{\partial q_1} \left(\frac{m}{h_1} \, v_1 \right) + \frac{\partial}{\partial q_2} \left(\frac{m}{h_2} \, v_2 \right) \right]_k \Delta q_3 \qquad (4.41)$$

In Eq. (4.4o) the normal velocity component at the point next to the
wall must also be prescribed together with the boundary condition for

q_3 =0. Here again, an approximation similar to the one chosen for the momentum equations must be used. By adjusting the stepsize in the vicinity of the wall the integration of the continuity equation can be initiated. The coefficients a_k and b_n are determined from Eqs.(4.15)-(4.18). They result from eliminating all second-order derivatives in these equations; since the coefficients change, depending on where they are evaluated they will not be given here. They are obtained without difficulty.

5. DISCUSSION OF NUMERICAL RESULTS

The accuracy of the numerical results obtained with the method described in the foregoing sections depends on several parameters. The most important ones are 1) the order of the truncation error, 2) the stepsize Δq, 3) the error of the iteration process for the normal velocity component ε_{v3}, 4) the error with which the edge conditions are approximated $_e$, 5) the accuracy of the finite-difference approximation of first-and second-order derivatives at the wall, and 6) indirectly the Reynolds number.

Fig. 7 shows a comparison of the shearing stress at the wall as calculated with the second-order solution and the Mehrstellen-method.The initial conditions were again taken from laminar flow and transition to turbulent flow was enforced by using an eddy-viscosity model in the solution. Several step-sizes were employed in the calculation and the influence of the step-size is clearly reflected in the results. The wave-like oscillations,which appear for large step sizes are caused by the error ε_δ; such a behaviour is typical for turbulent flows. The outer-edge conditions must be satisfied to an appropriate degree of accuracy or else the effect of the error ε_δ, is transmitted throughout the entire boundary layer.

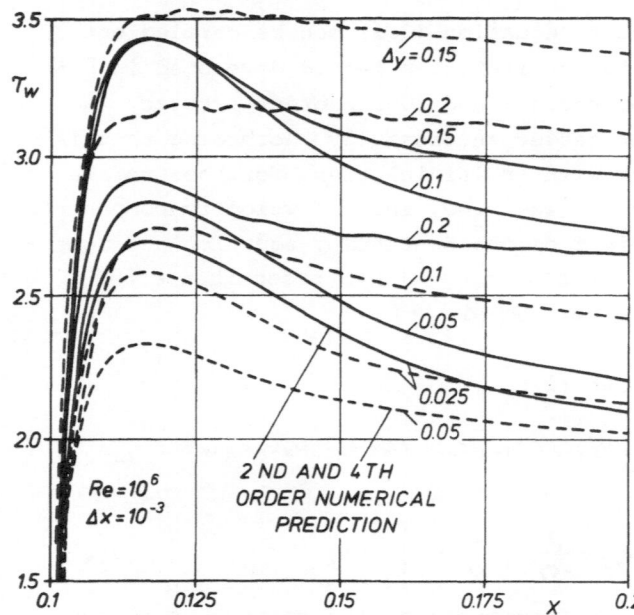

Fig.7 Effect of step size on shearing stress distribution at the wall

Fig. 8 shows the effect

of the error ε_{d} on the effective turbulent viscosity. It is seen that a small change of from 10^{-5} to 5.10^{-4} change the maximum value from 125 to 85. Unless a convention is introduced, through which the edge is properly defined, this error may result in noticable changes in the velocity profiles. This result, again, shows that the prediction of turbulent flows can greatly be influenced by the numerical accuracy.

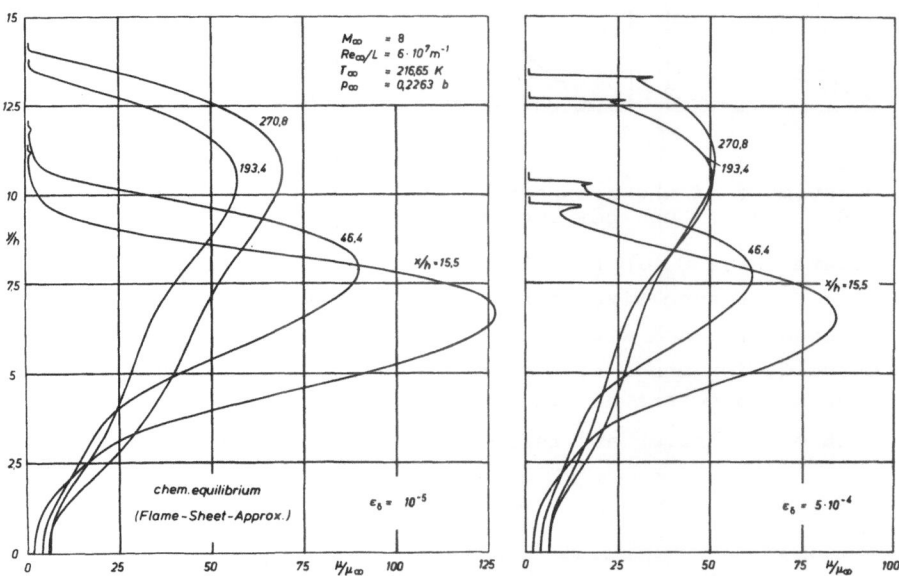

Fig.8   Effect of the accuracy with which the boundary layer is calculated on the profiles for the turbulent viscosity.

It was shown in Fig. 2 that the effective turbulent viscosity is large and exhibits steep gradients on both sides of its maximun value. The accuracy of the second- and fourth-order solution (Mehrstellen-method) can be improved substantially if Eq. (2.1) is divided by $\mu = 1 + \varepsilon$ and $\partial (\ln \mu)/\partial q_3$ is approximated by a finite-difference formula. The improvement of the accuracy can be seen in Fig.9, where the error of τ_w is plotted versus the step size h for the direction normal to the wall. The curve C_1 gives the results for matrix A_3 written in the form of Eq. (2.4), while the curve C_3 gives the same results, when the derivative is taken of $\ln \mu$.
The value of the shearing stress depends also on the form of the fini-te -difference approximation used for the point on the wall. If the end-point formula (either second- or fourth-order) is replaced by the

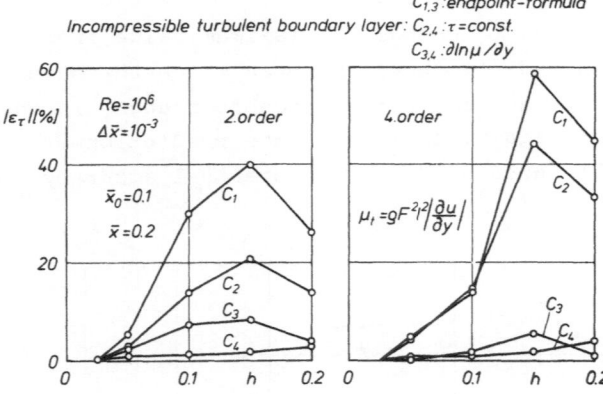

assumption that τ remains constant for the first step of integration for the direction normal to the wall -as outlined in the preceding section- and if τ_W is calculated from the resulting expansion, τ_W is less sensitive to an increase of the step size (curve C_2).

Fig.9 Improvement of the accuracy for the prediction of the shearing stress

The two approximations just described provide a marked increase in accuracy (compare curve C_4 in Fig. 5). There is only a small deviation of the results for a step size of h = 0,2 (less than five percent) compared to those obtained with a step size of h = 0,025. The decrease of the curve C_1 for large step sizes is caused by a cross-over of the over-all error to negative values. It should be kept in mind that all errors mentioned earlier are reflected in the results given in Fig.9.

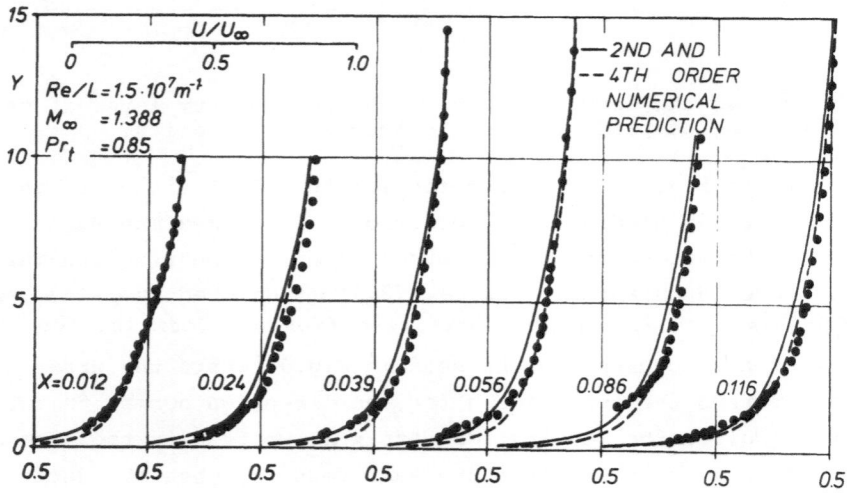

Fig. 1o Comparison of the Merstellen-method and 2nd order solution with experimental data. (Velocity Profiles.)

Finally, in Fig 1o and 11 the Mehrstellen-method is compared with the second-order solution and axperimental data. The better agreement of the Mehrstellen-method with the measurement clearly demonstrates the necessity of accurate numerical solutions, which, no doubt, will be indispensible for future analysis of turbulent flows. This is briefly pointed out in the next section.

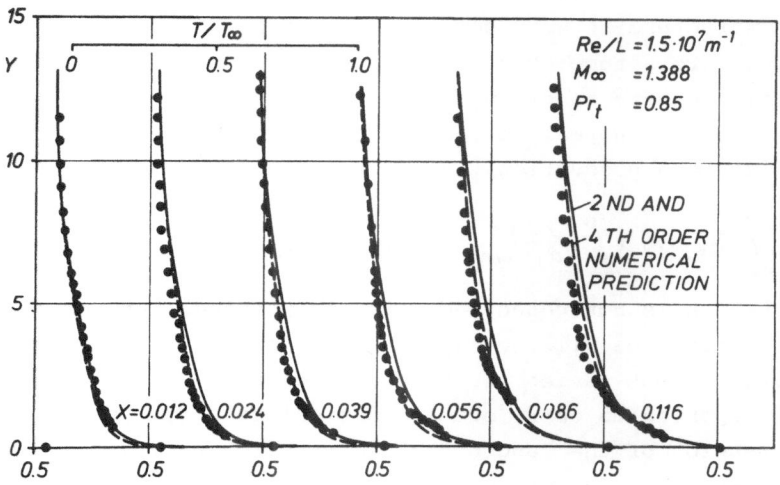

Fig. 11 Comparison of the Mehrstellen-method
and 2nd order solution with experimental
data. (Temperature Profiles)

6. REMARKS ON HIGHER-ORDER CLOSURE OF THE TRANSPORT EQUATIONS

As pointed out in section 2, the momentum equations (2.1) may be made determinate by a first-order model as, for example, by Eq.(2.13). For simple flows, such an approximation is sufficient and reasonable agreement with experimental data can be obtained. For more complex flows the system of equations (2.1) must be supplemented by differential equations for the Reynolds stresses and the kinetic fluctuation energy. These additional equations require naturally more assumptions than the first-order closure. Expressions for the turbulent dissipation, the pressure-shear velocity correlation term, the kinetic energy diffusion and the shear diffusion must be introduced. Several approximations presently being used are discussed in [26] for two-dimensional incompressible high Reynolds number flows. With the one exception described in [27], the differential equation for the Reynolds stresses and the kinetic energy are parabolic in nature and numerical solutions described in the preceeding sections can be employed. If the turbulent length scale L, which is the characteristic length of the eddies carrying the energy, and appears in the expression for the rate of dissipation, is held constant, only relatively simple flows can be predicted, as is shown in [26]. These short-comings due to over-simplifying the turbulence model may be remedied, if a transport equation for the length scale is introduced. Such an equation is given in [26]; although its structure is very complicated, it is again parabolic in nature. The closure assumptions which have to be introduced for the production term require an expansion of the derivative of the mean flow in the direction normal to the wall in a Taylor series. In the notation of [26] the production term is written in the following way:

$$\frac{3}{16}\left[\left(\frac{\partial U}{\partial y}\right)_y \int_{-\infty}^{\infty} R_{21} \, dr_y + \int_{-\infty}^{\infty} \left(\frac{\partial U}{\partial y}\right)_{y+r_y} R_{12} \, dr_y\right] = \overline{uv}\left[L_{12} \frac{\partial U}{\partial y} + \sum_{n=2}^{\infty} L_{12,n}^n \frac{\partial^n U}{\partial y^n}\right] \quad (6.1)$$

As this discussion is not concerned with the physical interpretation of the various terms we will not give the definitions of the integrals of Eq. (6.1) but advise the reader to consult Ref. [26]. Therein attempts are reported to carry out a numerical solution in which a two-term expression of the length scale production term was included in the transport equation. The solution proved to be unstable and according to Ref. [26] the instability had to be attributed to the third-order derivative. This may be due to various reasons as, for example, explicit formulation, the way of decoupling of the equations or the difference approcimations, which would require more than three net points. If, however, the instability is caused by this term alone, a convenient procedure can be devised in order to restrict the number of necessary net points to three and eliminate the third-order derivate from the truncated series in Eq. (6.1). Let the corresponding two-term expansion be designated by P_2 such that

$$P_2 = \overline{uv}\left[L_{12} \frac{\partial U}{\partial y} + L_{12,2}^2 \frac{\partial^2 U}{\partial y^2} + L_{12,3}^3 \frac{\partial^3 U}{\partial y^3}\right] \quad (6.2)$$

and write Eq. (4.17) for two-dimensional flows (in cartesian coordinates), i.e.

$$U \frac{\partial U}{\partial x} + V \frac{\partial U}{\partial y} + \frac{\partial p}{\partial x} - \frac{\partial^2 U}{\partial y^2} = -\frac{\partial \overline{uv}}{\partial y} \quad (6.3)$$

then, by inserting the continuity equation there results

$$-U \frac{\partial V}{\partial y} + V \frac{\partial U}{\partial y} + \frac{\partial p}{\partial x} - \frac{\partial^2 U}{\partial y^2} = -\frac{\partial \overline{uv}}{\partial y} \quad (6.4)$$

The third-order derivative can be expressed in terms of second-order derivatives

$$\frac{\partial^3 U}{\partial y^3} = -U \frac{\partial^2 V}{\partial y^2} + V \frac{\partial^2 U}{\partial y^2} + \frac{\partial^2 \overline{uv}}{\partial y^2}$$

(6.5)

and finally the two-term expansion P_2 becomes

$$P_2 = \overline{uv} \left[L_{12} \frac{\partial U}{\partial y} + \left(L^2_{12,2} + L^3_{12,3,V} \right) \frac{\partial^2 U}{\partial y^2} - L^3_{12,3} \, U \frac{\partial^2 V}{\partial y^2} + L^3_{12,3} \frac{\partial^2 \overline{uv}}{\partial y^2} \right]$$

(6.7)

The right hand side of the last equation can then be discretized in the usual manner if the equations are written in implicit formulation or explicitly with a three-level scheme in the x-direction .

6. CONCLUSIONS

Prandtl's boundary-layer equations were discussed for the analysis of two-and three-dimensional incompressible laminar and turbulent flows. Only boundary sheets were considered and curvature effects normal to the wall were neglected. The governing equations, written for non-orthogonal curvilinear coordinates, were put into a form suitable for implicit finite-difference integration. The numerical stability of the difference equations was investigated by freezing the coefficients and applying the von-Neumann analysis. The solution was found to be stable if the finite-difference representation of the convective terms is advective. Lower-order instabilities may arise when the curvature terms are large.

It was then shown how the order of the truncation error can be reduced by applying the Mehrstellen-method. The resulting-fourth order solution needs only three net-points for the direction normal to the wall if the coefficient matrices do not contain derivatives for the same direction. It was shown that for special cases this difficulty can be avoided by a suitable transformation. For the general case, however, five point formulae must be used to approximate the coefficients with proper accuracy. The difference equations are still tri-diagonal in form and rest on three net-points only.Compared to the second order solution the Mehrstellen-method enforces collocation at these three points and thereby provides a better accuracy.

The method was applied to several flow fields. The additional alge -

braic relations require about ten percent more program statements than
the second-order solution does. Results obtained so far indicate that
the new method is superior to second-order integration as it gives
either better accuracy or smaller calculation times if the coeffi -
cient matrices are properly normalized. Further improvements are possible. Finally it is shown how third-order terms arising in the transport equation for the length scale can be eliminated by introducing
a recursion of the momentum equation.

REFERENCES

[1] Krause, E., Hirschel, E.H., and Kordulla, W., Fourth Order
    "Mehrstellen"-Integration for Three-Dimensional Turbulent Boun-
    dary Layers. AIA-Computational Fluid Dynamics Conference,
    Palm Springs, Cal., 19-2o July, 1973,Conference Proceedings.

[2] Krause, E., Numerical Treatment of Boundary-Layer Problems.
    AGARD LS 64, 1973, Brussels.

[3] Bradshaw, P., Calculation of Three-Dimensional Turbulent Boun-
    dary Layers. J. Fluid Mech. (1971), Vol. 46, Part3, pp.417-445.

[4] Wesseling, P., Lindhout, J.P.F., Three-dimensional incompressi-
    ble turbulent boundary layers: comparison between calculations
    and experiments. Paper presented at the EUROMECH Colloquium 33
    "Three-dimensional turbulent boundary layers", 25 to 27 September
    1972, Berlin.

[5] Nash, J.F., Patel, V.C., Three-Dimensional Turbulent Boundary
    Layers, SBC Technical Books, 1972.

[6] Van den Berg, B., The Law of the Wall in Two-and Three-Dimen-
    sional Turbulent Boundary Layers. NLR TR 72111U, 1973.

[7] East, L.F., Measurements of the turbulent boundary layer on a
    slender wing. Paper presented at the EUROMECH Colloquium 33,1972,
    Berlin.

[8] Fannelop, T.K., A simple finite difference procedure for solving
    the three-dimensional laminar and turbulent boundary-layer equa-
    tions. Paper presented at the EUROMECH Colloquium 33,1972,Berlin.

[9] East, Jr., J.L., Pierce, F.J., Explicit Numerical Solution of
    the Three-Dimensional Incompressible Turbulent Boundary-Layer

Equations. AIAA Journal, Vol. 1o, No. 9, (1972), pp.1216-1223.

[1o] Klinksiek, W.F., and Pierce, F.J., A Finite-Difference Solution of the Two- and Three-Dimensional Incompressible Turbulent Boundary Layer Equations. Transactions of the ASME. Journal of Fluid Engineering Vol. 95, Series 1, No. 3, September 1973.

[11] Krause, E., Mehrstellenverfahren zur Integration der Grenzschicht-gleichungen, DLR Mitt. 71-13 (1971),S. 1o9-138.

[12] Krause, E., Hirschel, E.H., Kordulla, W., Finite difference so-lutions for three-dimensional turbulent boundary layers. Paper presented at the EUROMECH Colloquium 33, 1972, Berlin.

[13] Hocks, W., Korschelt, D., Küster, H., Peters,N., Arbeitsbe-richt der Projektgruppe "Turbulente dreidimensionale Grenzschich-ten", Teil I: Das numerische Verfahren.Inst. f. Thermo- und Fluiddynamik, Technische Universität Berlin (1972).

[14] Krause, E., Hirschel, E.H., Bothmann, Th., Numerische Stabilität dreidimensionaler Grenzschichten, ZAMM Sonderheft 48 (1968),T 2o5.

[15] Krause, E., Hirschel, E.H., Bothmann, Th., Die numerische Inte-gration der Bewegungsgleichungen dreidimensionaler laminarer kom-pressibler Grenzschichten. Fachtagung Aerodynamik, Berlin 1968, DGLR-Fachbuchreihe Bd. 3, Braunschweig (1969).

[16] Krause, E., Comment on Solution of a Three-Dimensional Boundary-Layer Flow with Separation, AIAA Journal, Vol.7, p. 575.

[17] Richtmyer, R.D., Morton, K.W., Difference Methods for Initial Value Problems, Interscience Publishers Inc., New York (1967), Second Edition.

[18] Kordulla, W., An Improved Calculation Method for Compressible Turbulent Boundary Layers. Paper presented at the EUROMECH 43 Colloquium " Heat transfer in turbulent boundary layer with variable fluid properties", 14 to 16 May 1973, Göttingen.

[19] Sells, C.C.L., Two-dimensional Laminar Compressible Boundary Layer Programme for a Perfect Gas. RAE TR 66243, Aug. 1966.

[2o] Keller, H.B., Cebeci, T., Accurate Numerical Methods for Boundary Layer Flows, I. Two-dimensional Laminar Flows. Pro-ceedings of the Second International Conference on Numerical

Methods in Fluid Dynamics, 197o, Berkeley,   Lecture Notes in
Physics No. 8, Springer, 1971.

[21]   Collatz, L., The numerical treatment of differential equations.
       Springer, 196o, Vol. 6o, 2nd printing of 3rd edition 1966.

[22]   Falk, S., Eine Variante zum Differenzenverfahren. ZAMM Vol. 45,
       1965, Sonderheft T 32.

[23]   Wirz, H.J., Eine Erweiterung des Verfahrens der Zwischenschritte
       auf allgemeinere parabolische und elliptische Differentialglei-
       chungen. ZAMM 52, 1972, S. 329-336.

[24]   Kordulla, W., Helium -und Wasserstoff-Wandstrahlen in atmos-
       phärischen Überschallgrenzschichten. Doctoral Dissertation.
       Aerodynamisches Institut, Aachen, 1974

[25]   Krause, E., Numerical Solution of the Boundary Layer Equations,
       AIAA Journal. Vol. 5 No.7 (1967) pp. 1231-1237.

[26]   Rotta, J.C., Recent Attempts to Develop a Generally Applicable
       Calculation Method for Turbulent Shear Layers. AGARD Conference
       Proceedings No. 93 on Turbulent Shear Flows North Atlantic
       Treaty Organisation. September 1971.

[27]   Bradshaw, P., Calculation of Three-Dimensional Turbulent
       Boundary Layers. J. Fluid Mech. (1971), Vol. 46, Part 3 pp.
       417-445.

COMPUTATION OF THREE-DIMENSIONAL,

INVISCID SUPERSONIC FLOWS

Paul Kutler

Ames Research Center, NASA

Table of Contents

Table of Contents (Continued)

List of Figures

List of Figures (Continued)

List of Figures (Continued)

COMPUTATION OF THREE-DIMENSIONAL, INVISCID SUPERSONIC FLOWS

by

Paul Kutler

Computational Fluid Dynamics Branch
Ames Research Center, NASA
Moffett Field, California 94035

1. INTRODUCTION

The numerical computation of three-dimensional, inviscid flow fields for either perfect or real gases about supersonic or hypersonic airplanes and reentry spacecraft can be of considerable importance to the vehicle designer. The continual increase in complexity of such prototype aerospace vehicles (e.g., the space shuttle) requires thousands of hours of costly wind-tunnel time (approximately $1000 per hour) to provide the aerodynamic flow simulations necessary for their development and design. Even with this investment of time, it is necessary to accurately scale the wind-tunnel results to real flight conditions, which is not always possible. Consequently, computer simulations offer an inexpensive supplement to the experimental data and, in certain instances, can completely eliminate the need for experimental testing. This paper describes some numerical procedures capable of determining the complicated supersonic flow fields about wings, bodies, and wing-body combinations.

The flow field about such configurations can contain multiple shocks, expansion waves, and slip surfaces. Basically, two philosophically different approaches can be used to compute such a flow field. The "shock-capturing technique" is inherently capable of predicting the location and strength of all flow discontinuities and their interaction without knowledge of their presence. The "sharp-shock technique" attempts to treat all known shock waves as sharp discontinuities by predicting their motion and applying the Rankine-Hugoniot equations across them. The shock-capturing technique has the single big advantage of being easy to apply while at the same time yielding an accurate solution. The flow discontinuities, rather than appearing as discrete jumps, are spread over several mesh intervals; however, can be precisely located within that region.

Probably all supersonic-flow problems involving multiple shock waves and their intersections could be solved using a sharp-shock procedure. To allow for all possible types of interactions and reflections, however, would be a formidable task. Consider, for example, the flow field generated by a planar oblique shock passing over a cone at angle of attack in supersonic flight (Fig. 1). The resulting flow field is shown in Fig. 2. This time-dependent problem can lead to a variety of possible shock patterns depending on the incident shock-wave inclination angle and strength. This particular problem, however, is well suited for solution by the shock-capturing

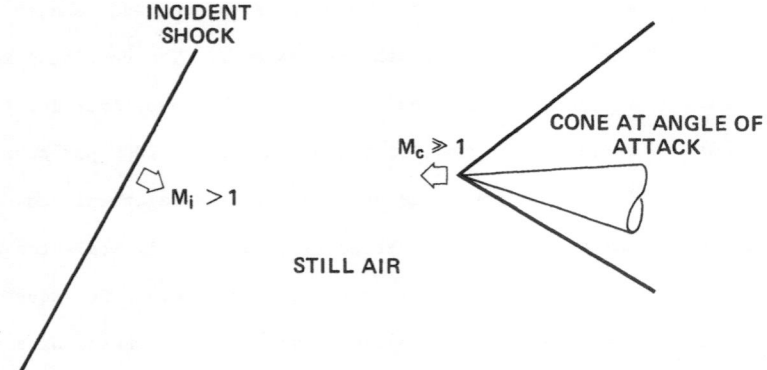

Fig. 1. Three-dimensional interfering shock problem; to determine unsteady flow field interactions.

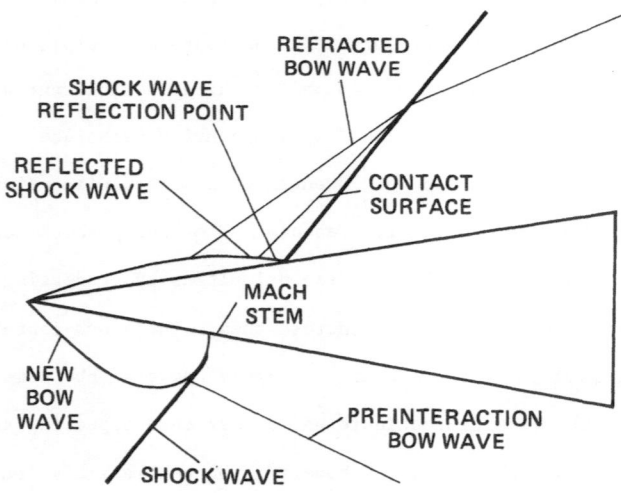

Fig. 2. Typical shock wave and contact surface structure for interfering shock problem.

approach because of its ability to accurately predict the location and intensity of all shock waves without any special treatment. Examples of the equivalent two-dimensional problem are presented later.

In applying the shock-capturing technique, the appropriate hyperbolic equations that govern inviscid flows are written in conservation law form; that is, the dependent conservative variables (ρu, ρuv, $p+\rho u^2$, etc.) are composed of the product and sum of the individual state (nonconservative) variables (p, ρ, u, v, etc.) and yield a set of partial differential equations (pde's) whose coefficients are unity. They are then integrated using a finite-difference scheme of the desired accuracy with the appropriate boundary conditions being applied at the extremities of the computational domain.

In the following sections, the pde's for both unsteady and steady flows are presented. First-, second-, and third-order finite-difference schemes are discussed briefly. Boundary condition procedures used in the numerical computations are also reviewed and discussed. The computational solutions discussed here, which have all been presented or published elsewhere, include supersonic flow about cones, delta wings, conical wing-body combinations, internal corners, blunt bodies, and three-dimensional wing-body configurations.

2. GAS-DYNAMIC EQUATIONS

2.1 Euler Equations

The basic equations of gasdynamics include those for the conservation of mass, conservation of species, conservation of momentum, and conservation of energy as well as an equation of state. These nonlinear partial differential equations are written in terms of four independent variables; time, plus three space dimensions. Under the assumptions of an inviscid, non-heat-conducting gas in local thermochemical equilibrium, these equations are written in the following conservation law form for a generalized orthogonal coordinate system:

$$\frac{\partial U}{\partial t} + \frac{\partial E}{\partial x_1} + \frac{\partial F}{\partial x_2} + \frac{\partial G}{\partial x_3} + H = 0 \tag{2.1}$$

where

$$
U = h_1 h_2 h_3 \begin{pmatrix} \rho \\ \rho u \\ \rho v \\ \rho w \\ e \end{pmatrix} , \quad
E = h_2 h_3 \begin{pmatrix} \rho u \\ p+\rho u^2 \\ \rho uv \\ \rho uw \\ (e+p)u \end{pmatrix} , \quad
F = h_1 h_3 \begin{pmatrix} \rho v \\ \rho uv \\ p+\rho v^2 \\ \rho vw \\ (e+p)v \end{pmatrix} , \quad
G = h_1 h_2 \begin{pmatrix} \rho w \\ \rho uw \\ \rho vw \\ p+\rho w^2 \\ (e+p)w \end{pmatrix}
$$

$$
H = \begin{pmatrix}
0 \\[4pt]
\rho uvh_3 \dfrac{\partial h_1}{\partial x_2} + \rho uwh_2 \dfrac{\partial h_1}{\partial x_3} - (p+\rho v^2)h_3 \dfrac{\partial h_2}{\partial x_1} - (p+\rho w^2)h_2 \dfrac{\partial h_3}{\partial x_1} \\[10pt]
\rho vwh_1 \dfrac{\partial h_2}{\partial x_3} + \rho uvh_3 \dfrac{\partial h_2}{\partial x_1} - (p+\rho w^2)h_1 \dfrac{\partial h_3}{\partial x_2} - (p+\rho u^2)h_3 \dfrac{\partial h_1}{\partial x_2} \\[10pt]
\rho uwh_2 \dfrac{\partial h_3}{\partial x_1} + \rho vwh_1 \dfrac{\partial h_3}{\partial x_2} - (p+\rho u^2)h_2 \dfrac{\partial h_1}{\partial x_3} - (p+\rho v^2)h_1 \dfrac{\partial h_2}{\partial x_3} \\[10pt]
0
\end{pmatrix}
$$

where t and x_i are the independent variables and h_i are the metric coefficients (or arcual derivatives - $h_i = \partial S_i / \partial x_i$, where S_i is the arc length measured in the x_i direction with the remaining two independent variables fixed). For example, the metric coefficients for a spherical coordinate (x_1, x_2, $x_3 = r$, θ, ϕ) system (Fig. 3) are

$$
\begin{aligned}
h_1 &= \partial S_1 / \partial r = 1 \\
h_2 &= \partial S_2 / \partial \theta = r \sin \phi \\
h_3 &= \partial S_3 / \partial \phi = r
\end{aligned}
$$

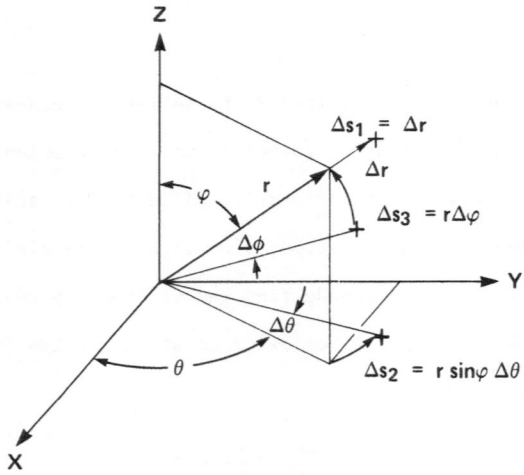

Fig. 3. Spherical coordinate system (r, θ, ϕ).

In Eq. (2.1), p represents the pressure; ρ, the density; u, v, and w, the velocity components in the x_i directions, respectively; and e, the total energy per unit volume. The energy e is related to p, ρ, u, v, and w by

$$e = \rho\left[h(p,\rho) + \frac{u^2+v^2+w^2}{2}\right] - p \qquad (2.2)$$

where $h(p,\rho)$ is the state equation for static enthalpy. The specific formulation for h depends, in particular, on whether the gas is assumed to be perfect or everywhere in local thermodynamic equilibrium. For a perfect gas, $h(p,\rho)$ is simply related to pressure and density ($h = (p/\rho)[\gamma/(\gamma-1)]$), and total energy is thus given by

$$e = \frac{p}{\gamma-1} + \rho\,\frac{u^2+v^2+w^2}{2} \qquad (2.3)$$

For a real gas, no such simple explicit functional relationship exists. The conventional procedure[1,2] for evaluating real gas state relations is to use a combination table lookup and curve-fitting procedure.

For steady flow, the differential form of energy equation (2.1) can be replaced by the integrated form, yielding the equation for total enthalpy:

$$H_t = h(p,\rho) + \frac{u^2+v^2+w^2}{2} = \text{constant} : \text{real gas} \qquad (2.4a)$$

$$H_t = \frac{p}{\rho}\,\frac{\gamma}{\gamma-1} + \frac{u^2+v^2+w^2}{2} = \text{constant} : \text{perfect gas} \qquad (2.4b)$$

2.2 Decoding Conservative Variables

It is well known that Eq. (2.1) is hyperbolic for all flow regimes. It is also known that, for steady flow, Eq. (2.1) is hyperbolic with respect to the x_1, x_2, or x_3 coordinate if the velocity component in that direction is greater than the local speed of sound. When one solves such hyperbolic equations by finite-difference procedures, prior to each integration step, the physical flow variables p, ρ, u, v, w, and e must be obtained from the components u_i of U for unsteady problems (p, ρ, u, v, and w from e_i of E for steady problems (x_1-hyperbolic)) to form the other conservative variables (E, F, and G, unsteady; F and G, steady).

For the unsteady case,

$$\left.\begin{array}{l} \rho = u_1 \\ u = u_2/u_1 \\ v = u_3/u_1 \\ w = u_4/u_1 \\ e = u_5 \end{array}\right\} \text{ real or ideal gas} \tag{2.5}$$

The pressure can be obtained directly from Eq. (2.3) for an ideal gas and iteratively or possibly directly in conjunction with gas property tables from Eq. (2.2) for a real gas.

For the steady case, the decoding procedure necessitates the solution of five simultaneous, nonlinear equations consisting of Eq. (2.4) together with the four elements e_i. The velocity components v and w are given by

$$v = e_3/e_1 \, , \quad w = e_4/e_1 \tag{2.6a}$$

If the e_i along with the above relations are used to eliminate the explicit dependence of p, ρ, v, and w from Eq. (2.4a), the following implicit expression for the velocity u is obtained:

$$D(u) = u^2/2 + h[p(u),\rho(u)] - \Gamma/2 = 0 \tag{2.7a}$$

where

$$\left.\begin{array}{l} p(u) = (e_2 - e_1 u)/h_2 h_3 \\ \rho(u) = e_1/h_2 h_3 u \\ \Gamma = 2H_t - (e_3^2 + e_4^2)/e_1^2 \end{array}\right\} \tag{2.7b}$$

The decoding procedure is now reduced to a problem of root finding—that is, the x_1—velocity component u that satisfies Eq. (2.7a). Two roots exist, one corresponding to subsonic flow, the other to supersonic flow. We seek the supersonic root because the flow in the x_1 direction is assumed to be supersonic, which resulted in the x_1 hyperbolicity of our equations. The procedure for solving Eq. (2.7a) depends on whether a perfect or real gas is being considered and, consequently, on the function $h(p,\rho)$.

For an ideal gas,

$$h(p,\rho) = \frac{\gamma}{\gamma-1} \frac{p}{\rho}$$

and, when combined with Eq. (2.7a), yields a quadratic equation that can be solved, resulting in an algebraic expression for the supersonic velocity u:

$$u = [-B + (B^2-4AC)^{1/2}]/2A \qquad (2.6b)$$

where

$$A = \frac{\gamma+1}{2\gamma}$$

$$B = -e_2/e_1$$

$$C = \frac{\gamma-1}{2\gamma} (2H_t - v^2 - w^2)$$

To find the roots of Eq. (2.7a) for a real gas, a root-finding algorithm is used in conjunction with gas property tables. According to Kutler et al.,[3] the successive linear interpolation scheme described by Dekker[4] was found to be very efficient and required, on the average, about seven iterations to find the desired supersonic root.

2.3 Eigenvalues of Governing Equations

The solution of initial value problems using finite difference procedures requires an integration step size. The value of this step size (discussed in a later section) is a function of the eigenvalues of the coefficient matrices of the governing pde's when written in the form:

$$\frac{\partial U}{\partial t} + P \frac{\partial U}{\partial x_1} + Q \frac{\partial U}{\partial x_2} + R \frac{\partial U}{\partial x_3} + S = 0 \; ; \quad \text{unsteady} \qquad (2.8a)$$

$$\frac{\partial E}{\partial x_1} + J \frac{\partial E}{\partial x_2} + K \frac{\partial E}{\partial x_3} + L = 0 \; ; \quad \text{steady } (x_1\text{-hyperbolic}) \qquad (2.8b)$$

where P, Q, R, J, and K are the Jacobian matrices $\partial E/\partial U$, $\partial G/\partial U$, $\partial F/\partial E$, and $\partial G/\partial E$, respectively.

Equations (2.8) are difficult to derive algebraically, and once the derivation is complete, it is even more difficult to determine their eigenvalues. An easier way to obtain the same result is to rewrite Eq. (2.8) as

$$\frac{\partial U}{\partial t} + A \frac{\partial U}{\partial x_1} + B \frac{\partial U}{\partial x_2} + C \frac{\partial U}{\partial x_3} + D = 0 \; ; \quad \text{unsteady} \qquad (2.9a)$$

$$\frac{\partial U}{\partial x_1} + M \frac{\partial U}{\partial x_2} + N \frac{\partial U}{\partial x_3} + A^{-1}D = 0 \; ; \quad \text{steady} \qquad (2.9b)$$

where $U = [u,v,w,p,\rho]^t$, $M = A^{-1}B$, and $N = A^{-1}C$.

The eigenvalues of the coefficient matrices A, B, C, M, and N are

$$\sigma^A_{1,2} = (u\pm c)/h_1 \,, \quad \sigma^A_{3,4,5} = u/h_1$$

$$\sigma^B_{1,2} = (v\pm c)/h_2 \,, \quad \sigma^B_{3,4,5} = v/h_1 \left.\vphantom{\begin{matrix}1\\1\\1\end{matrix}}\right\} \text{unsteady} \qquad (2.10a)$$

$$\sigma^C_{1,2} = (w\pm c)/h_3 \,, \quad \sigma^C_{3,4,5} = w/h_1$$

$$\sigma^M_{1,2} = \frac{h_1}{h_2}\left[\frac{uv\pm c\sqrt{u^2+v^2-c^2}}{u^2-c^2}\right] \,, \quad \sigma^M_{3,4,5} = \frac{v}{u}\frac{h_1}{h_3}$$

$$\left.\vphantom{\begin{matrix}1\\1\end{matrix}}\right\} \text{steady} \qquad (2.10b)$$

$$\sigma^N_{1,2} = \frac{h_1}{h_3}\left[\frac{uw\pm c\sqrt{u^2+w^2-c^2}}{u^2-c^2}\right] \,, \quad \sigma^N_{3,4,5} = \frac{w}{u}\frac{h_1}{h_2}$$

The σ terms in Eq. (2.10a) can be recognized as the slopes of the characteristics and streamlines in the t-x_1, t-x_2, and t-x_3 planes, while the σ terms in Eq. (2.10b) are the slopes in the x_1-x_2 and x_1-x_3 planes. Note that any secondary transformation of the independent variables of Eq. (2.1) will logically change the eigenvalues of the governing equations (as shown in a later example).

2.4 Conservative Versus Nonconservative Variables

As mentioned in the Introduction, shock-capturing techniques are used to determine the supersonic flow fields presented here. As Lax[5] points out, it is necessary to formulate the governing equations, therefore, in conservation law form and integrate them using a conservative finite-difference procedure to guarantee that any discontinuities in the flow will be processed correctly, for example, shock waves will be of the correct intensity and in the correct position. It was demonstrated by Lax,[5] Gary,[6] and Abbett[7] that use of the nonconservative form of the equations can produce significant errors in the speed of the shock wave, while Longley[8] showed that the conservation form yielded the correct shock speed for a variety of finite-difference schemes.

The shock waves captured using such a procedure, although correctly positioned and of the proper strength, generally will not appear as sharp discontinuities, but will be spread over several mesh intervals. For second- and higher-order, finite-difference schemes,[9] the solution will display post- and precursor oscillations of the dependent variables in the vicinity of the shock. The simple reason that these discrepancies occur is that the dependent variables being differenced, although conservative, can be discontinuous across the shock, and the finite-difference scheme

approximates the solution by passing a polynomial of some degree through the discontinuity. (The degree equals the order of the finite-difference scheme.) This is analogous to approximating a discontinuity by the first few terms of a Fourier series
(Fig. 4).

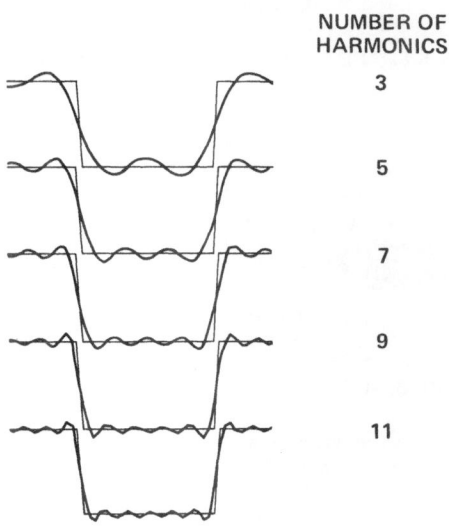

NUMBER OF
HARMONICS

3

5

7

9

11

Fig. 4. Approximation square wave using first several harmonics.

Spreading of the shock and oscillations near the shock wave will not always occur
with a shock-capturing approach. Consider, for example, the steady supersonic flow
over a wedge (Figs. 5(a) and (b)). The flow properties (nonconservative variables)
on either side of the wedge shock are constant. For polar coordinates [$(x_1, x_2 = r, \theta)$;
$h_1 = 1$, $h_2 = r$], the conservative variables F of Eq. (2.1) are continuous across the
shock (Fig. 5(b)) since $F_{\text{free stream}} = F_{\text{wedge}}$, which are simply the Rankine-Hugoniot
equations for a normal shock. Therefore, any conservative finite-difference approximation of $\partial F/\partial \theta$ in the free stream, across the shock, or in the shock layer will be
zero, and will not produce oscillations of the dependent variables. Because of this,
the shock will span only one mesh interval.

Consider the steady supersonic flow over a cone, again in polar coordinates
(Figs. 5(c) and (d)). The flow properties in the shock layer are continuous but not
constant, while the conservative variable F is continuous everywhere, but with a
discontinuous first derivative ($\partial F/\partial \theta$) at the shock ($F_{\text{free stream}} = F_{\text{cone}}$ at shock).

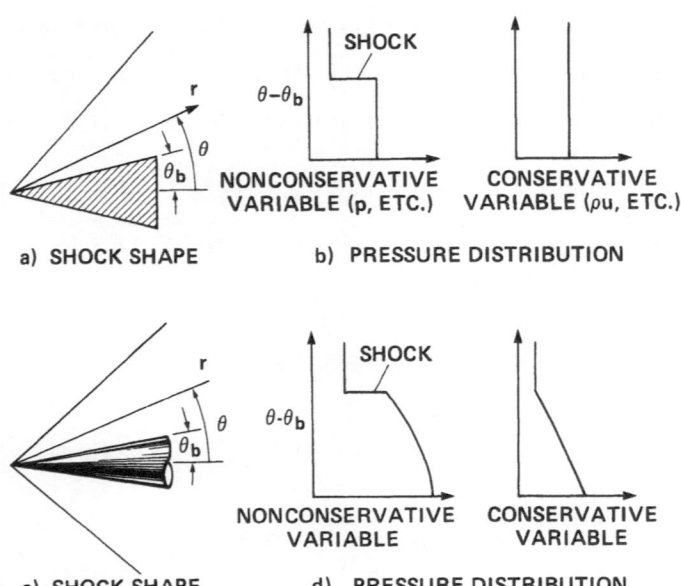

Fig. 5.   Conservative variable distribution through shock generated
by a wedge (a and b) or cone (c and d).

Therefore, differencing across the shock in this case will generate oscillations in
the vicinity of the shock.

In both examples mentioned above, one of the coordinate lines (θ = const) was
aligned with the shock wave, which resulted in the continuously varying F conservative variables across the shock. If the coordinate system is chosen so that the shock
is not parallel to one of the coordinate lines, then none of the conservative variables will be continuous across the shock, which will induce added oscillations at the
shock. Therefore, the more skewed the shock is with respect to the coordinate lines,
and the more the flow variables change near the shock, the greater the oscillations
will be in the vicinity of the shock. Examples presented later show that although
captured shocks exhibit oscillations, the remaining flow field is still accurately
predicted.

As noted, the time-dependent Euler equations (Eq. (2.1)) are hyperbolic and generally can be used to solve any unsteady or steady three-dimensional problem using
finite-difference procedures. However, it would be inappropriate to solve for the
steady supersonic flow over a pointed ogive, for example, using the unsteady equations

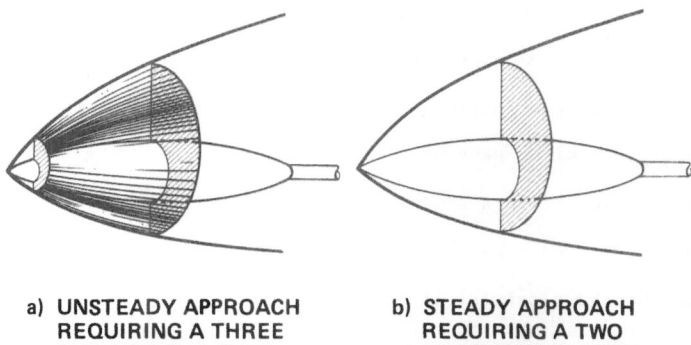

**a) UNSTEADY APPROACH
REQUIRING A THREE
DIMENSIONAL GRID**

**b) STEADY APPROACH
REQUIRING A TWO
DIMENSIONAL GRID**

Fig. 6. Supersonic flow over a pointed ogive.

that require a three-dimensional grid (Fig. 6(a)) when the steady equations are hyperbolic in the longitudinal direction and can be solved with a marching procedure requiring only a two-dimensional grid (Fig. 6(b)).

As another example, consider the flow over a three-dimensional, blunt-nosed configuration in which the last plane of data must be normal to the body axis and in the
supersonic flow region. Again, the unsteady equations could be used to solve the entire problem (Fig. 7(a)). However, a more efficient procedure is to discretize only
the embedded three-dimensional subsonic region for solution by the unsteady approach
as did Rizzi and Inouye[10] (Fig. 7(b)) and then use a steady marching approach from
the warped unsteady boundary to the required axis normal plane (Rizzi et al.[11]).

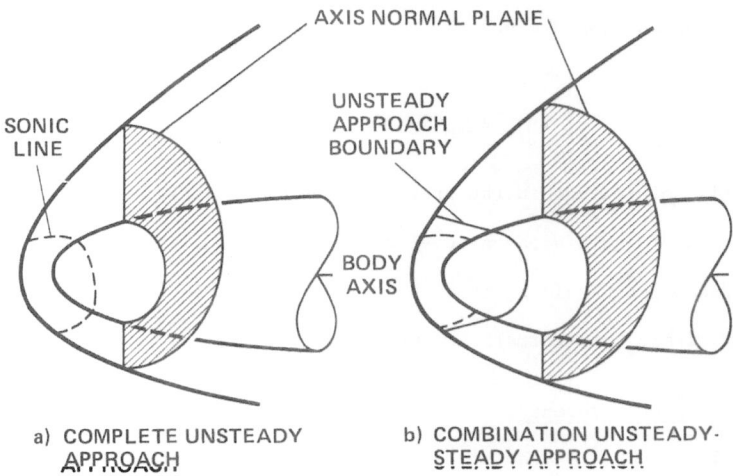

a) COMPLETE UNSTEADY
APPROACH

b) COMBINATION UNSTEADY-
STEADY APPROACH

Fig. 7. Supersonic blunt-body flow.

When one solves supersonic flow problems involving three-dimensional geometries, it is quite possible that embedded regions of subsonic flow may exist at other places in the field than near the blunt nose. In Fig. 8, for example, the supersonic flow over a space shuttle orbiter generates both a bow shock and wing shock which, when they intersect, can result in a small region of locally subsonic flow. Determination of the flow in a region such as this requires a local solution of the unsteady equations, which may not be a trivial matter.

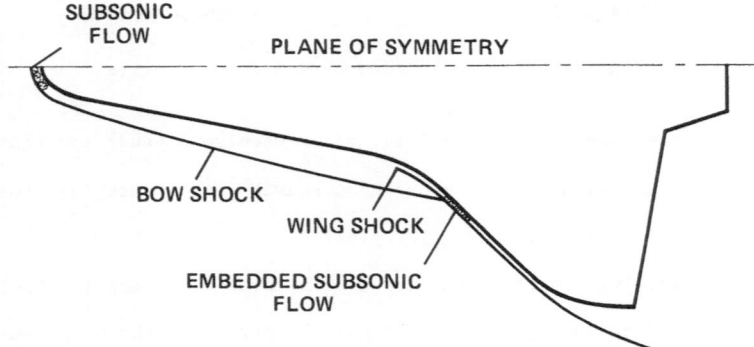

Fig. 8. Planform shock pattern for space shuttle orbiter.

Note that although the inclusion of the time derivative in the governing pde's affords us the ability to calculate transient flows, some workers[10,12] are interested only in the steady-state solution. In such instances, one should strive to obtain the solution without the rigors required to maintain a truly time-accurate procedure if this speeds convergence.

3. COORDINATE SYSTEMS

As briefly pointed out in the previous section, a judicious choice of the independent variable can have pronounced effects on the resulting numerical solution. Thus some forethought should go into the choice of coordinate system and subsequent independent variable transformations that are to be used.

3.1 Basic Orthogonal Systems

Table 1 lists three of the most commonly and one not so commonly used orthogonal coordinate systems along with their metric coefficients. Some explanation should be given to better understand the surface-oriented coordinate system (Fig. 9). The co-ordinates μ and ξ lie on the surface, with μ being the distance measured along the surface from an origin ξ the distance along the surface from the $\xi = 0$ line, normal to $\mu = $ const lines. The third coordinate η is the distance measured normal to the surface. The metric coefficients for this system involve the radii of curvature R^μ and R^ξ of the surface in the μ and ξ directions. Note that this system is somewhat difficult to work with, but slight variations of it are particularly useful.

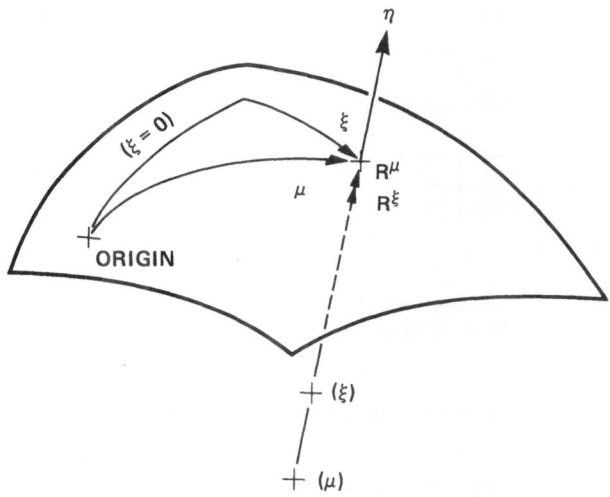

Fig. 9. Surface-oriented coordinate system.

The last three coordinate systems listed in Table 1 contain singularities due to the transformation from a Cartesian system that has no singularities in a finite region; for example, spherical coordinates (Fig. 3) are singular when $\phi = 0$ or $r = 0$. In most instances when these various coordinate systems are used, computation near the singularities is not required. However, when it is, a simple application of

Table 1.

Metric Coefficients for Some Orthogonal Coordinate Systems

Coordinates	x_i	h_i
Cartesian	x, y, z	1, 1, 1
Cylindrical	z, r, ϕ	1, 1, r
Spherical	r, θ, ϕ	1, r sin ϕ, r
Surface oriented (Fig. 9)	μ, η, ξ	$1+\eta/R^\mu$, 1, $1+\eta/R^\xi$

l'Hospital's rule yields a new set of governing equations to be used at the singular points (see, for example, Bohachevsky and Mates[13]).

3.2 Surface Alignment Transformations

Once the basic orthogonal coordinate system is selected, subsequent independent variable transformations can be made to align various surfaces with a particular coordinate and/or concentrate grid points at a given location or in a particular area of the discretized flow region. One of the goals underlying the choice of the coordinate system or transformation should be to align the surface of the body with a coordinate surface (Fig. 10) This eliminates the necessity of unequally spaced grid points at

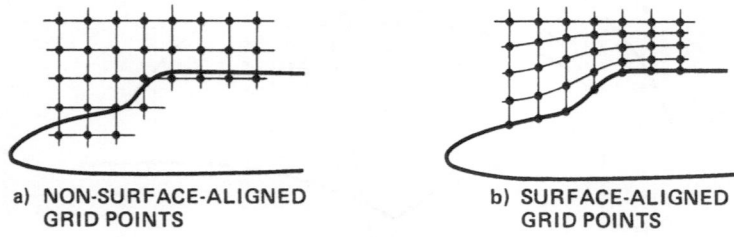

a) NON-SURFACE-ALIGNED
   GRID POINTS

b) SURFACE-ALIGNED
   GRID POINTS

Fig. 10. Result of transformation to align body with a coordinate direction.

the body (Fig. 10(a)) and the subsequent formation of instabilities in the numerical calculation (see, for example, Burstein[14]). For some problems, the choice of coordinate system is such that one of the coordinates is already coincident with the surface of the body and need not be transformed. Consider, for example, the following problems and their coordinate systems as shown in Fig. 11:

1. Wedge flow (Cartesian)

2. Conical wing-body combination (spherical)

3. Two-dimensional or axisymmetric pointed ogive (surface oriented)

    ($h_1 = 1 + y/R^x$, $h_2 = 1$, $h_3 = y \cos \theta_b + r_b$, where θ_b is the local body angle and r_b is the cylindrical body radius).

4. Two-dimensional blunt body (polar)

In most simple, supersonic flow problems, a single shock separates the free stream from the disturbed flow region surrounding the body. In obtaining a solution to the blunt-body problem, Morretti and Abbett[12] introduced a nonorthogonal coordi-

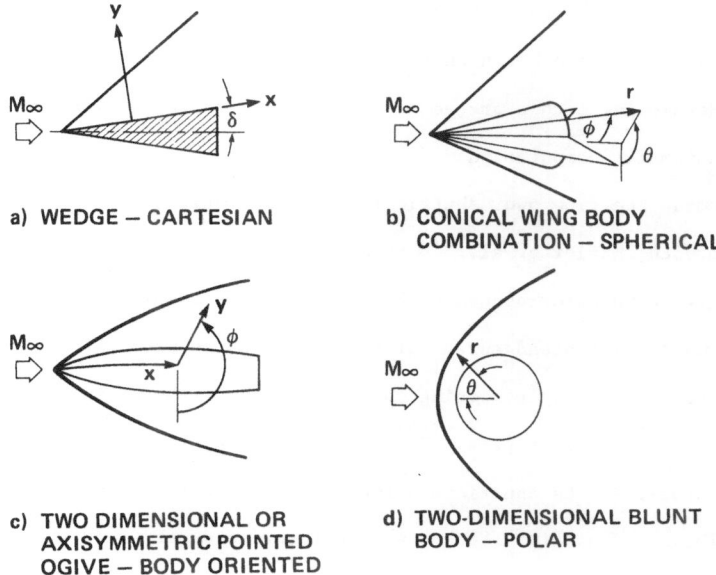

a) WEDGE — CARTESIAN

b) CONICAL WING BODY
   COMBINATION — SPHERICAL

c) TWO DIMENSIONAL OR
   AXISYMMETRIC POINTED
   OGIVE — BODY ORIENTED

d) TWO-DIMENSIONAL BLUNT
   BODY — POLAR

Fig. 11. Configurations and coordinate systems for coordinate
alignment with body.

nate transformation that normalized the distance between the body and such a shock

(which was treated as a sharp discontinuity). Both surfaces become coordinate sur-

faces as shown in Fig. 12. A similar procedure was developed by Morretti et al.[15]

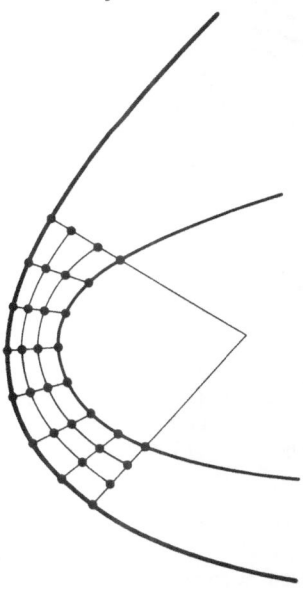

Fig. 12. Body and shock coordinate alignment.

and Marconi and Salas[16] for more complicated supersonic flows involving multiple
shock waves, in which each known shock coincided with a coordinate line. The logic
involved in developing such a procedure is, however, quite complicated and rather cumbersome for automatic computation.

To determine the flow over shuttle-like configurations using a complete shock-
capturing approach, Kutler et al.[17] normalized the equations between the body and
an analytically known outer boundary (Fig. 13(a)) that completely encompassed the outermost shock wave. The disadvantage of such a procedure was that the peripheral shock
location was not known. By combining the attributes of the sharp shock and shock-
capturing procedures, Kutler et al.[3] developed a computer code for three-dimensional
bodies that normalized the equations between the body and peripheral or outermost
shock (Fig. 13(b)). Interior shock waves that formed due to such things as canopy,
wing, or recompression waves were captured, while the peripheral shock was treated as
a discontinuity. Since only special treatment of the finite-difference procedure is
required at the body and shock, this particular formulation is rather well suited for
automatic computation, especially on the new computers being developed that make use
of parallel processing (such as the ILLIAC IV designed by Burroughs Corp.).

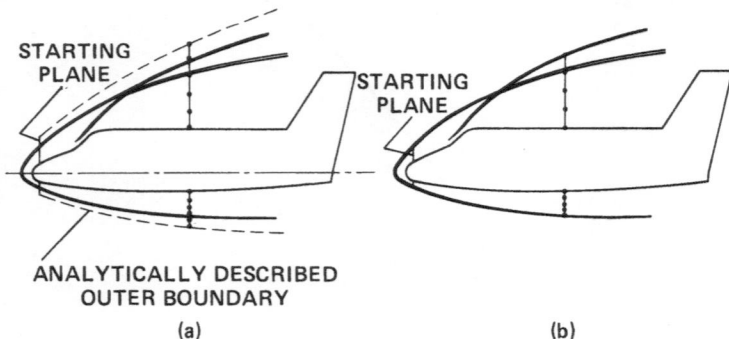

Fig. 13. Three-dimensional coordinate alignment;
(a) normalization between body and outer
boundary, and (b) normalization between
body and peripheral shock.

3.3 Self-Similar Transformations

Some problems encountered in supersonic inviscid flow are self-similar; that is,
the flow field is invariant with respect to a particular independent variable. One of

the prime examples is the wide range of conical flow problems. This self-similar
feature of certain problems can be used as the basis for solution and can be insti-
tuted easily using a nonorthogonal coordinate transformation. Take, for example, the
flow about a supersonic edge delta wing (Fig. 14). The basic coordinate system is
selected as Cartesian with the origin at the vertex of the wing. The governing equations are then transformed according to the conical transformation:

$$\zeta = x , \quad \eta = y/x , \quad \xi = z/x \qquad (3.1)$$

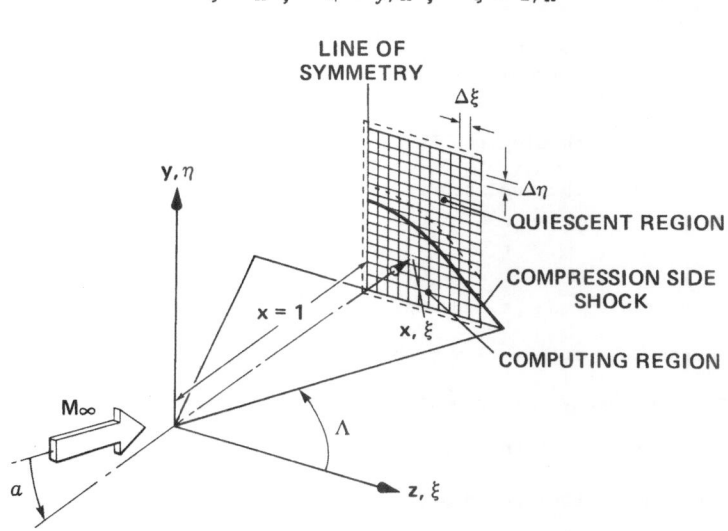

Fig. 14. Coordinate system and mesh description for planar delta wing.

In the η,ξ plane (ξ-derivatives zero), the governing equations are mixed elliptic-
hyperbolic, but if the ζ-derivative terms are included, the resulting set of equations is hyperbolic with respect to ζ. These equations can thus be solved iteratively using finite-difference procedures until the ζ-derivative terms become zero, which
implies the establishment of a conical flow field.

Another interesting self-similar problem is posed by the flow field generated by
passing a planar oblique shock over a wedge or cone in supersonic flight. The nonex-
istence of a characteristic length associated with the body and the fact that the incident wave is planar are the basis for the self-similarity with respect to time.
Consider the two-dimensional problem shown in Fig. 15 where a wedge traveling at Mach
number M_w is struck by an oblique shock wave moving in the opposite direction at

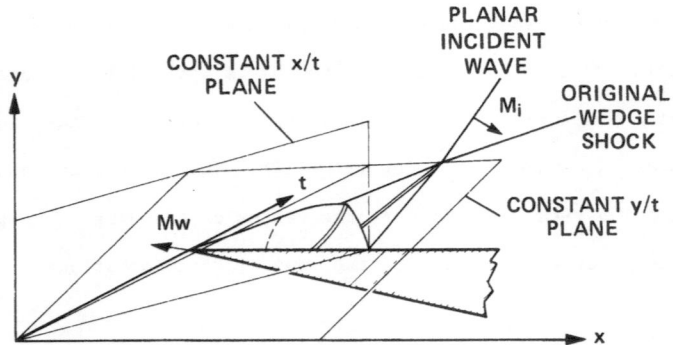

Fig. 15. Two-dimensional interfering shock problem with
self-similar planes.

Mach number M_i. By transforming the Euler equations in Cartesian coordinates according to the following nonorthogonal transformation,

$$\tau = t \ , \quad \xi = x/t \ , \quad \eta = y/t \tag{3.2}$$

the time-dependent problem can be made self-similar. In effect, the unsteady problem
has now been reduced to a steady-flow problem.

As mentioned previously, it is desirable when using the shock-capturing approach
to align the shock with one of the coordinate directions. For the previous example,
it is possible through another coordinate transformation to align the incident planar
shock, the original wedge shock, and the new wedge shock with certain coordinates by
use of the following transformation:

$$\tau = t \ , \quad \xi = \frac{x-b(t,y)}{t} \ , \quad \eta = y/x \tag{3.3}$$

where $b(t,y) = y/\tan \lambda + b_o t$. This transformation yields the network shown in Fig.
16. As opposed to discretizing the desired flow region with a rectangular grid using
Eq. (3.2) (Fig. 16), this transformation results in a more efficient use of the
points, that is, more points appear in the unknown flow regions than in the known, in
addition to its shock alignment properties.

It was mentioned earlier that the basic coordinate system can contain singular
points. It is also possible to create other singularities by the subsequent nonorthogonal transformation. For example, in the transformation given by Eq. (3.3), singularities occur at $t = 0$ and $x = 0$. If computation is required in regions near these

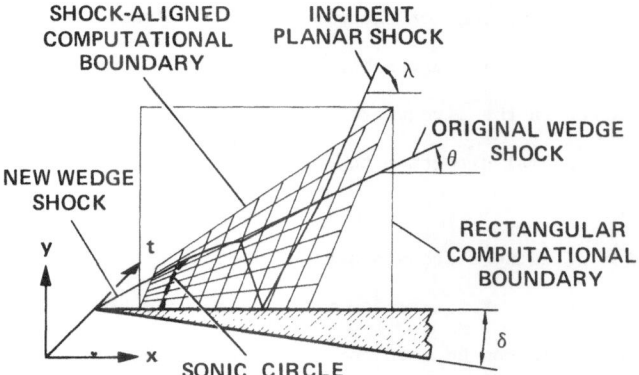

Fig. 16. Rectangular and shock-aligned computation boundaries
for interfering shock problem.

singularities, it will generally result in a reduced overall step size, an increase in
the number of iterations required for convergence, and thus a greater amount of computer time. However, in most instances, it is believed that the resulting enhanced
accuracy of the numerical solution under the transformation will warrant the additional increase in computer time.

3.4 Point Concentration Transformations

    As mentioned earlier, the other purpose of a coordinate transformation is to
cluster or concentrate points in a particular region where the dependent variables are
changing rapidly. Basically, there are two procedures by which clustering can be in-
stituted: (1) by physically inserting points where desired, and (2) by an analytic
transformation of the independent variables. Both Blottner and Roache[18] and Kalnay
de Rivas[19] agree that the latter approach is best and can result in a significant
improvement in accuracy over the former.

    The physical insertion of points approach has been successfully applied by Chu
and Powers,[20] however, in their method-of-characteristics technique for three-
dimensional flows. On the other hand, using a finite-difference procedure, Kutler et
al.,[3] in determining the flow over three-dimensional winged bodies, used the simple
transformation:

$$\eta(\phi) = \tan^{-1}(\kappa \tan \phi) \tag{3.4}$$

(where κ controls the degree of clustering) to group points in the circumferential direction near the leading edge of the wing (ϕ = 90°); good results were obtained. In determining the flow field over a blunt delta body, Thomas et al.[21] used a clustering transformation in the circumferential direction to group points near the leading edge and in the radial direction to capture the thinning entropy layer generated by the blunt nose.

The mechanics of using a transformation as opposed to a manual change of the mesh is somewhat simpler. For example, with an analytic transformation, the same finite-difference scheme can be used for the entire grid, whereas for the mesh change the finite-difference scheme must change when transitioning from one mesh to the next.

The coordinate systems mentioned and the nonorthogonal transformations discussed in this section are by no means an exhaustive discussion of the existing possibilities. Rather, a few have been mentioned to describe the importance of the proper choice in formulating and simplifying a given problem, and, more basic, how this choice facilitates and improves the solution obtained.

4. FINITE-DIFFERENCE SCHEMES

Our main interest in computing three-dimensional supersonic flow fields in an Eulerian mesh is to accurately predict the location and intensity of all shock waves along with the continuous portions of the disturbed region. At most, only one shock wave will be treated as a sharp discontinuity. All others therefore must be captured by the finite-difference solution. Thus we rely heavily on the ability of the finite-difference algorithm in conjunction with the boundary conditions to correctly describe multiple shocked flows.

It is not the purpose of this section to survey all existing finite-difference schemes or even a small number of them. Only three explicit schemes containing no explicit "artificial viscosity" terms are considered: the first-order scheme of Lax,[5] the second-order scheme of MacCormack[22] (a Lax-Wendroff variant), and the third-order scheme of Warming et al.[23] (a Rusanov[24] or Burstein and Mirin[25] variant). In discussing the properties of finite-difference schemes, we use the following terms, which are based on simple linear partial differential equations (see Richtmyer and Morton[26]):

1. <u>Properly posed initial value problem</u>. The solution of the governing pde exists, it is unique, and it depends continuously on the initial data.

2. <u>Truncation error</u>. A measure of how well a solution of the pde satisfies the finite-difference equation.

3. <u>Consistency</u>. The difference equation is said to be a consistent approximation of the pde if the truncation error approaches zero as the step size (Δt) approaches zero.

4. <u>Accuracy</u>. The order of the truncation error is the formal accuracy of the finite-difference scheme.

5. <u>Convergence of iterations</u>. In solving a problem iteratively, as the number of iterations approaches infinity, the difference between the solution at the nth iterate and the solution at infinity approaches zero for a fixed step size.

6. <u>Convergence</u>. The solution of the difference equation approaches the solution of the pde as the step size approaches zero. (Lax's Equivalence Theorem:[26] for a properly posed intial value problem, the consistency of approximation plus the stability of the scheme is a necessary and sufficient condition for convergence.)

7. <u>Stability</u>. Depends solely on the difference scheme and not the pde being solved, requires that the solution be bounded as the step size approaches zero.

8. <u>Modified equation</u>.[27] The pde that is actually solved numerically. It is derived by expanding each term of the finite-difference equation into a Taylor series and then eliminating time derivatives higher than first order (including mixed time and space derivatives) by repeated use of the expanded equation.

9. <u>Dissipation</u>. The ability of the finite-difference scheme to damp high-frequency terms. The effect of dissipation is indicated by the even-order spatial derivatives of the modified equation.

10. <u>Dispersion</u>. The ability of the finite-difference scheme to propagate waves of different wavelengths at the same speed. The effect of dispersion is indicated by the odd-order spatial derivative of the modified equation.

11. <u>Diffusion</u>. A property of the finite-difference scheme that results in a spreading or smearing out of a discontinuity, generally due to both dissipative and dispersive effects.

The analysis of finite-difference schemes is generally constrained to simple linear pde's of the form:

$$\frac{\partial u}{\partial t} + c\,\frac{\partial u}{\partial x} = 0 \qquad (4.1a)$$

or for a system of equations,

$$\frac{\partial U}{\partial t} + A\,\frac{\partial U}{\partial x} \qquad (4.1b)$$

where A is a constant matrix with eigenvalues σ_i. But this convective equation is of interest since it carries the main features of the inviscid fluid-flow equations. Application of the conclusions drawn for the linear equations can be generalized to the nonlinear modified form of Burgers'[28] equation:

$$\frac{\partial u}{\partial t} + u\,\frac{\partial u}{\partial x} = 0 \qquad \text{nonconservative} \qquad (4.2a)$$

or

$$\frac{\partial u}{\partial t} + \frac{\partial(u^2/2)}{\partial x} = 0 \qquad \text{conservative} \qquad (4.2b)$$

which is a more accurate model of the inviscid equations.

In describing the three finite-difference schemes, we use the modified equation approach developed by Warming and Hyett[27] for the analysis of such schemes. The finite-difference schemes themselves are presented in a form applicable for use in Eq. (2.1).

4.1 First-Order Scheme

Lax[5] proposed a finite-difference scheme for calculating time-dependent, one-dimensional, compressible fluid flows containing strong shocks. His single-step scheme, which uses central differences for the spatial derivatives and a forward difference for the time derivative, when applied to Eq. (2.1), becomes

$$
\begin{aligned}
U_{i,j,k}^{n+1} = {}& \frac{1}{6}\left(U_{i+1,j,k}^{n} + U_{i-1,j,k}^{n} + U_{i,j+1,k}^{n} + U_{i,j-1,k}^{n} + U_{i,j,k+1}^{n} + U_{i,j,k-1}^{n} \right) \\
& - \frac{\Delta t}{2\Delta x_1}\left(E_{i+1,j,k}^{n} - E_{i-1,j,k}^{n} \right) - \frac{\Delta t}{2\Delta x_2}\left(F_{i,j+1,k}^{n} - F_{i,j-1,k}^{n} \right) \\
& - \frac{\Delta t}{2\Delta x_3}\left(F_{i,j,k+1}^{n} - F_{i,j,k-1}^{n} \right)
\end{aligned} \qquad (4.3)
$$

where

$$U_{i,j,k}^n = U(n\Delta t, \ i\Delta x_1, \ j\Delta x_2, \ k\Delta x_3)$$

$$E_{i,j,k}^n = E\left(U_{i,j,k}^n, \ n\Delta t, \ i\Delta x_1, \ j\Delta x_2, \ k\Delta x_3\right), \ \text{etc.}$$

The modified equation for Lax's scheme applied to Eq. (4.1a) is

$$\frac{\partial u}{\partial t} + c\,\frac{\partial u}{\partial x} + \left(\frac{c^2\Delta t}{2} - \frac{\Delta x^2}{2\Delta t}\right)\frac{\partial^2 u}{\partial x^2} + \left(-\frac{1}{3}\,c\Delta x^2 + \frac{c^3\Delta t^2}{3}\,\frac{\partial^3 u}{\partial x^3}\right)$$

$$+ \left(\frac{1}{12}\,\frac{\Delta x^4}{\Delta t} - \frac{1}{3}\,c^2\Delta x^2\Delta t + \frac{c^4\Delta t^3}{4}\right)\frac{\partial^4 u}{\partial x^4} + \cdots = 0 \qquad (4.4)$$

The coefficient of the lowest-order error term is

$$\frac{c^2\Delta t}{2} - \frac{\Delta x^2}{2\Delta t}$$

so the scheme is first-order accurate in Δx and Δt. The scheme can be inconsistent since, as Δt and Δx approach zero (e.g., in the ratio $\Delta x^2/\Delta t$), this error term does not go to zero. The stability bound is shown to be

$$\nu = \frac{c\Delta t}{\Delta x} < 1 \qquad (4.5)$$

where, for Eq. (4.1b), c is replaced by $|\sigma_{max}|$, where $\sigma_{max} = \max(\sigma_i)$. The parameter ν is referred to as the "Courant number." The dissipation of the scheme (lowest-order even derivative of error terms) is of order 2, while the dispersion (lowest-order odd derivative) is of order 3. Phase errors generated using Lax's scheme lead to the exact solution.

4.2 Second-Order Scheme

MacCormack,[22] in studying hypervelocity impact cratering, developed a noncentered, two-step, finite-difference scheme. Basically, the scheme uses one-sided differences (forward or backward, hence the term "noncentered") to replace the spatial derivatives as opposed to the centered differences of Lax's[5] or Richtmyer's[26] scheme.

For the two-dimensional plus time version of Eq. (2.1), MacCormack's scheme is written

$$U_{i,j}^{(1)} = U_{i,j}^n - \alpha_1 \left\{ \frac{\Delta t}{\Delta x_1} \left[(1-\epsilon_1) E_{i+1,j}^n - (1-2\epsilon_1) E_{i,j}^n - \epsilon_1 E_{i-1,j}^n \right] \right.$$

$$\left. + \frac{\Delta t}{\Delta x_2} \left[(1-\epsilon_2) F_{i,j+1}^n - (1-2\epsilon_2) F_{i,j}^n - \epsilon_2 F_{i,j-1}^n \right] + \Delta t H_{i,j}^n \right\} \qquad (4.6a)$$

$$U_{i,j}^{n+1} = \frac{1}{2} \left(U_{i,j}^n + U_{i,j}^{(1)} \right) - w_1 \left\{ \frac{\Delta t}{\Delta x_1} \left[\epsilon_1 E_{i+1,j}^{(1)} + (1-2\epsilon_1) E_{i,j}^{(1)} + (\epsilon_1-1) E_{i-1,j}^{(1)} \right] \right.$$

$$\left. + \frac{\Delta t}{\Delta x_2} \left[\epsilon_2 F_{i,j+1}^{(1)} + (1-2\epsilon_2) F_{i,j}^{(1)} + (\epsilon_2-1) F_{i,j-1}^{(1)} \right] + \Delta t H_{i,j}^n \right\} \qquad (4.6b)$$

where $\alpha_1 = 1$, $w_1 = 1/2$, $U_{i,j}^n$ and $E_{i,j}^n$ are as defined before, and

$$E_{i,j}^{(1)} = E^{(1)} \left[U_{i,j}^{(1)}, \ (n+\alpha_1)\Delta t, \ i\Delta x_1, \ j\Delta x_2 \right]$$

This scheme allows four possible variations for replacing the space derivatives in the predictor and corrector steps:

$$\left. \begin{array}{llll} \text{I:} & \epsilon_1 = 0 \ , & \epsilon_2 = 0 \\ \text{II:} & \epsilon_1 = 1 \ , & \epsilon_2 = 1 \\ \text{III:} & \epsilon_1 = 0 \ , & \epsilon_2 = 1 \\ \text{IV:} & \epsilon_1 = 1 \ , & \epsilon_2 = 0 \end{array} \right\} \qquad (4.7)$$

MacCormack[22] suggests that in applying this scheme to general flow problems the four variations (I-IV) be cyclically permuted to obtain the most unbiased results. The generalization to three dimensions plus time is straightforward.

The modified equation corresponding to MacCormack's scheme (which reduces to the Lax-Wendroff[29] scheme for linear equations) is

$$\frac{\partial u}{\partial t} + c \frac{\partial u}{\partial x} + \frac{c}{6} (\Delta x^2 - c^2 \Delta t^2) \frac{\partial^3 u}{\partial x^3} + \frac{c^2 \Delta t}{8} (\Delta x^2 - c^2 \Delta t^2) \frac{\partial^4 u}{\partial x^4} + \cdots = 0 \qquad (4.8)$$

Therefore, the scheme is uniformly second-order accurate in both time and space. The scheme is consistent, and the stability bound is the same as for Lax's scheme, namely,

$$\nu = c \frac{\Delta t}{\Delta x} < 1$$

According to Lax's equivalence theorem, the scheme is convergent.

The dissipation of the scheme is of order 4 while the dispersion is of order 3. Phase errors generated using MacCormack's scheme will be predominantly lagging.

4.3 Third-Order Scheme

Warming et al.[23] derived a third-order scheme that is a variant of Rusanov's[24] and Burstein and Mirin's[25] scheme. It is a three-step method that uses the first two steps of MacCormack's method. The algorithm can be written as

$$U_{i,j}^{(1)} = \text{same as right side of (4.6a) with } \alpha_1 = 2/3 \qquad (4.9a)$$

$$U_{i,j}^{(2)} = \text{same as right side of (4.6b) with } \alpha_1 = 2/3 \text{ and } w_1 = 1/3 \qquad (4.9b)$$

$$
\begin{aligned}
U_{i,j}^{n+1} = U_{i,j}^n &- \left[(\omega_1)_{i+1/2,j}^n /24 \right] \left[U_{i+2,j}^n -3U_{i+1,j}^n +3U_{i,j}^n -U_{i-1,j}^n \right] \\
&+ \left[(\omega_1)_{i-1/2,j}^n /24 \right] \left[U_{i+1,j}^n -3U_{i,j}^n +3U_{i-1,j}^n -U_{i-2,j}^n \right] \\
&- \left[(\omega_2)_{i,j+1/2}^n /24 \right] \left[U_{i,j+2}^n -3U_{i,j+1}^n +3U_{i,j}^n -U_{i,j-1}^n \right] \\
&+ \left[(\omega_2)_{i,j-1/2}^n /24 \right] \left[U_{i,j+1}^n -3U_{i,j}^n +3U_{i,j-1}^n -U_{i,j-2}^n \right] \\
&- \frac{1}{24} \left[(\Delta t/\Delta x_1) \left(-2E_{i+2,j}^n +7E_{i+1,j}^n -7E_{i-1,j}^n +2E_{i-2,j}^n \right) \right. \\
&\left. + (\Delta t/\Delta x_2) \left(-2F_{i,j+2}^n +7F_{i,j+1}^n -7F_{i,j-1}^n +2F_{i,j-2}^n \right) + 6\Delta t H_{i,j}^n \right] \\
&- \frac{3}{8} \left[(\Delta t/\Delta x_1) \left(E_{i+1,j}^{(2)} -E_{i-1,j}^{(2)} \right) + (\Delta t/\Delta x_2) \left(F_{i,j+1}^{(2)} -F_{i,j-1}^{(2)} \right) + 2\Delta t H_{i,j}^{(2)} \right] \quad (4.9c)
\end{aligned}
$$

It should be emphasized here that $U_t^{(1)}$, $U^{(2)}$, $F^{(1)}$, etc., are evaluated at the t coordinate $[n+(2/3)]\Delta t$. The free-parameter terms of Eq. (4.9c), that is, the brackets with multiplicative coefficients ω_1 and ω_2 have been differenced conservatively in both the x_1 and x_2 directions, and formulas for them are given later. Since these terms represent a fourth-order difference operator,[23] they do not affect the third-order accuracy of the algorithm. Any of the four possible variations of MacCormack's method are applicable for the first two steps (Eqs. (4.9a) and (4.9b)), and each is consistent with the third-order accuracy of the final step.

The modified equation corresponding to this difference algorithm is

$$\frac{\partial u}{\partial t} + c \frac{\partial u}{\partial x} + \frac{\Delta x^4}{24\Delta t} (\omega -4\nu^2 +\nu^4) \frac{\partial^4 u}{\partial x^4} + \frac{c\Delta x^4}{120} [-5\omega +(4\nu^2 +1)(4-\nu^2)] \frac{\partial^5 u}{\partial x^5} + \cdots = 0 \quad (4.10)$$

where $\nu = c\Delta t/\Delta x$. Therefore, the scheme is uniformly third-order accurate in both time and space. It is consistent and stable if[24]

$$4\nu^2 - \nu^4 < \omega \leq 3 \text{ and } |\nu| < 1 \qquad (4.11)$$

The third-order method can have either a leading or lagging phase error, depending on the choice of the free parameter ω.[23] The difficulty is in choosing the correct value for the free parameter. The modified equation for this scheme, however, affords some insight into the proper choice. If one wishes to minimize the dissipation, then ω is chosen so that the lowest-order, even derivative coefficient of the modified equation (Eq. (4.10)) is zero or

$$\omega = 4\nu^2 - \nu^4 \qquad (4.12)$$

To minimize the dispersion, the coefficients of the lowest-order odd derivative is set to zero; that is,

$$\omega = (4\nu^2 + 1)(4 - \nu^2)/5 \qquad (4.13)$$

Warming et al.[23] and Kutler et al.[17] were successful in using the free parameter ω to enhance the shock-capturing ability of the third-order scheme by allowing it to vary from grid point to grid point. Examples presented later demonstrate the variable free-parameter effects.

Again, note that the analysis of finite-difference schemes is based on the use of a linear pde; however, Hirt[30] has argued that certain stability criteria are applicable to nonlinear equations with variable coefficients. In addition, since no complete rigorous nonlinear analysis exists, we must rely on the theory developed for the linear case and account for the nonlinear effects by conservative estimates of the integration step size.

4.4 Solution of $u_t + (u^2/2)_x = 0$

To determine the relative merits or deficiencies of a particular finite-difference algorithm for solving the gasdynamic equations, the scheme can be applied to the modified Burger's equation, Eq. (4.2b), which is representative of the inviscid Euler equations. Because of its simplicity, the computer coding is easier and the core requirements and computer time are minimal. Kutler[31] and Warming et al.[23] performed such studies, and some of their results are presented here for the three schemes discussed in this section.

Of prime interest is how well the finite-difference scheme can predict and follow discontinuities in the flow. Therefore, the initial condition for Eq. (4.2b) is chosen as a step discontinuity that is supposed to simulate a shock moving from left to right. Solutions are obtained for different step sizes or Courant numbers (Eq. (4.5)) to study the effects of dispersion, dissipation, and diffusion.

The results obtained using Lax's first-order method, Eq. (4.3), for Courant numbers of 1.0 and 0.5 are shown in Fig. 17. There is no overshoot or undershoot for either case; for $\nu = 1.0$, the discontinuity is spread over three mesh intervals, while for $\nu = 0.5$ it is spread over approximately six. This diffusion of the discontinuity indicates that Lax's method is highly dissipative. However, the location of the discontinuity is predicted quite well by this scheme.

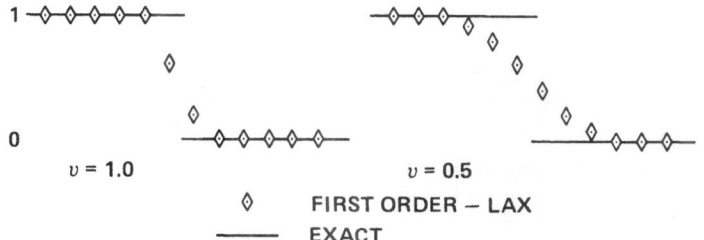

Fig. 17. First-order solution of Eq. (4.2b) using Lax's scheme for Courant numbers of 1.0 and 0.5.

MacCormack's second-order scheme has been termed "preferential" by Kutler,[31] which means that the solution obtained will be more favorable, depending on the version (Eq. (4.7)) of MacCormack's scheme used. In particular, for discontinuities that propagate in the direction of increasing i (in the present case, to the right), the preferred version of MacCormack's scheme is I, in which forward space differences are used in the predictor and backward space differences are used in the corrector. This preferential behavior is shown in Fig. 18 by the results for Version I (Fig. 18(a)) and version II (Fig. 18(b)) for the right-moving discontinuity. For $\nu = 1.0$, with version I (the preferred scheme), there is no overshoot or undershoot and the discontinuity is only spread over two mesh intervals, whereas for version II (the nonpreferred scheme) there is a slight overshoot. For a Courant number of 0.5, both versions exhibit oscillations following the discontinuity (a result of dispersion errors), with version II having the larger amplitudes. The discontinuities are still

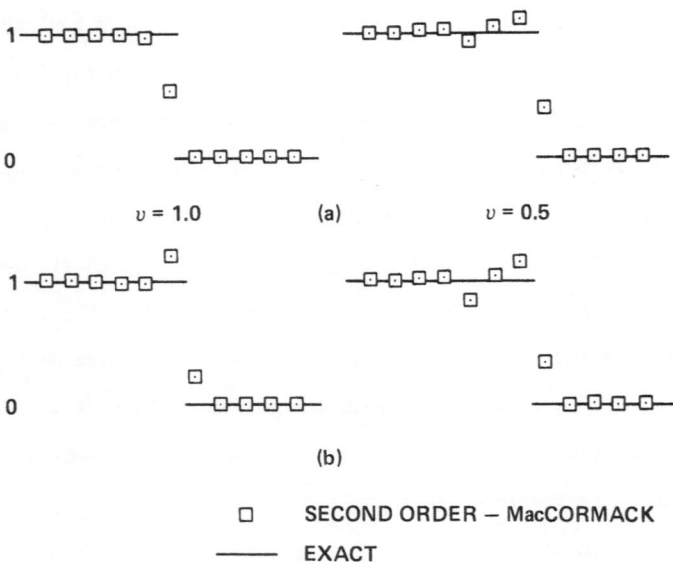

Fig. 18. Second-order solution of Eq. (4.2b) using MacCormack's scheme for Courant numbers of 1.0 and 0.5; (a) version I, and (b) version II.

only spread over two mesh intervals, indicating the improvement in the diffusion properties over Lax's scheme.

If versions I and II are permuted as suggested by MacCormack,[22] this results in the solutions in Figs. 19(a) and 19(d) for Courant numbers of 0.9 and 0.1. The dis-

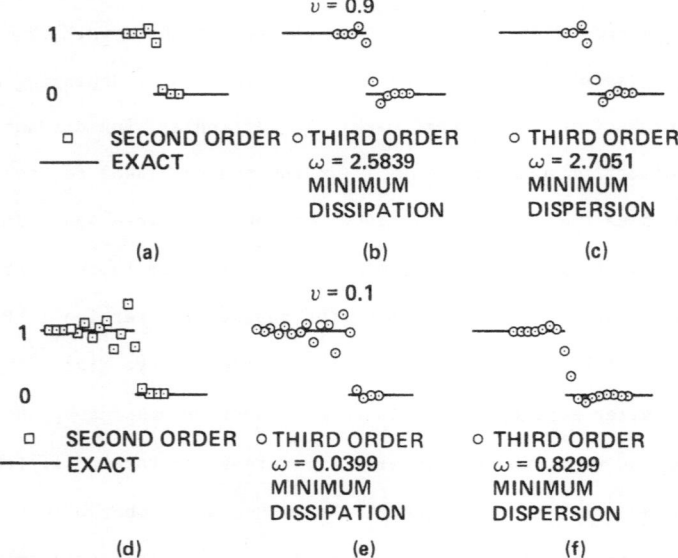

Fig. 19. Comparison of second- (versions I and II permuted) and third-order solutions of Eq. (4.2b) for Courant numbers of 0.9 and 0.1.

continuity is captured as well as that using either version I or II, but the overshoot is greater than version I and less than version II. For $\nu = 0.1$, the dispersion errors are predominant and result in oscillations behind the discontinuity.

The analysis governing the properties of finite-difference schemes, including the free parameter of the third-order scheme, was based on a linear equation and, strictly speaking, is invalid for nonlinear problems. However it provides a guide of what to expect, and for the third-order method, what value of the free parameter to use in the nonlinear case. Figure 19 shows the results obtained for Eq. (4.2b) using the third-order method with values of ω for minimum dissipation and dispersion and four Courant numbers of 0.9 and 0.1. These results are compared with the second-order permuted MacCormack scheme. For the large Courant number, there is not much difference between the second- and third-order solutions. However, for a Courant number of 0.1, the third-order method using ω for minimum dispersion results in the best solution for the captured discontinuity.

The previous examples have shown that the resolution of discontinuities computed by the third-order method is rather sensitive to the choice of the parameter ω. In complicated flow patterns with multiple shocks, an optimum choice of ω at some spatial point will probably result in poor shock-capturing ability in other regions of the flow field if ω is constant throughout the flow. Figure 20(a) depicts a weak shock computed in the presence of a strong shock for the nonlinear Eq. (4.2b). The

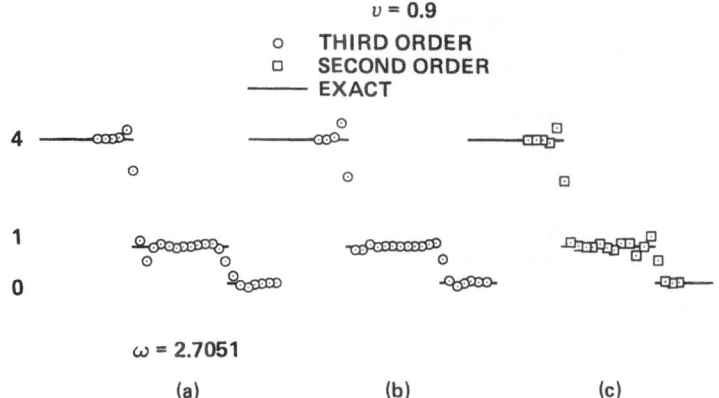

Fig. 20. Calculation of a weak shock in the presence of a strong shock;
(a) third-order method with a constant ω,
(b) third-order method with a variable ω, and
(c) second-order method (versions I and II permuted).

numerical solution was computed for a Courant number $\nu = |\sigma|_{max} \Delta t / \Delta x = 0.9 \Delta t / \Delta x = 0.225$, and with $\omega = \omega(0.9) = 2.7051$ calculated from Eq. (4.13). The local value of the Courant number for the weak shock is $\nu = 0.225$ and, consequently, the value of ω used is too high for this lower Courant number, resulting in a rather smeared shock.

It is possible to make ω a local function of the Courant number ν rather than keep it constant throughout the computational mesh.[22] Figure 20(b) shows the same double-shock system as in Fig. 20(a), except ω was assumed to be a variable calculated locally for minimum dispersion from Eq. (4.13). The result, while not showing dramatic improvement, indicates that the weak shock is less smeared since the jump that takes essentially three mesh intervals in Fig. 20(a) has been reduced to two mesh intervals in Fig. 20(b). For comparison, Fig. 20(c) illustrates the double-shock solution using the second-order method.

With regard to programming logic and basic core storage requirements, Lax's method is the easiest to program and requires minimum storage, while the third-order method is the most difficult and requires the most core storage. The choice of the finite-difference scheme to be used, therefore, will depend strongly on the problem to be solved and the computer used to solve it.

5. BOUNDARY CONDITIONS

Probably the single most important aspect in the successful application of any numerical technique to the solution of fluid-flow problems is the proper treatment of the impermeable and permeable boundaries that encompass the computational plane. An impermeable boundary is one across which no mass can flow, such as the solid surface of a body, a plane of symmetry, or a contact discontinuity, and along which therefore the tangency condition must be satisfied. A permeable boundary, such as a shock wave or inflow or outflow boundary, allows mass to flow through its surface. Of all these boundaries, the impermeable surface of a solid body has probably received the most attention by the numerical modelist, and rightfully so. It is this boundary, in conjunction with the free-stream conditions, that generates the behavior of the surrounding flow field.

5.1 Impermeable Boundaries

A wide variety of body boundary condition schemes exists, some of which are discussed by Abbett,[32,33] Moretti,[34] and Roache.[35] Three of these procedures (used in the examples presented later) are discussed, namely, (1) the reflection technique, (2) the explicit one-sided derivative technique, and (3) the simple wave corrector technique of Abbett.[33] All these schemes are easy to apply in that they attempt to treat the boundary points as regular points of the interior, and when used properly, they yield quite accurate results.

The "reflection principle" or "reflection technique" is only truly applicable for simulating the slip condition on a plane or flat wall in which one of the coordinates lies along the wall. It forces all nonconservative flow variables other than the normal velocity component to behave as even functions with respect to the wall. And it forces the normal velocity to behave as an odd function, thus yielding a zero normal velocity at the wall. If the conservative variables are used, all but the normal momentum component behave as odd functions since they are multiplied by the normal velocity component.

Two approaches can be used in applying the reflection technique (Fig. 21, areas I and II): (1) image point approach (area I), which requires a sequence of grid points one Δy below or inside the body and thus allows the wall points to be treated as regular interior points of the flow or (2) nonimage point approach (area II), which

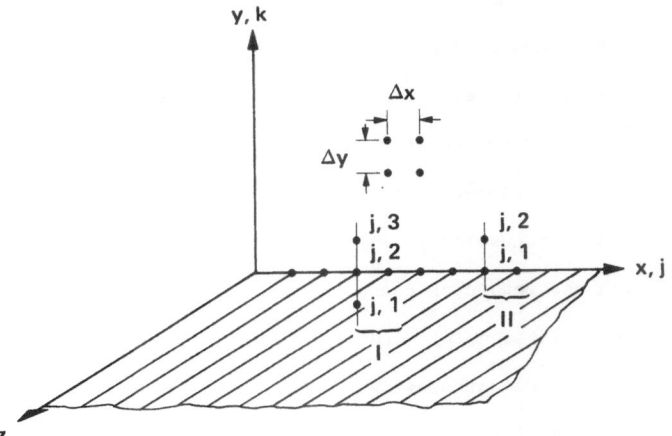

Fig. 21. Reflection technique for flat surface in x-z plane; (I) image point approach and (II) nonimage point approach.

requires no subwall points but requires a simple change in the finite-difference scheme at the body.

Thus, in applying the reflection technique using the image point approach with conservative variables (see F in Eq. (2.1)), the values of $F_{(1)}$, $F_{(2)}$, $F_{(4)}$, and $F_{(5)}$ at points (j,1) are set equal to the negative of the values at points (j,3), while $F_{(3)}$ at (j,1) is set equal to the value at (j,3). In using the nonimage point approach, as with MacCormack's scheme, when a backward difference is required at the wall, it is computed using the conservative variables at points (j,1) and (j,2) only (see area II); that is, $\partial F/\partial y|_{back} = F_{(1,2,4,5)_{j,1}} + F_{(1,2,4,5)_{j,2}}$ and $\partial F/\partial y|_{back} = F_{(3)_{j,1}} - F_{(3)_{j,2}}$.

The reflection technique, although exact only for planar bodies, has been used for nonplanar bodies with moderate success by a number of investigators, including Bohachevsky and Mates,[13] Bohachevsky and Rubin,[36] Kutler,[31] and Kutler and Lomax.[37] The accuracy is directly proportional to the curvature of the body, the number of grid points used in the direction normal to the body, and how well behaved the flow variables are near the body.

A rather simple and easily implemented procedure for modeling the tangency condition is to use the normal interior point finite-difference scheme with one-sided differences at the body, along with the fact that the normal velocity component is zero. For example, in using version I of MacCormack's scheme, Eq. (3.10), the predictor step at the body is the same as for the interior. The subsequent corrector step, however, is modified so that the required backward difference in the body normal direction is replaced by a forward difference. Following the corrector, the normal velocity component, which will normally not be zero, is set to zero. Again, the success of this scheme depends on the body geometry, number of points, and flow variable behavior.

Abbett,[32] in his discussion of various surface boundary condition procedures, made comparisons with the method of characteristics for both the reflection technique and explicit one-sided differencing technique for simple compression and expansion flows—both of which were computed using a marching procedure. Figure 22 is a sketch of the comparison which depicts a convergent oscillatory behavior of the two approximate procedures. The noteworthy characteristic of this curve is that as x increases

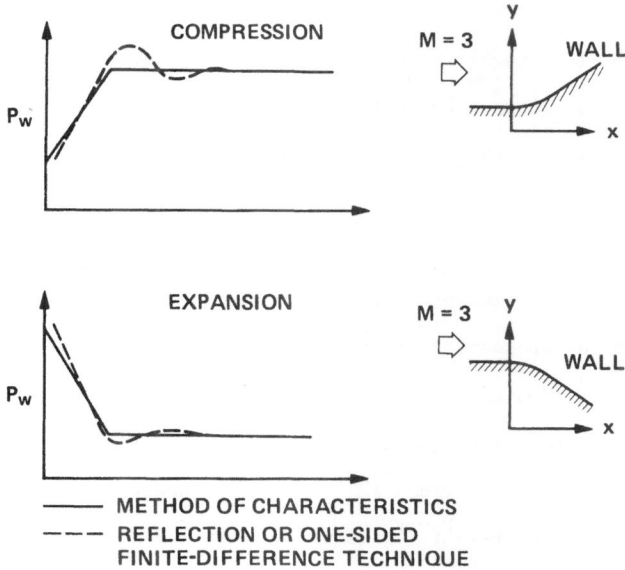

Fig. 22.   Comparison sketch of method of characteristics and reflection
or one-sided, finite-difference techniques.

both techniques approach the method of characteristics solution, which implies that if
an iterative procedure is used as opposed to a marching procedure, these boundary condition techniques will yield acceptable results. Examples of self-similar flow problems involving multiple shock waves using only an iterative approach are presented
later; these demonstrate solutions obtained using the two boundary condition procedures, which incidentally are capable of properly treating shocks impinging on the
body.

It should be concluded from the above discussion that reflection and one-sided,
finite-difference boundary condition techniques are best used for iterative solutions
and that, in using a marching procedure for three-dimensional supersonic flows, a more
accurate boundary condition technique is required. Again, many procedures are available; some are quite accurate but difficult to apply, while others are inaccurate but
easy to apply.

Abbett's[32] concept of the simple-wave corrector technique is selected because
it possesses the attributes of being both accurate and easy to apply. The details of
the scheme are presented by Kutler et al.,[3,17] but are repeated here for complete-
ness. The three-dimensional supersonic flow-field program is based on a cylindrical

coordinate system (see Table 1) marching in the z direction and uses version I of MacCormack's scheme (Eq. (4.6)) to solve the governing equations.

This boundary condition procedure is basically a predictor-corrector technique in that the flow variables at the surface are first predicted using finite-difference algorithms and then corrected using a simple compression or expansion wave to satisfy the tangency condition. The first step of MacCormack's scheme, Eq. (4.6a), is applied at the body to yield the first predicted conservative variable $U^{(1)}$. It is followed by Eq. (4.6b) with the backward difference in the radial direction replaced by a forward difference and yields the "second predicted" conservative variable $U^{(2)}$. These variables are then decoded (see discussion following Eq. (2.4)) to obtain $p^{(2)}$, $\rho^{(2)}$, $u^{(2)}$, $v^{(2)}$, and $w^{(2)}$ at the station $z^{n+1} = z + \Delta z$. Generally, the resulting velocity vector $\vec{q}^{(2)} = u^{(2)}\hat{i}_z + v^{(2)}\hat{i}_r + w^{(2)}\hat{i}_\phi$ will not satisfy the surface tangency condition $\vec{q} \cdot \vec{n}_b = 0$, where $\vec{n}_b$ is the outward unit normal vector to the body and, in fact, will be rotated out of the surface tangent plane by a small angle $\Delta\theta$. This angle can be determined from

$$\Delta\theta = \sin^{-1}\left[\vec{q}^{(2)} \cdot \vec{n}_b / q^{(2)}\right] \tag{5.1}$$

where $q^{(2)}$ is the magnitude of $\vec{q}^{(2)}$ and the unit normal $\vec{n}_b$ can be calculated from

$$\vec{n}_b = \frac{\nabla f_b}{|\nabla f_b|} = \frac{-r_{b_z}\hat{i}_z - \hat{i}_r - \left[r_{b_\phi}/r_b\right]\hat{i}_\phi}{\left[r_{b_z}^2 + 1 + \left[r_{b_\phi}/r_b\right]^2\right]^{1/2}} \tag{5.2}$$

where $f_b = r_b - r_b(z,\phi) = 0$ describes the body surface. If $\Delta\theta$ is positive, then an expansion wave is necessary for the rotation of $\vec{q}^{(2)}$ and if $\Delta\theta$ is negative, a compression wave is required. The corrected value of the static pressure is found from the integral relation[38] for the Prandtl-Meyer turning angle $\nu(p; H_t, s)$, which depends on pressure and has the total enthalpy and entropy as parameters. The corrected value of pressure is found by solving

$$\nu(p^{n+1}; H_t, s) = \nu(p^{(2)}; H_t, s) + \Delta\theta \tag{5.3}$$

for the pressure p^{n+1}. In this equation, $\Delta\theta$ is given by Eq. (5.1). If $\Delta\theta$ is sufficiently small, Eq. (5.3) can be inverted and solved analytically for p^{n+1} only for a perfect gas to yield

$$\frac{p^{n+1}}{p^{(2)}} = 1 - \frac{\gamma(M^{(2)})^2}{[(M^{(2)})^2 - 1]^{1/2}} \Delta\theta + \gamma(M^{(2)})$$

$$\times \left\{\frac{(\gamma+1)(M^{(2)})^4 - 4[(M^{(2)})^2 - 1]}{4[(M^{(2)})^2 - 1]^2} (\Delta\theta)^2\right\} + 0[(\Delta\theta)^3] \qquad (5.4)$$

where

$$M^{(2)} = \frac{q^{(2)}}{c^{(2)}} \quad \text{and} \quad c^{(2)} = \left[\frac{\gamma p^{(2)}}{\rho^{(2)}}\right]^{1/2}$$

For a real gas, Eq. (5.3) can be inverted by the use of a table lookup method. The isentropic flow assumption requires that the table be generated only once at the very beginning of a flow-field calculation when the entropy on the body stream surface is known. The table elements are pressure and Prandtl-Meyer turning angle ν. The procedure for generating the table is described by Hayes and Probstein.[38]

The pressure p^{n+1} in Eq. (5.3) is determined by first finding $\nu(p^{(2)}; H_t, s)$ from the table with the predicted pressure $p^{(2)}$ as the argument. The angle $\Delta\theta$ given by Eq. (5.1) is then added to the result and the desired value for the corrected pressure p^{n+1} is then found from the same table with $\nu(p^{(2)}; H_t, s) + \Delta\theta$ as the argument. The remaining flow variables ρ^{n+1}, u^{n+1}, v^{n+1}, and w^{n+1} are then determined as follows.

From the starting solution, all surface flow variables are known along with the value of entropy at the body, and since entropy is assumed to be constant over the entire body, that constant can be determined from

$$\left.\frac{p^n}{(\rho^n)^\gamma}\right|_{\text{starting plane}} = \text{const} \quad \text{(nonconical flow)} \qquad (5.5a)$$

When a cone solution is generated, the surface entropy during the iteration procedure is calculated from

$$\left.\frac{p^n}{(\rho^n)^\gamma}\right|_{\text{windward shock}} = \text{const} \quad \text{(conical flow)} \qquad (5.5b)$$

since the entropy throughout the shock layer in the windward plane of symmetry is known to be constant.

If constant entropy is assumed along the body, the density ρ^{n+1} can therefore be calculated from

$$\rho^{n+1} = (p^{n+1}/\text{const})^{1/\gamma} \tag{5.6}$$

The velocity magnitude q^{n+1} can then be determined from the energy equation (2.4) as

$$q^{n+1} = \left[2 \left(H_t - \frac{p^{n+1}}{\rho^{n+1}} \frac{\gamma}{\gamma-1} \right) \right]^{1/2}$$

Finally, it is necessary to determine the individual components of $\vec{q}^{n+1}$. Since $\vec{q}^{(2)}$ was rotated through an angle $\Delta\theta$ in the plane of $\vec{n}_b$ and $\vec{q}^{(2)}$, it follows that the direction of the final velocity $\vec{q}^{n+1}$ must be in the direction of the vector $\vec{q}^{(2)} - \left(\vec{q}^{(2)} \cdot \vec{n}_b \right) \vec{n}_b$. Thus,

$$\vec{q}^{n+1} = q^{n+1} \vec{n}_t \tag{5.7}$$

where

$$\vec{n}_t = \frac{\vec{q}^{(2)} - \left(\vec{q}^{(2)} \cdot \vec{n}_b \right) \vec{n}_b}{\left| \vec{q}^{(2)} - \left(\vec{q}^{(2)} \cdot \vec{n}_b \right) \vec{n}_b \right|} \tag{5.8}$$

The velocity components obtained from Eq. (5.7) are

$$u^{n+1} = q^{n+1} (u^{(2)} + r_{b_z} M/N)/L \tag{5.9a}$$

$$v^{n+1} = q^{n+1} (v^{(2)} - M/N)/L \tag{5.9b}$$

$$w^{n+1} = q^{n+1} [w^{(2)} + (r_{b_\phi}/r_b)(M/N)]/L \tag{5.9c}$$

where

$$M = (-u^{(2)} r_{b_z} + v^{(2)} - w^{(2)} r_{b_\phi}/r_b)/N$$

$$N = [r_{b_z}^2 + 1 + (r_{b_\phi}/r_b)^2]^{1/2}$$

$$L = \left[u^{(2)} + r_{b_z} M/N + (v^{(2)} - M/N)^2 + \left(w^{(2)} + \frac{r_{b_\phi}}{r_b} \frac{M}{N} \right)^2 \right]^{1/2}$$

and $r_{b_z} = \partial r_b/\partial z$, etc.

A comparison of this boundary condition scheme for the simple two-dimensional compression and expansion flows mentioned earlier showed virtually no differences with the method of characteristics. The three-dimensional version just described relies on the two predictor steps to yield an accurate flow description in the ϕ direction.

Results presented later establish the accuracy of this procedure.

The impermeable boundary created by a plane of symmetry is simply simulated numerically by applying the refelection technique along it. This is an exact boundary condition treatment and therefore does not contribute to the numerical error of a solution.

5.2 Permeable Boundaries

As with the surface boundary condition, there is a variety of schemes to treat the permeable boundary created by a shock wave. Abbett[39] discusses and compares the most widely used techniques. Based on personal investigations and the conclusions of Abbett, a modified version of Thomas' scheme[40] (pressure approach) is adopted here for the supersonic flow-field calculations presented that used a sharp-shock approach. The procedure is presented for both the two-dimensional unsteady (which can easily be generalized to three dimensions) and the three-dimensional steady cases. In both cases, one of the constant coordinate lines is assumed to lie along the shock wave.

The pressure approach method of Thomas, originally derived for the steady flow nonconservative equations, is relatively simple to implement and yields quite accurate results. The Rankine-Hugoniot relations can be written as a function of the pressure downstream of the shock wave. This pressure is predicted by the first step of MacCormack's method (version I), Eq. (4.6a), from the decoded conservative variables, and through the Rankine-Hugoniot relations, the other predicted variables can be found. The pressure is then recomputed by the corrector step, Eq. (4.6b), and the rest of the procedure is repeated. The details for the unsteady case assuming a perfect gas follow.

Figure 23 depicts a two-dimensional unsteady shock with a shock velocity q_s in a polar coordinate system whose surface is described by

$$r_s = r_s(t,\theta) \tag{5.10}$$

where $r_{s_t} = \partial r_s / \partial t = q_{s_r}$ and $r_{s_\theta} = \partial r_s / \partial \theta$. From geometrical considerations, the shock angle β is

$$\beta = \frac{\pi}{2} - \theta + \tan^{-1}(r_{s_\theta}/r_s) \tag{5.11}$$

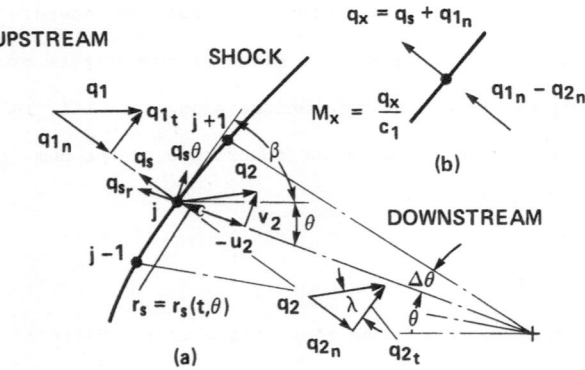

Fig. 23. Unsteady shock for two-dimensional supersonic flow in polar coordinates (t, r, θ).

The pressure behind the shock wave is known from the finite-difference scheme; therefore, the shock Mach number M_x (Fig. 23(b)) is given by

$$M_x = \{[(\gamma+1)p_2/p_1 + (\gamma-1)]2\gamma\}^{1/2} \tag{5.12}$$

The density behind the shock is given by

$$\rho_2 = \rho_1(\gamma+1)M_x^2/[(\gamma-1)M_x^2 + 2] \tag{5.13}$$

The upstream velocity components normal and tangential to the shock are

$$\left.\begin{aligned} q_{1_n} &= q_1 \sin \beta \\ q_{1_t} &= q_1 \cos \beta \end{aligned}\right\} \tag{5.14}$$

From kinematical relations, the shock velocity is given by

$$q_s = M_x c_1 - q_{1_n} \tag{5.15}$$

where $c_1 = (\gamma p_1/\rho_1)^{1/2}$ and thus

$$q_{s_r} = r_{s_t} = q_s \sin(\beta+\theta) \tag{5.16}$$

The normal velocity behind the shock is given by

$$q_{2_n} = c_2 \left[\frac{(\gamma-1)M_x^2 + 2}{2\gamma M_x^2 - (\gamma-1)}\right]^{1/2} - q_s \tag{5.17}$$

where $c_2 = (\gamma p_2/\rho_2)^{1/2}$.

The velocity components in the r and θ directions are therefore

$$u_2 = -q_2 \cos \mu \left.\begin{array}{c}\\\\\end{array}\right\} \tag{5.18}$$
$$v_2 = q_2 \sin \mu$$

where $q_2 = (q_{2_n}^2 + q_{1_t}^2)^{1/2}$ and $\mu = \beta + \theta - \lambda$, where $\lambda = \tan^{-1}(q_{2_n}/q_{1_t})$.

The shock wave is propagated using the following Euler predictor/modified Euler corrector in conjunction with the predictor-corrector philosophy of computing interior points:

$$r_s^{(1)} = r_s^n + q_{s_r}^n \Delta t \quad \text{(predictor)} \tag{5.19a}$$

$$r_s^{n+1} = r_s^n + \frac{1}{2}\left[q_{s_r}^n + q_{s_r}^{(1)}\right]\Delta t \quad \text{(corrector)} \tag{5.19b}$$

where Δt is the time step. The partial derivative r_{s_θ} required in the Rankine-Hugoniot equations is found numerically (Fig. 23(a)) using the second-order, central-difference formula:

$$r_{s_\theta} = (r_{s_{j+1}} - r_{s_{j-1}})/2\Delta\theta \tag{5.20}$$

The above equations are used as follows. Initially, p_2, ρ_2, u_2, v_2, r_s, r_{s_θ}, and r_{s_t} are known along the shock at time step n. The pressure at the shock $p_2^{(1)}$ is predicted using Eq. (4.6a) with a backward difference in the radial direction. The shock wave is then moved using Eq. (5.19a), and $r_{s_\theta}^{(1)}$ is calculated from Eq. (5.20). From Eq. (5.11), $\beta^{(1)}$ can then be found, followed by the upstream velocity components given by Eq. (5.14). The density, shock velocity, and velocity components behind the shock can be found from Eqs. (5.13), (5.16), and (5.18), respectively. The same pro-cedure is used in the corrector step except that Eq. (4.6b) is used to correct the pressure and Eq. (5.19b) is used to correct the shock position. Results obtained using this procedure at the shock for the supersonic flow over a two-dimensional blunt body are presented later.

To apply a sharp-shock procedure in steady three-dimensional supersonic flow of either an ideal or real gas for cylindrical coordinates (z, r, ϕ), the pressure ap-proach can again be used. Consider the three-dimensional shock shown in Fig. 24 for which the governing equations are integrated in the longitudinal direction z. The free-stream velocity vector $\vec{q}_1$ at an angle of attack α is given by

$$\vec{q}_1 = (u_1, v_1, w_1) = q_1 \cos \alpha \hat{i}_z - q_1 \sin \alpha \cos \phi \hat{i}_r + q_1 \sin \alpha \sin \phi \hat{i}_\phi \tag{5.21}$$

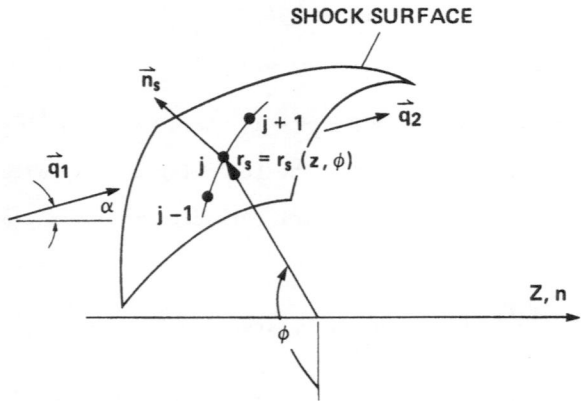

Fig. 24.   Steady shock for three-dimensional supersonic flow in
cylindrical coordinates (z, r, ϕ).

Two quantities that appear in the following equations are inherently dependent on
equations of state, the upstream component of velocity normal to the shock surface $\tilde{u}_1$
and the downstream density ρ_2:

$$\tilde{u}_1 = \tilde{u}_1(p_2; p_1, \rho_1) \tag{5.22}$$

$$\rho_2 = \rho_2(p_2; p_1, \rho_1) \tag{5.23}$$

Here the quantities upstream and downstream of the shock-wave discontinuity are denoted by subscripts 1 and 2, respectively. The procedure used to evaluate Eqs. (5.22)
and (5.23), which depends on whether a real or perfect gas is considered, is covered
later. For the present it is assumed that such relations exist.

The vector components of velocity normal and tangential to the shock wave are
given by

$$\vec{u}_1 = -(\vec{q} \cdot \vec{n}_s)\vec{n}_s \tag{5.24}$$

$$\vec{v}_1 = \vec{q}_1 - \vec{u}_1 \tag{5.25}$$

where $\vec{n}_s$ denotes the outward unit normal to the shock surface. The formula for the
shock surface normal is identical with that for the body surface normal, Eq. (5.2),
except that subscript b is replaced by subscript s.

The magnitude of the normal velocity component $\tilde{u}_1$ is found as a function of
shock surface derivatives by use of Eq. (5.24). The resulting expression is then inverted to yield the following representation for the derivative r_{s_z}:

$$r_{s_z} = u_1 \Omega + \tilde{u}_1 \sqrt{\Omega^2 + [1 + (r_{s_\phi}/r_s)^2]/(u_1^2 - \tilde{u}_1^2)} \qquad (5.26)$$

where

$$\Omega = \frac{v_1 - w_1(r_{s_\phi}/r_s)}{(u_1^2 - \tilde{u}_1^2)}$$

The downstream velocity $\vec{q}_2$ is given by either

$$\vec{q}_2 = u_2 \hat{i}_z + v_2 \hat{i}_r + w_2 \hat{i}_\phi \qquad (5.27a)$$

or

$$\vec{q}_2 = \vec{\tilde{u}}_2 + \vec{\tilde{v}}_2 \qquad (5.28a)$$

The tangential component $\vec{\tilde{v}}_2$ is conserved across the shock wave and hence its value is the same as $\vec{\tilde{v}}_1$. This variable is eliminated from Eq. (5.28b) by use of Eq. (5.25) and, in addition, mass conservation across the shock wave is introduced in the form $\tilde{u}_2 = \tilde{u}_1 \rho_1/\rho_2$ to eliminate $\tilde{u}_2$. The result is

$$\vec{q}_2 = \vec{q}_1 - r_{s_z} a \hat{i}_z + a \hat{i}_r - \frac{r_{s_\phi}}{r_s} a \hat{i}_\phi \qquad (5.29)$$

where

$$a = \frac{|\tilde{u}_1|(1 - \rho_1/\rho_2)}{\sqrt{r_{s_z}^2 + 1 + (r_{s_\phi}/r_s)^2}}$$

As in the unsteady case, the shock wave is moved using the following Euler predictor/modified Euler corrector:

$$r_s^{(1)} = r_s^n + r_{s_z}^n \Delta z \quad \text{(predictor)} \qquad (5.30a)$$

$$r_s^{n+1} = r_s^n + \frac{1}{2}(r_{s_z}^n + r_{s_z}^{(1)})\Delta z \quad \text{(corrector)} \qquad (5.30b)$$

where Δz is the integration step size. The quantity r_{s_ϕ} is evaluated according to the formula (Fig. 23):

$$r_{s_\phi} = (r_{s_{j+1}} - r_{s_{j-1}})/2\Delta\phi \qquad (5.31)$$

Equations (5.19) and (5.20) for an ideal gas can be written explicitly as

$$\tilde{u}_1(p_{2_j} p_1, \rho_1) = \frac{p_1}{2\rho_1}\left[(\gamma+1)\frac{p_2}{p_1} + (\gamma-1)\right] \qquad (5.32)$$

$$\rho_2(p_2, p_1, \rho_1) = \rho_1 \frac{\frac{p_2}{p_1} + \frac{\gamma-1}{\gamma+1}}{1 + \frac{\gamma-1}{\gamma+1} \frac{p_2}{p_1}} \tag{5.33}$$

Equivalent analytic representations for a real gas are not available. However, a table-lookup scheme can also be adapted here as in the body boundary condition scheme. At the beginning of the flow-field calculation after p_1 and ρ_1 are specified, a pair of tables is generated that contains the upstream normal velocity $\tilde{u}_1$ and the downstream density ρ_2 as elements with the pressure p_2 as the argument. The table lookup procedure can then simultaneously return values $\tilde{u}_1$ and ρ_2 for a given pressure p_2. The procedure used to generate the table is similar to that described by Vincenti and Kruger[41] (p. 179) in their discussion on steady shock waves.

The procedure used to apply the above equations for a steady three-dimensional shock is basically the same as that used in the unsteady example presented earlier and therefore is not repeated here. Again, results presented later use this procedure and demonstrate the obtainable accuracy.

In the computation of supersonic flow problems, two other types of permeable boundaries are encountered: the outflow or downstream boundary and the inflow or upstream boundary. Figure 25 presents examples of both types.

The computational region for determining the supersonic flow about a blunt body (see Fig. 25(a)) is bounded by the body, bow shock, plane of symmetry, and the line $\theta = \theta_{max}$. This line is a supersonic outflow boundary and must be chosen downstream of the sonic line, that is, q_θ along θ_{max} must be greater than the local speed of sound

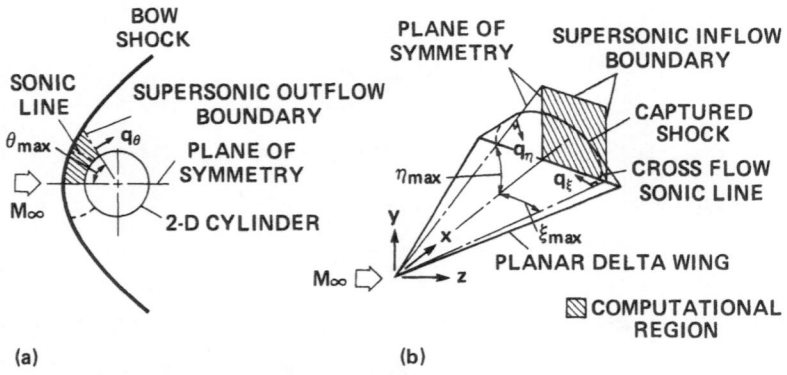

Fig. 25. Supersonic outflow and inflow computational boundaries; (a) blunt body, and (b) planar delta wing-compression side.

in order that the time-dependent problem be well posed. A variety of techniques exists for determining the dependent variables along such a boundary, but the simple linear extrapolation, or equivalently one-sided differencing, has been found to work well and is usually used. Roache[35] discusses other procedures for accomplishing the same thing.

Supersonic inflow boundaries are quite common in the examples presented later. Two such boundaries occur in the determination of the conical supersonic flow over the compression side of a planar delta wing (Fig. 25(b)). The computational boundary for conical coordinates in this case consists of the body, a plane of symmetry, and the two supersonic inflow boundaries along the lines $\eta_{max} = (y/x)_{max}$ and $\xi_{max} = (z/x)_{max}$. For this problem to be well posed, the fluid velocity q_ξ and q_η normal to these boundaries and at every point along them must be greater than the local speed of sound. The exact flow conditions along this type of boundary must be known at each step during the integration procedure; in general, they are usually constant. In Fig. 25(b), the dependent variables along the line η_{max} assume the values of the free stream, while along the line ξ_{max} they assume values of both the free stream and "sliding wedge" flow.

The boundary condition procedures discussed in this section are not meant to be the best possible for solving supersonic flow problems. But when used properly, however, they yield solutions of more than acceptable quality.

6. INITIAL CONDITIONS, STEP SIZE, AND CONVERGENCE

To initiate a given calculation, the values of the dependent variables at each grid point in the computational plane must be specified. This initialization procedure depends on whether a marching technique or an iterative approach is being used to obtain the solution.

For an iterative approach in which one is interested only in the converged solution, the initial data, in most instances, can be a rather rough approximation of what the actual solution is believed to be. In using a complete shock-capturing approach, for example, it is a common practice to set the flow variables at the interior points equal to the free-stream variables. In doing this, imposition of the tangency condition at the body creates either a compression or expansion wave that propagates into the field and eventually settles down to its correct location as the solution converges.

Initialization for the blunt-body problem in which the bow shock is treated as a discontinuity is somewhat more complicated. In addition to initializing the dependent variables at the interior points, one must estimate the shock position and slope. Usually, this simply requires that the shock be specified analytically, thus yielding its slope and the flow variables behind it. In conjunction with a simple approximation for the body variables for example, modified Newtonian flow followed by a linear interpolation for the flow variables between the body and shock, the initialization procedure is complete.

The initial data required when using a marching procedure in which the solution at each step is of interest are usually determined from some other source and supplied to the marching code. For example, when solving for the supersonic flow over a three-dimensional, blunt-nose configuration (Fig. 13(b)), a blunt-body computer program supplies the data in the starting plane. This includes all flow variables between the body and the shock, the shock position, and shock slopes.

Once the initialization procedure is complete, a step size must be selected that is small enough to be commensurate with the stability bound, but large enough to finish the computation with a minimum of computer time. Use of the largest possible step size for hyperbolic equations ensures that the finite-difference scheme is as nearly compatible with the method of characteristics or the "perfect shift condition" (see Kutler and Lomax[37] or Warming et al.[23]) as possible. To determine the range of values the step·size should have, many authors rely on amplification matrix theory because of its ability to predict a stability condition conservatively and quickly. The method is based on a locally linear analysis of the governing partial differential equations coupled with a discrete harmonic analysis of the linear-difference scheme. In Section 4, the modified equation approach was used to get a stability bound. Both theories yield the same result.

The partial differential equation system for the linear analysis is given by Eq. (2.9a) for the unsteady case and by Eq. (2.9b) for the steady case. The stability theories, at least for (t, x_1) space in unsteady flow and (x_1, x_2) space in steady flow, require that

$$\frac{\Delta t}{\Delta x} \leq \frac{1}{|(\sigma_\ell^A)_{max}|} \quad ; \quad \text{unsteady} \tag{6.1a}$$

or

$$\frac{\Delta x_1}{\Delta x_2} \leq \frac{1}{|(\sigma_\ell^M)_{max}|} \quad ; \quad \text{steady} \tag{6.1b}$$

$$\ell = 1, 3$$

where $(\sigma_\ell^A)_{max}$ and $(\sigma_\ell^M)_{max}$ are the maximums of the local maximum eigenvalues (Eq. (2.10a) and (2.10b)) of all the points at a particular constant t or x_1 plane, respectively. Similar relations can be obtained for the (t, x_2), (t, x_3), and (x_1, x_3) spaces:

$$\frac{\Delta t}{\Delta x_2} \leq \frac{1}{|(\sigma_\ell^B)_{max}|}$$

$$\frac{\Delta t}{\Delta x_3} \leq \frac{1}{|(\sigma_\ell^C)_{max}|} \quad \text{unsteady} \qquad \ell = 1, 3 \tag{6.2a}$$

$$\frac{\Delta x_1}{\Delta x_3} \leq \frac{1}{|(\sigma_\ell^N)_{max}|} \quad \text{steady} \tag{6.2b}$$

This planar analysis has been shown[3,17,31,42] to give a good bound on the step size in multidimensional problems if the right-hand size of Eqs. (6.1) and (6.2) are multiplied by a constant $C \leq 1$, which can be varied during the computation and is usually assigned a value of approximately 0.9.

For multidimensional problems in unsteady flow, for example, Δx_1, Δx_2, and Δx_3 are known. It is therefore necessary to determine the step size from the minimum Δt predicted by Eqs. (6.1a) and (6.2a). This ensures that all inequalities are satisfied. This same procedure is followed for multidimensional steady flows. In flow fields where the dependent variables are changing rapidly, the step size should be recalculated after each complete integration step to eliminate the possibility of an instability.

Independent variable transformations change the structure of the eigenvalues and hence the step size. Consider, for example, the following transformation applied to the steady flow equations (Eq. (2.9b)):

$$\zeta = x_1$$
$$\eta = \eta(x_1, x_2, x_3) \qquad\qquad (6.3)$$
$$\xi = \xi(x_1, x_2, x_3)$$

Applied to Eq. (2.9b), this yields

$$U_\zeta + PU_\eta + QU_\xi + A^{-1}D = 0 \qquad\qquad (6.4)$$

where

$$P = (\eta_{x_1} + M\eta_{x_2} + N\eta_{x_3})$$
$$Q = \xi_{x_1} + M\xi_{x_2} + N\xi_{x_3}$$

The eigenvalues of the matrices P and Q can be determined and then used in Eqs. (6.1b) and (6.2b) to calculate the integration step size. An example of this effect is presented later.

The convergence of an iterative solution is recognized when the flow variables, to the accuracy of the method being used, are not changing. All calculations presented in the following section were obtained using a cathode ray display tube (CRT) linked with the computer. This allowed the solution to be displayed after each integration step so that convergence could be judged.

7. NUMERICAL RESULTS

This section presents the numerical results for a number of different problems ranging from simple wedge flow to multiple-shocked three-dimensional flows. The majority of these computations were performed on an IBM 360/67 computer system linked with an IBM 2250 cathode ray display tube, which allowed on-line interaction with the computer while simultaneously displaying the results.

7.1 Two-Dimensional Wedge and Axisymmetric Cone Flow

The pedagogical problems of supersonic flow over a wedge and a cone at zero angle of attack are by no means difficult to solve. They are used here to demonstrate the effects of the shock-capturing, finite-difference approach on solutions of the inviscid Euler equations in comparison with solution of the modified Burger's equation discussed earlier. Results are presented using the first-order method of Lax and the second-order method of MacCormack.

To solve the wedge problem, a Cartesian coordinate system is used in which the x coordinate lies along the surface of the wedge and the y coordinate is normal to the surface (Fig. 26(a)). The reflection principle is applied at the body, and the flow variables are initially set equal to the free-stream variables. The governing equations are integrated from $x = 1$ to $x = 2$ and then stepped back to $x = 1$. This procedure is repeated until the transients have been damped. The shock wave during the integration from 1.0 to 2.0 propagates upward (or forward) in the computational line as did the simulated shocks when solving the modified Burger equation, thus allowing for an easy comparison of the two solutions.

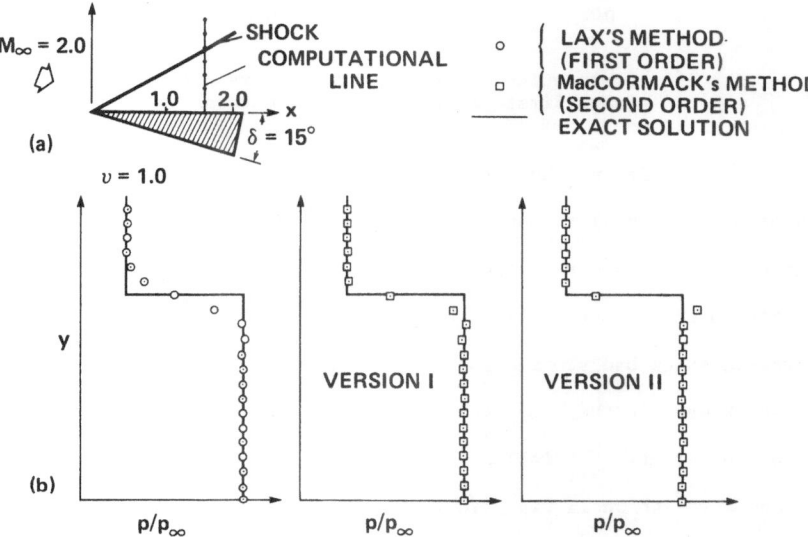

Fig. 26. Computation of simple supersonic wedge flow using first- and second-order methods.

The wedge solutions are shown in Fig. 26(b). Using Lax's method, the captured shock is spread over approximately four mesh intervals, and its location agrees quite well with the exact shock location. This same behavior was observed in Fig. 17 for Burger's equation. Using version I of MacCormack's scheme (the preferred version for this problem), the shock is spread over two intervals and has no noticeable overshoots or undershoots as does the solution obtained using version II (the nonpreferred version.

A spherical coordinate system is used in solving the cone at zero angle of attack

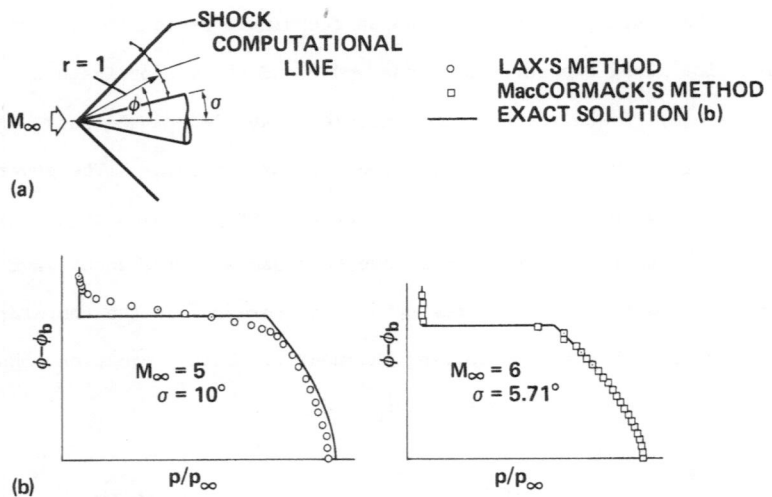

Fig. 27.   Computation of axisymmetric supersonic cone flow
using first- and second-order methods.

problem (Fig. 27(a)). Again, the reflection principle is used at the body, and the

flow variables are set equal to the free stream initially. The governing equations

are integrated iteratively with respect to r until converged. Solutions obtained

using Lax's method and MacCormack's method are shown in Fig. 27(b). The first-order

scheme smears the shock badly while the second-order scheme yields a rather sharp

shock. The shock wave in the computational line for the near-converged solution does

not move. The conservative variables in the ϕ direction therefore are continuous

through the shock (as shown in Fig. 5(d)), but their first derivative is discontinu-

ous. When differencing across the shock in this case, it is this discontinuous behav-

ior that results in the minor oscillations of the second-order solution.

The conclusions to be drawn from this simple study are that (1) the modified form

of Burger's equation can be used to represent the inviscid Euler equations for numeri-

cal studies, (2) the numerical solutions obtained using MacCormack's second-order

scheme are far superior in their shock-capturing ability than the solutions obtained

using Lax's scheme, and (3) shock alignment with one of the coordinate directions

yields captured shock waves that are better defined.

7.2 Planar Delta Wing

The supersonic flow field surrounding a lifting delta wing with supersonic leading edges has been the object of many numerical, theoretical, and experimental studies in the past. The purpose here is first to determine the flow fields on both the compression and expansion sides of a planar delta wing and, second, to combine these two solutions at the trailing edge to form an initial data plane that can be integrated downstream to yield the inviscid flow in the wake of the delta wing.

Compression side numerical calculations have been obtained by such investigators as Fowell,[43] South and Klunker,[44] Babaev,[45] Beeman and Powers,[46] and Voskresenskii.[47] The flow field on this side of the delta wing poses no great problem for the sharp-shock techniques used by these investigators, and their results are comparable.

Expansion side calculations have been attempted by Fowell,[43] Babaev,[48] and Beeman and Powers.[45] This flow field is somewhat more complicated than the compression side since, at some distance inboard of the leading edge, an embedded shock wave is produced because of the deceleration of the supersonic crossflow velocities.

In the past, no attempt has been made to determine the complete inviscid flow field in the wake of a lifting delta wing. Oswatitsch and Sun[49] used the method of characteristics in conjunction with a limiting procedure to handle the trailing-edge expansion to determine the influence of the near-field flow in the far-field wave formation beneath the delta wing. Their main conclusion was that the bow shock, due to attenuation by the trailing-edge expansion wave, terminates a finite distance from the wing. Roe,[50] in a subsequent experiment, found that the bow shock did not disappear, but the wave pattern beneath the planar delta wing developed into the classical "N" wave.

The near-field flow behind the delta wing is complicated not only by the existing compression and expansion side-flow fields, which contain various combinations of shock and expansion waves, but also by the shock and expansion waves generated at the trailing edge. This class of problems, which would be difficult to solve using a complete sharp-shock approach, is a good test for the usefulness of the shock-capturing technique.

To solve this problem, a Cartesian coordinate system is used, and the wing with respect to this system is oriented as shown in Fig. 14. An independent variable transformation to conical coordinates, normalized with respect to the tangent of the wing angle β ($\zeta = x$, $\eta = y/x \tan \beta$, $\xi = z/x \tan \beta$) is performed on Eq. (2.1). The resulting equation is rearranged in conservation law form, and since it is hyperbolic with respect to ζ, it is integrated cyclically from $\zeta = 1$ to $\zeta = 1 + \Delta\zeta$ until conical flow is established. Plane-of-symmetry boundary conditions are applied at the left-hand boundary (Fig. 14) of the mesh since the flow field is symmetrical to it. The uppermost part of the mesh is chosen so that it remains entirely in the free stream, and the boundary conditions there are fixed accordingly. Exact boundary conditions are applied at the right-hand side since these can be computed from the oblique-shock or Prandtl-Meyer relations which apply along supersonic leading edges. At the body, the reflection principle, which is exact in this case, is used. For both the compression and expansion side calculations, initial conditions are supplied by using the values of the flow variables in the free stream.

After the complete flow field is determined on both sides of the delta wing, the two solutions are combined at the trailing edge to form an initial data plane that can then be numerically integrated downstream through the wake. In this case, the only boundary condition imposed is that the free stream exist along all edges of the mesh where symmetry does not apply.

Computations were performed (see Kutler and Lomax[42]) to determine the flow field over the compression side of a 50° swept, planar delta wing. Figure 28 shows the results of those computations for angles of attack of 5°, 10°, and 15° at Mach 4. The semispan pressure distributions are compared with the results of South and Klunker,[44] who used a method of lines technique. South, in turn, made comparisons with Voskresenskii[47] and found good agreement, and with Babaev[45] and found poor agreement. The present results substantiate the results of South and Voskresenskii. Figure 29 compares the shock shapes for the above flow conditions. There is very little disagreement with South's results. The shock shape obtained using the shock-capturing technique was located using a weighted gradient technique coded into the program.

The surface-pressure coefficient on the expansion side of a 45° swept delta wing in Mach 3 flow for 4°, 8°, and 12° angle of attack is shown in Fig. 30. The results are compared with a method-of-characteristics technique devised by Beeman and Pow-

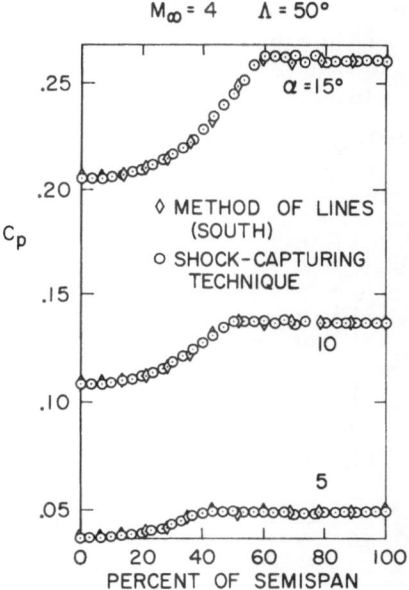

Fig. 28. Spanwise pressure distribution on compression side of planar
delta wing; M = 4, Λ = 50°, = 5°, 10°, and 15°.

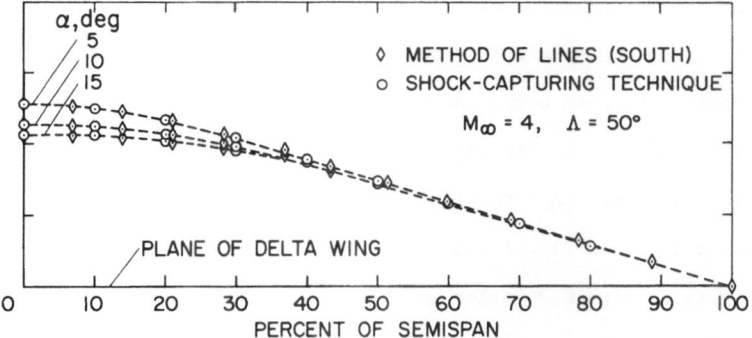

Fig. 29. Compression side-shock shapes; M = 4, Λ = 50°,
α = 5°, 10°, and 15°.

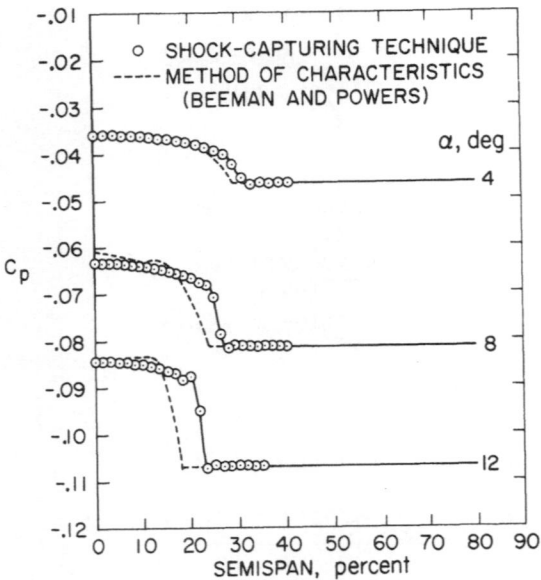

Fig. 30.   Spanwise pressure distribution on expansion side of planar
delta wing; M = 3, Λ = 45°, α = 4°, 8°, and 12°.

ers,[46] who chose to neglect the weak crossflow shock, assuming instead an isentropic compression. The disagreement in the location of the embedded shock is believed to be the result of their assumption. However, the pressure distributions on either side of the shock are in good agreement. The grid size used for these problems consisted of 34 points normal to the body and 26 points parallel to the body, and computation times were on the order of 10 min on the IBM 360/67.

Bannink and Nebbeling[51] obtained very good experimental results on the expansion side of a 44.7° swept delta wing at 12° angle of attack in Mach 2.94 flow. The identical case was solved numerically and plots of the spanwise pitot-pressure distribution at various heights above the wing are compared with the experimental results in Fig. 31. The locations of the crossflow shock and conical sonic line are compared in Fig. 32. The agreement in both figures is remarkable.

The complete inviscid flow field in the wake of a 55° swept, planar delta wing at 1.8° angle of attack in Mach 2.7 flow was determined after the compression and expansion side flow fields were determined. The results of this calculation clearly show the formation of the trailing-edge shock wave on the leeward side and expansion wave on the windward side. The formation of the secondary recompression shock on the wind-

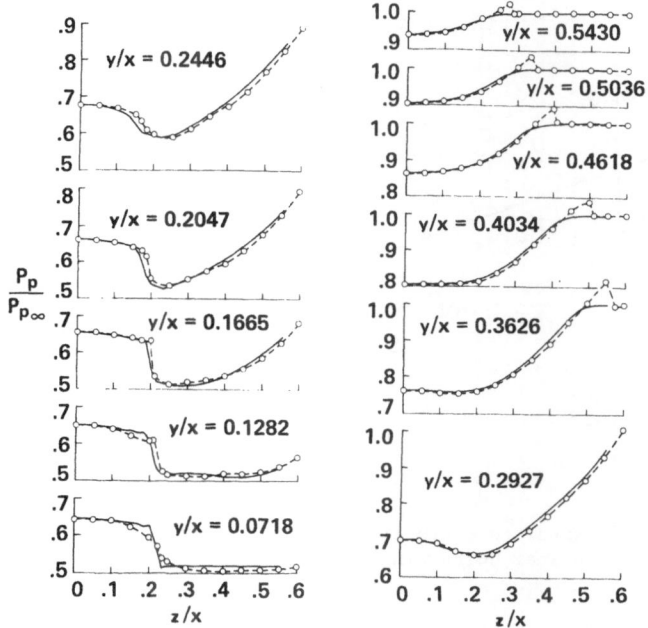

Fig. 31. Spanwise pitot pressure distribution on planar delta
wing; M = 2.94, Λ = 44.7°, α = 12°.

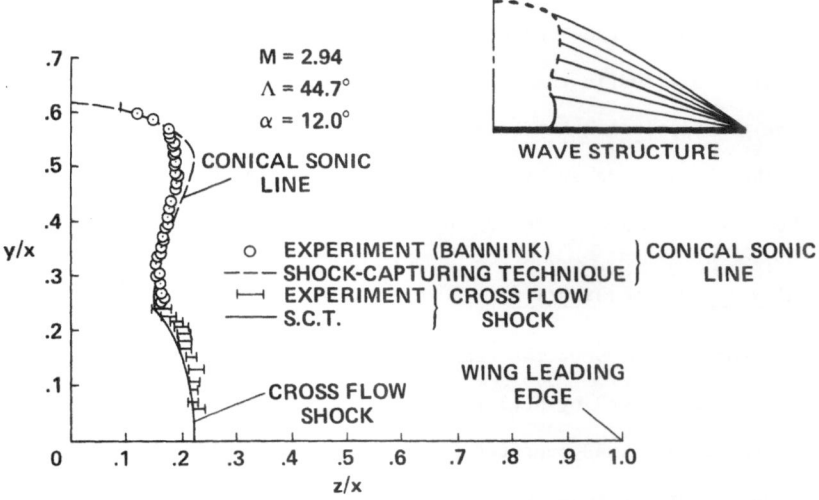

Fig. 32. Comparison with experiment of shock wave and conical sonic
line for planar delta wing; M = 2.94, Λ = 44.7°, α = 12°.

CENTERLINE SONIC BOOM SIGNATURE
FOR PLANAR DELTA WING

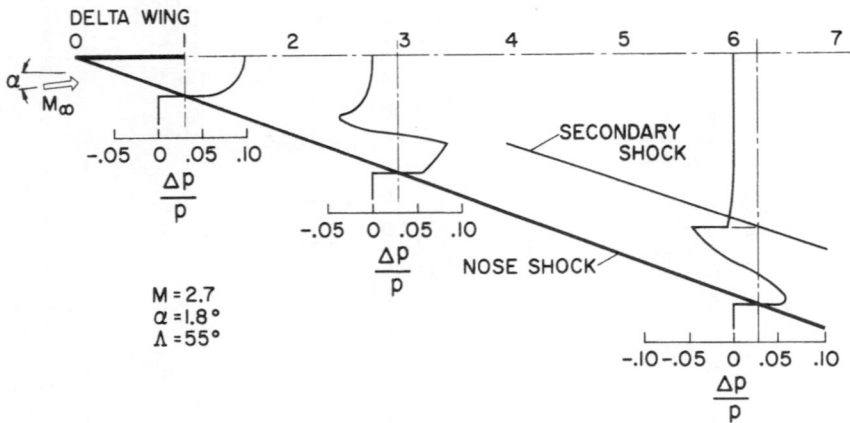

Fig. 33. Windward centerline pressure distributions beneath
planar delta wing; M = 2.7, Λ = 55°, α = 1.8°.

ward side is also clearly recognizable. Figure 33 shows the plane-of-symmetry, compression side-shock locations, and normal pressure distributions between the plane of the wing and the free stream at various stations downstream. The pressure curves shown were faired through the numerical data, and the actual shock waves did not appear as sharp discontinuities but were spread over two mesh intervals.

The results obtained in this wake study did not reveal the disappearance of the bow shock for the distance behind the wing for which the equations were solved. The results did show a rapid decay of the disturbed flow region away from the plane of symmetry, but always indicated an N wave beneath the wing in the plane of symmetry.

7.3 Conical Wing-Body Combination

The supersonic flow about the compression side of a circular cone/planar delta wing combination was solved by Kutler and Lomax[42] using the shock-capturing technique. The results obtained compared well with existing experimental data. Results are now presented for a similar wing-body configuration for both the compression and expansion side-flow regions.

Consider the quarter-body in Fig. 34. The basic coordinate system is again Cartesian, and the governing equations are transformed to conical coordinates. They are

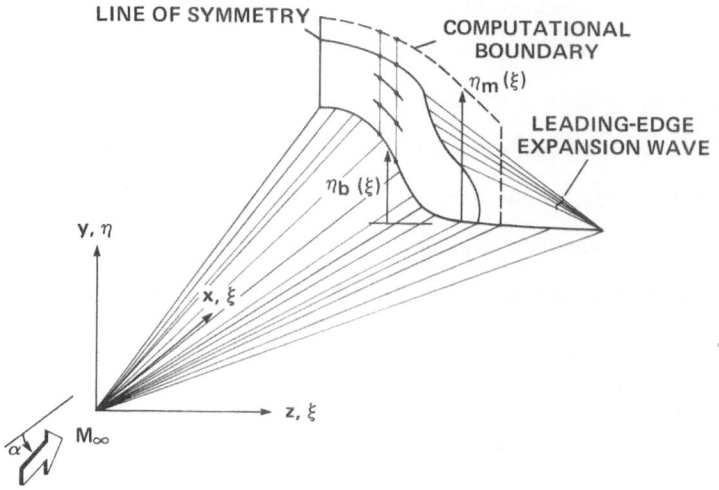

Fig. 34. Coordinate system for conical wing-body combination.

transformed again to normalize the distance between the body given by $\eta_b = \eta_b(\xi)$ and the upper computational boundary given by $\eta_m = \eta_m(\xi)$ as follows:

$$\left.\begin{aligned} \zeta &= \zeta \\ \eta' &= \frac{\eta - \eta_b(\xi)}{\eta_m(\xi) - \eta_b(\xi)} \\ \xi &= \xi \end{aligned}\right\} \tag{7.1}$$

The resulting equation is then rearranged in conservation law form and integrated with respect to ζ to establish conical flow.

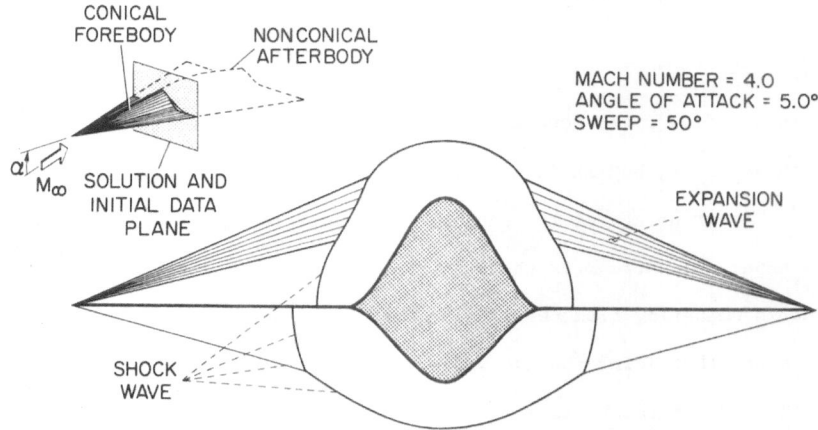

Fig. 35. Computed location of shock and expansion waves around wing-body combination; $M = 4$, $\Lambda = 50°$, $\alpha = 5°$.

The results of such a computation for a cosine shaped body and planar delta wing at 5° angle of attack in Mach 4 flow are shown in Figs. 35 and 36. The computed shock and expansion wave patterns are shown in Fig. 35. On the compression side, there exists a slip surface (not shown) emanating from the triple point and extending to the crossflow stagnation point on the body. The surface-pressure distribution for both sides of the vehicle is shown in Fig. 36. The discontinuous behavior in the curves results from the embedded shocks striking the body.

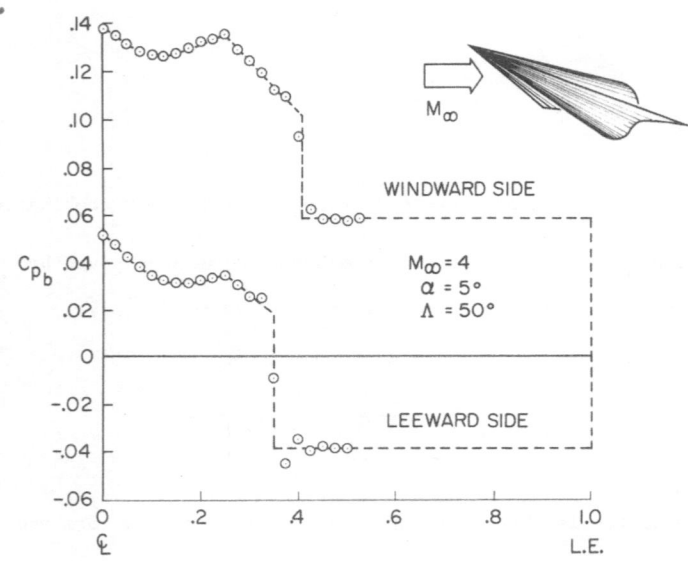

Fig. 36. Surface-pressure distribution on wing-body combination; M = 4, Λ = 50°, α = 5°.

7.4 Internal Corner Flow

The supersonic flow field generated by an internal corner contains multiple shocks and slip surfaces, but it is easily determined using the shock-capturing approach.[52,53] The structure of the conical flow generated by two intersecting wedges immersed in a supersonic stream is shown in Fig. 37. The shock structure consists of the planar shocks emanating from the leading edge of each wedge (with angles δ_1 and δ_2), a corner shock that joins the two wedge shocks, and two embedded shocks that stretch from the body to their respective triple points. Stretching between each of triple points and the axial corner is a slip surface or inviscid shear layer. A vor-

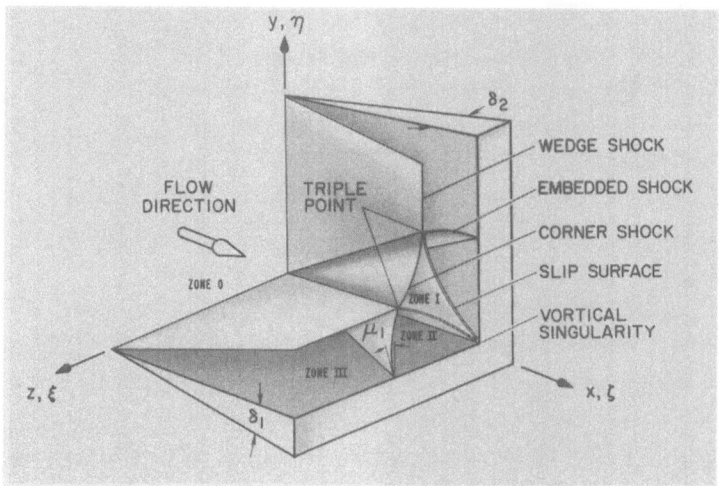

Fig. 37. Coordinate system and typical wave structure.

tical singularity (a point with multiple values of entropy) exists at the axial corner
of the two wedges. It is generated as a result of the flow passing through different
points of the curved embedded and corner shocks and converging at the axial corner.

The regions bounded by the shocks and slip surfaces are denoted as zones 0, I,
II, and III (as illustrated in Fig. 37). Zone 0 corresponds to the free stream, and
zone III contains simple wedge flow; both have constant flow properties. The flow in
zones I and II is rotational because of the convex corner shock and concave embedded
shock (viewed from corner). The conical crossflow velocity (the component that results when the total velocity is projected on a sphere whose center is at the origin
of the coordinate system) is supersonic in zone III and subsonic in zones I and II.
Therefore, the conical problem is mixed elliptic-hyperbolic. If the three-dimensional
steady-flow equations (Eq. (2.1)) in Cartesian coordinates are transformed into non-
orthogonal, conical coordinates ($\zeta = x$, $\eta = y/x$, and $\xi = z/x$), they are made hyperbolic with respect to ζ.

A rectangular region that completely encompasses the shock structure in the corner is discretized and results in the computational plane shown in Fig. 38. In the

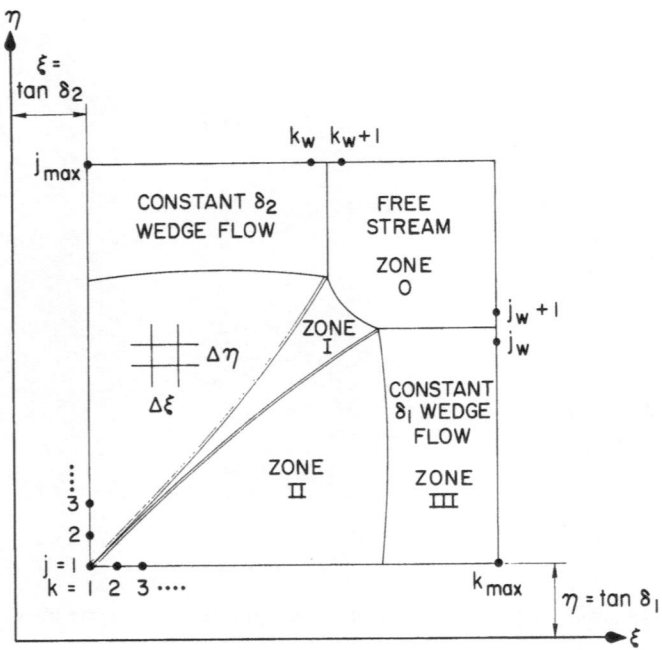

Fig. 38. Computational plane for internal corner flow.

η direction, the lower boundary is the wedge surface, $\eta = \tan \delta_1$ where j = 1 (along with the tangency condition is satisfied using one-sided finite differences), while the upper boundary, $j = j_{max}$, is chosen to fall in a region of known flow properties (well outside the wedge-flow Mach cone from the corner). An analogous procedure is used in the ξ direction. All interior points are assigned values of the free-stream variables initially.

The second-order-accurate, predictor-corrector scheme of MacCormack is used to integrate the governing equations iteratively until the term E_ζ is zero, which implies the establishment of a conical flow field. Shock waves and slip surfaces that should exist form automatically and are correctly positioned within the computational network of points.

The most recent published experimental data obtained for the corner flow problem are by West and Korkegi.[54] They tested an equal wedge angle ($\delta_1 = \delta_2 = 9.49°$) configuration in Mach 2.98 flow over a Reynolds number range from 0.4×10^6 to 60×10^6, which included laminar, transitional, and turbulent boundary layers. A numerical so-

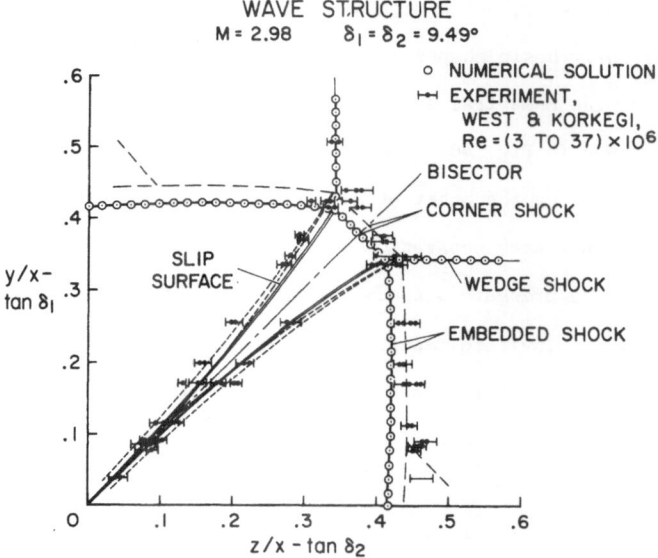

Fig. 39. Comparison of numerical and experimental shock patterns;
$M = 2.98$, $\delta_1 = \delta_2 = 9.49°$.

lution* for the corresponding inviscid case was obtained, and the shock wave and slip

surface structure are compared with the high Reynolds number experiment in Fig. 39.

The inviscid embedded shock is slightly concave when viewed from the origin and

falls inside the location of the corresponding experimental shock. The corner shock,

which is slightly convex when viewed from the origin, also falls inside the experi-

mental shock. The position of the experimental and numerical wedge shocks agrees ex-

actly. It appears, therefore, that the displacement effects of the boundary layer in

the region bounded by the corner and embedded shocks result in an effective thickening

of the body, and this forces the shock structure outward.

The slip surface locations for this case can be found from plots of density and

are shown as the thin double line that stretches from the triple point to the origin

in Fig. 39. The slip surface is slightly curved and asymptotically approaches the bi-

sector near the origin. The experimental shear layer is also curved, but it appears

to merge before the origin is reached. Since the positions of the numerical and ex-

perimental triple points are different, the comparison between the inviscid slip sur-

---

*For the numerical results presented, a 30×30 rectangular grid was used. Each
computation, consisting of 400 iterations, required approximately 20 min of CPU time
on an IBM 360/67.

face and viscous shear layer, which originate at the triple points, is unfair. Qualitatively, however their basic shapes are the same.

A comparison of the numerical and experimental (turbulent boundary layer) surface pressures is shown in Fig. 40. The first pressure rise in the experimental data
(decreasing z/x) indicates the onset of separation. This is followed by a reduced
gradient region that indicates separation and again a rapid pressure rise that indicates reattachment. The pressure between the reattachment point and the origin is
greater than that of the inviscid result. This higher pressure indicates an apparent
thickening of the body in this region due to boundary-layer displacement effects.

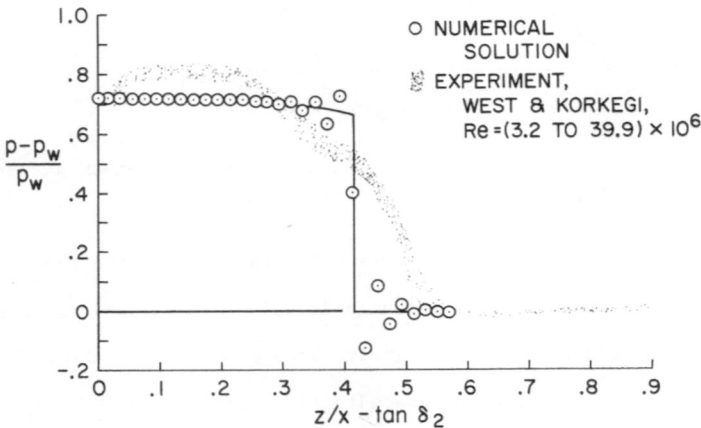

Fig. 40. Comparison of numerical and experimental surface-pressure
distributions; M = 2.98, $\delta_1 = \delta_2 = 9.49°$.

Figure 41 shows the distortion in the numerical and experimental wave structure
that results from an asymmetrical configuration with $\delta_1 = 3.5°$ and $\delta_2 = 12.2°$ in
M = 3.17 flow. The experiment was performed by Charwat and Redekeopp.[55] Both the
computed and experimental vertical embedded shocks are practically straight, while
both horizontal embedded shocks curve rapidly into the triple point. Again, the position of the experimental wave pattern is displaced outward when compared with the numerical result, but both exhibit the same asymmetrical behavior.

7.5 Two-Dimensional Blunt Body

The blunt-body problem has been solved by various investigators in various ways.
Our goal is not to devise a new approach but to determine the flow field that results

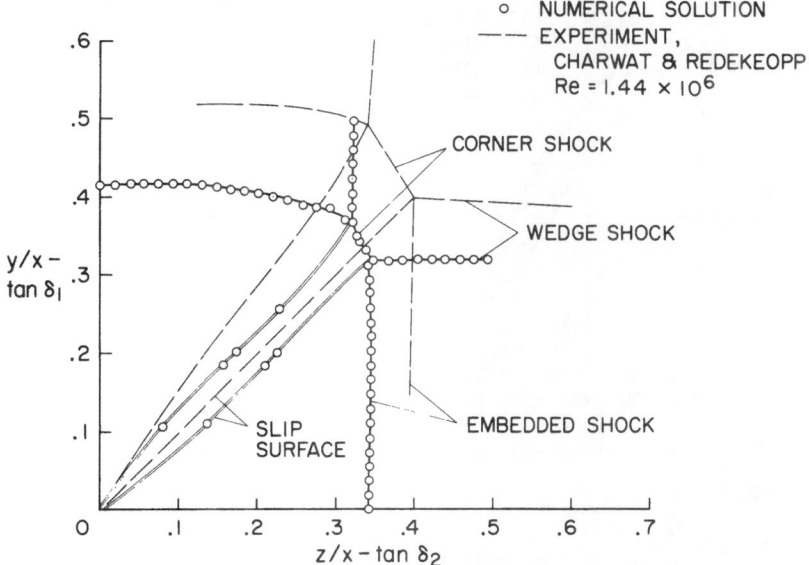

Fig. 41. Comparison of numerical and experimental shock patterns for
unequal wedge angles; M = 3.17, δ_1 = 3.5°, δ_2 = 12.2°.

when a planar shock wave intersects the bow shock generated by a blunt body. The

analysis and results presented here, however, are only for the two-dimensional, blunt-

body solution and do not include the effects of the impinging shock.

The basic orthogonal coordinate system chosen is polar (t, r, θ) (Fig. 11(d)).

The unsteady equations are then transformed to normalize the distance between the body

and shock, which is treated as a sharp discontinuity in this case. The resulting com-

putational plane is shown in Fig. 25(a). The unsteady shock relations discussed ear-

lier are applied at the shock boundary, and data at the supersonic outflow boundary

(θ = θ_{max}) are obtained using a second-order extrapolation technique. Plane-of-

symmetry conditions are applied along the line θ = 0, while the surface tangency con-

dition is satisfied at the body using the "normal momentum equation approach" dis-

cussed by MacCormack and Warming.[56]

The grid points are initialized according to a procedure discussed earlier, and

the time-dependent equations are integrated using MacCormack's scheme until a steady

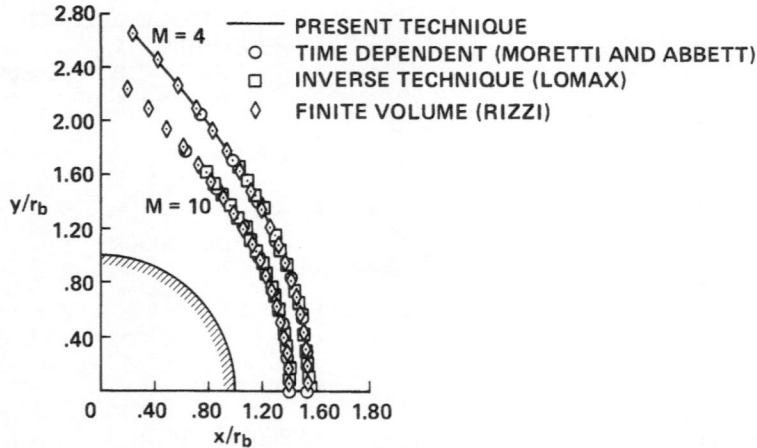

Fig. 42. Shock shape for two-dimensional blunt body; M = 4 and 10.

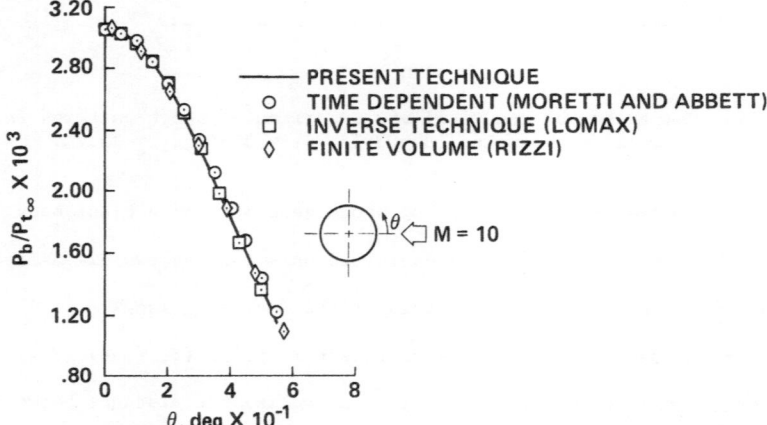

Fig. 43. Surface-pressure distribution for two-dimensional blunt body; M = 10.

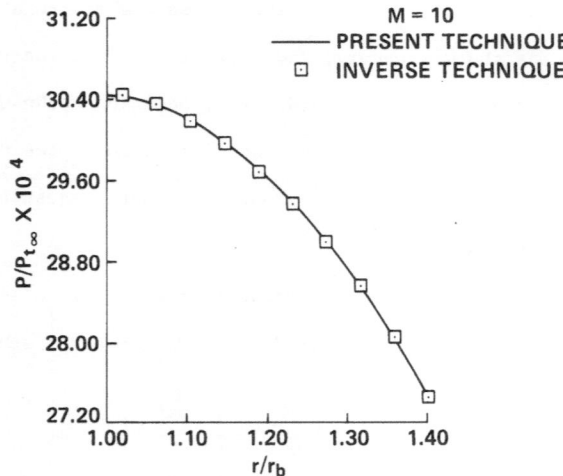

Fig. 44. Stagnation streamline pressure distribution for two-dimensional blunt body; M = 10.

state is reached. Results* of typical calculations are shown in Figs. 42-44. The shock shapes for Mach numbers 4 and 10 are shown in Fig. 42 and compared with the existing solutions of Morretti and Abbett,[12] Lomax and Inouye,[2] and Rizzi and Inouye.[10] The surface-pressure distribution for Mach 10 is shown in Fig. 43 along with the other numerical solutions. Figure 44 is a plot of the pressure distribution along the stagnation streamline, which is compared with the results from the inverse solution only. All comparisons are excellent and no numerical difficulties were encountered during the converging process.

7.6 Interfering Shock Problem

A rather interesting unsteady problem and a good test for the shock-capturing approach is the flow field that results when a moving planar shock wave interferes with the supersonic flow over a pointed cone (Fig. 1). The resulting shock wave and contact surface structure for a typical encounter is shown in Fig. 2. To develop the numerical procedure required to determine this three-dimensional flow field, the two-dimensional counterpart of this problem was first solved.[57] This analysis and the resulting solutions are now presented.

The basis for solving this unsteady problem is that it is self-similar—that is, as time increases, the wave structure expands linearly (assuming that at t = 0 the incident shock is at the wedge tip) and thus in constant x/t and y/t planes (Fig. 15), the flow field is invariant with time. This self-similarity is generated because there is no characteristic length associated with the body, that the incident shock wave is planar. Thus, if one applies a nonorthogonal coordinate transformation based on self-similarity to the unsteady gasdynamic equations, the unsteady problem is reduced to a steady problem.

From discussions in previous sections, it has been shown that shock waves, if they are aligned with one of the coordinate directions, will be captured with a minimum of post- and precursor oscillations. Because of the complicated shock structure in this problem, it is not possible to align all the shock waves. By an appropriate transformation of the independent variables, however, the incident planar shock can be

---

*This work was performed in conjunction with Mr. James Daywitt, Graduate Assistant at Iowa State University and Reese Sorenson of Ames Research Center.

aligned with one of the coordinate directions, and the original and new wedge shocks, since they are rays from the wedge vertex, can be aligned with the other coordinate direction. It was also mentioned previously that one should also strive to align the body with one of the coordinate directions. The independent variable transformation to satisfy these criteria, which is applied to the two-dimensional unsteady version of Eq. (2.1), is

$$\tau = t$$
$$\xi = [x - b(t, y)]/t \qquad (7.2)$$
$$\eta = y/a(x)$$

where

$$a(x) = x \tan \theta$$
$$b(t,y) = y/\tan \lambda + x_b t$$

where $x_b = x_{min}$ at $t = t_{initial}$.

The above transformation results in the shock-aligned computational plane shown in Fig. 16, where the left-hand boundary is chosen to be upstream of the sonic circle, the right-hand boundary is chosen to be downstream of the intersection point of the incident shock and original wedge shock, and the top boundary is chosen to be at an angle greater than the new wedge-shock angle. Along all these supersonic inflow boundaries, the flow conditions are known and held fixed during the entire computation. Along the lower boundary, which corresponds to the wedge surface, the tangency condition is satisfied by use of a one-sided, finite-difference procedure. MacCormack's scheme is used to advance the interior points in the $\tau = 1.0$ plane, which are initially set equal to free-stream values.

Merritt and Aronson[58] performed an experiment in which they obtained Schlieren photographs of the two-dimensional, shock-wave interactions generated by a 30° wedge at Mach 3 and a head-on ($\lambda = 60°$), $M_i = 2$, planar incident shock. This same case was calculated numerically and the results are shown in Figs. 45 and 46. Figure 45 is a density contour plot of the computational plane, which clearly shows all the shock waves and slip surfaces. The discontinuities in the flow that are not aligned with a coordinate direction are spread over approximately two mesh intervals, but their precise location can be pinpointed as denoted by the broad white dashed lines (shock waves) and the thin dotted lines (slip surface) superimposed on the contour plot. The

Fig. 45. Density contours for two-dimensional interfering shock
problem; M_w = 3.15, δ = 30°, M_i = 2, λ = 60°.

Fig. 46. Pressure contours for two-dimensional interfering shock
problem; M_w = 3.15, δ = 30°, M_i = 2, λ = 60°.

existence of a slip surface can be seen by comparing the pressure contour (Fig. 46)
with the density contour. Since pressure is continuous across the slip surface, the
constant-pressure lines do not build up where the slip surface exists.

The experimental wave structure for this case is shown in Fig. 47, and compares
well with the numerical results of Fig. 45. Note, however, that in the experiment the
original wedge shock, if extrapolated toward the wedge vertex, does not intersect it.
This indicates that there were some nonuniformities in the tunnel flow conditions.

7.7 Three-Dimensional Supersonic Flow

In this section, a numerical procedure (see Kutler et al.[3]) is discussed that

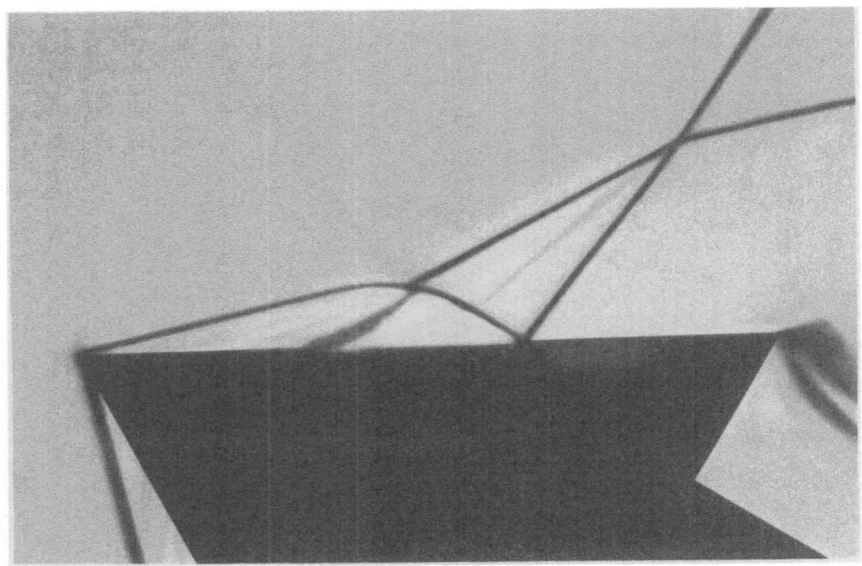

Fig. 47. Schlieren photograph of two-dimensional interfering shock
problem; $M_w = 3$, $\delta = 30°$, $M_i = 2$, $\lambda = 60°$.

is capable of determining the inviscid supersonic flow field for either pefrect or
real gases about three-dimensional wing-body configurations, in particular, vehicles
similar to the space shuttle orbiter. The complicated geometry of such a configuration traveling at supersonic velocities results in an intricate pattern of intersecting shocks, expansion waves, and slip surfaces—a natural problem for the shock-
capturing approach.

In this procedure, the outermost shock—that is, the shock that separates the
free stream from the disturbed region generated by the body—is treated as a sharp
discontinuity. For example, before the canopy or wing is reached, the outermost shock
is that generated by the nose of the vehicle. If the shock generated by the canopy
intersects the bow shock, the resulting coalesced shock then becomes the outermost
shock. When the shock generated off the wing leading edge intersects the bow shock,
it becomes the outermost shock and is then treated as a sharp discontinuity. Hence,
there are segments of the outermost shock that could consist of the original nose
shock, the canopy shock, and the wing leading-edge shock. The resulting flow field
beneath the body and outermost shock is treated in a shock-capturing fashion and
therefore allows for the correct formation of secondary internal shocks.

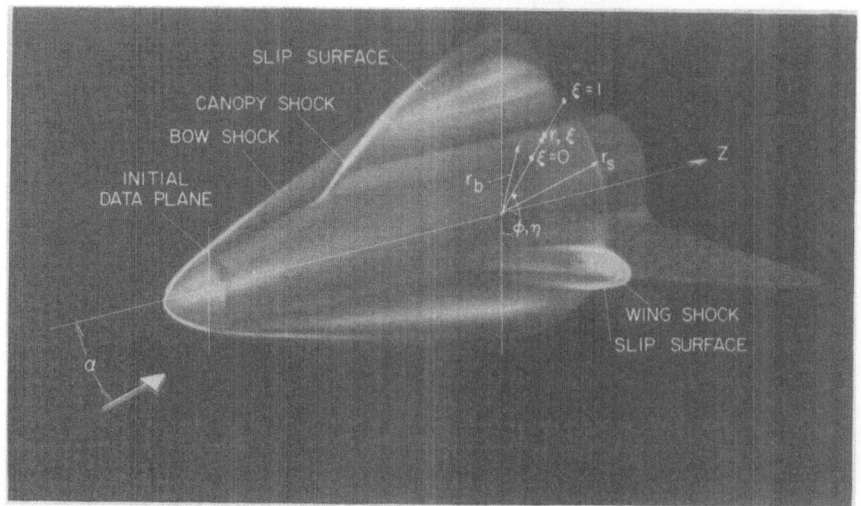

Fig. 48.   Coordinate system for three-dimensional configurations.

The basic coordinate system used for this problem is cylindrical; its orientation with respect to the body is shown in Fig. 48.   The vehicle body geometry and the location of the outer or peripheral shock surface are represented by functions of the form

$$r_b = r_b(z, \phi)$$
$$r_s = r_s(z, \phi)$$

(7.3)

The function r_b is known, and r_s is determined during the numerical computation. As is common practice in problems of this type, the distance between the body and peripheral shock is normalized (Fig. 49(a)) by a transformation of the radial variable r.   This yields a rectangular computational plane whose boundaries consist of the plane of symmetry and the body and shock surfaces as shown in Fig. 49(b).

In regions where the body curvature is large, the flow variables change rapidly in the meridional direction and to avoid degrading the rest of the solution, it is necessary to cluster points in such a region.   As was pointed out earlier, this is best accomplished by an independent variable transformation such as that given by Eq. (3.4), which allows for clustering only at the $\phi = 90°$ meridian.   Thomas et al.[21] used a transformation that clusters points about any meridian, given by

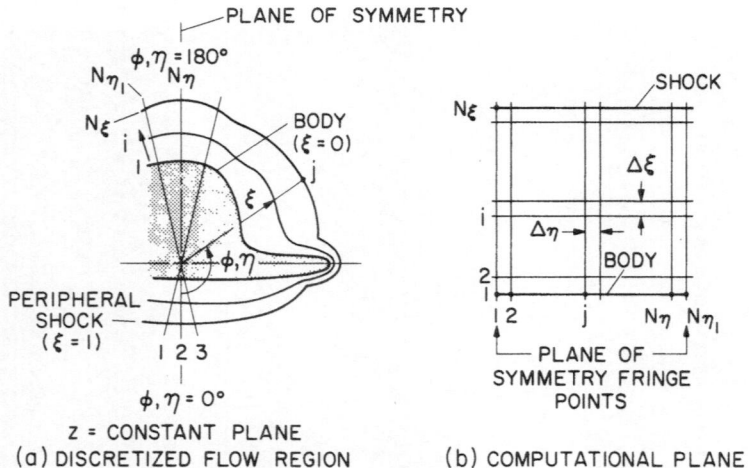

Fig. 49. Mesh description for three-dimensional problem.

where

$$\left. \begin{array}{l} \eta(\phi) = \pi \left\{ C + \dfrac{1}{\beta} \sinh^{-1}\left[\left(\dfrac{\phi}{\phi_o} - 1\right) \sinh(\beta C)\right]\right\} \qquad \beta > 0 \\[3mm] C = \dfrac{1}{2\beta} \ln\left[\dfrac{1 + (e^\beta - 1)\phi_o/\pi}{1 - (1 - e^{-\beta})\phi_o/\pi}\right] \\[3mm] \eta(\phi) = \phi \hspace{8.5cm} \beta = 0 \end{array} \right\} \qquad (7.4)$$

The angle ϕ_o is the point about which clustering occurs while β is the parameter that controls the degree of clustering. Either the above transformation or the one given by Eq. (3.4) is used in the results presented later.

Thus, the equations of the independent variable transformations are

$$\left. \begin{array}{l} \zeta = z \\[2mm] \xi(z,r,\phi) = [r - r_b(z,\phi)]/[r_s(z,\phi) - r_b(z,\phi)] \\[2mm] \eta = \eta(\phi) \end{array} \right\} \qquad (7.5)$$

This transformation is applied to the three-dimensional steady version of Eq. (2.1), and the resulting equation is then rearranged in conservation law form. The steady-flow energy equation for either a real or perfect gas is used.

The pressure approach discussed earlier is used to treat the peripheral shock wave as a sharp discontinuity while the simple-wave corrector scheme is used at the body to satisfy the surface tangency condition. Symmetry conditions are applied at the $\phi = 0°$ and $180°$ planes.

MacCormack's scheme is used to integrate the equations in the z direction from a given initial data plane downstream over the body. The computer code relies on an external program to provide the starting data if the body has a blunt nose, but it is capable of generating a solution for a pointed cone at angle of attack; results are presented later which describe the flow field surrounding circular cones at large angles of attack.

The function $r_b(z,\phi)$ that describes the body is a known analytic function. A typical body cross section is composed of various segments (Fig. 50) (such as straight lines, ellipses, etc.) governed by certain parameters that vary longitudinally as cubic polynomials.

SEGMENT	FUNCTION
① - ②	STRAIGHT LINE
② - ③	ELLIPSE
③ - ④	STRAIGHT LINE
④ - ⑤	CUBIC POLYNOMIAL
⑤ - ⑥	STRAIGHT LINE
⑥ - ⑦	ELLIPSE

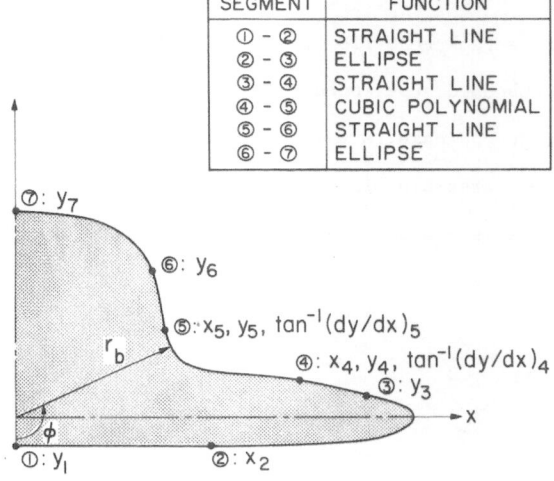

Fig. 50. Typical body cross section.

As mentioned earlier, to determine the integration step size, the eigenvalues of the matrices of Eq. (6.4) for this problem under the above transformation are

$$
\left.
\begin{aligned}
\sigma^P_{1,2} &= \frac{1}{\tilde{c}}\left[\tilde{a} + \frac{u(v+\tilde{b}w/r) \pm c[(v+\tilde{b}w/r)^2 + (u^2-c^2)(1+\tilde{b}^2/r^2)]^{1/2}}{u^2-c^2}\right] \\
\sigma^P_{3,4,5} &= \frac{1}{\tilde{c}}\,[\tilde{a} + (v+\tilde{b}w/r)/u]
\end{aligned}
\right\}
\qquad (7.6)
$$

where

$$
\tilde{a} = -r_{b_z} - \xi(r_{s_z}-r_{b_z})\ , \qquad \tilde{b} = -r_{b_\phi} - \xi(r_{s_\phi}-r_{b_\phi})
$$

$$
\tilde{c} = r_s - r_b
$$

and

$$\left.\begin{array}{l} \sigma^Q_{1,2} = \dfrac{1}{r}\left[\dfrac{uw \pm c(u^2+w^2-c^2)^{1/2}}{u^2-c^2}\right]\eta_\phi \\[12pt] \sigma^Q_{3,4,5} = (w/ur)\eta_\phi \end{array}\right\} \qquad (7.7)$$

The three-dimensional program can be used to determine the flow field about cones at large incidence (σ is the cone half-angle and α is the angle of attack) by a distance asymptotic approach. The conical flow problem, when solved in the crossflow plane, is elliptic for small angles of attack and mixed elliptic/hyperbolic for large angles of attack. The problem is made totally hyperbolic (with respect to the marching coordinate z) by treating it three dimensionally. The governing equations are thus integrated downstream over the cone until a conical flow field has been established.

Two large angle-of-attack cone flow solutions are presented. In the first case ($M = 3$, $\alpha/\sigma = 2$), the embedded supersonic crossflow is confined to a small region near the body. In the second case ($M = 7$, $\alpha/\sigma = 1.5$), this supersonic crossflow encompasses a large region stretching from the body to the shock.

The results of the first case are shown in Figs. 51 and 52. To obtain this so-

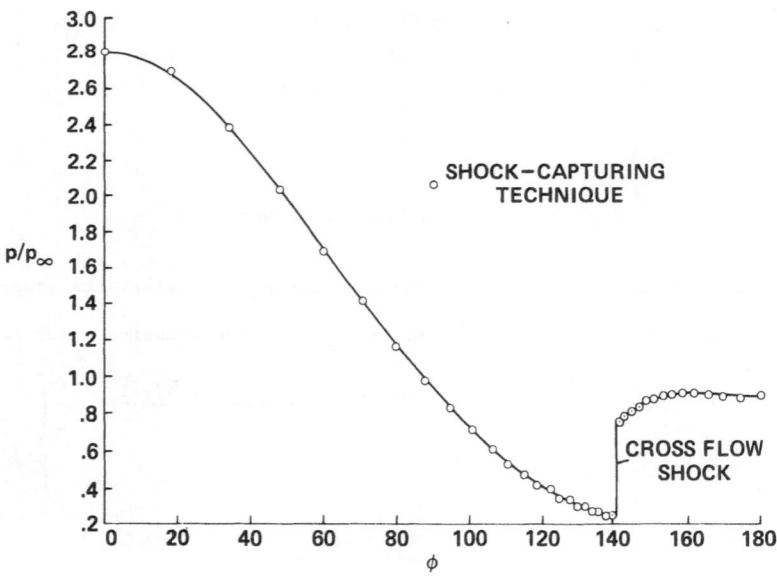

Fig. 51. Surface-pressure distribution for cone at angle of attack;
$M = 3$, $\alpha = 15°$, $\sigma = 7.5°$.

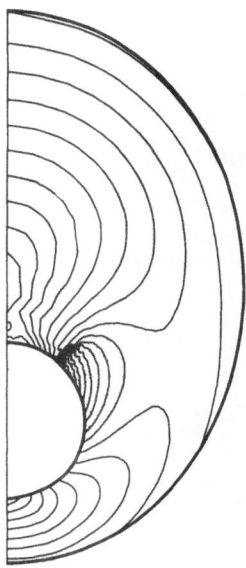

Fig. 52. Crossflow Mach number contour plot for cone at angle of attack;
M = 3, α = 15°, σ = 7.5°.

lution, a 22×37 point (r,ϕ) grid was used with clustering (β = 5 in Eq. (7.4)) at the

140° meridian. The surface pressure distribution for this case is shown in Fig. 51.

Near the ϕ = 140° plane, a crossflow shock exists that results from the deceleration

of the supersonic crossflow velocity. This shock was captured by the numerical

scheme. If the crossflow Mach number in front of this shock is known, the Rankine-

Hugoniot relations yield the conditions just downstream of it. The pressure from

such a calculation is plotted in Fig. 51, and it agrees well with the computed result.

Figure 52 is a crossflow Mach number contour plot of the conical flow field. The

shaded portion is the embedded supersonic crossflow region, and the coalescence of

contours is the crossflow shock. For this case, the vortical singularity in the lee-

ward plane of symmetry lifted from the body. The program encountered no numerical

difficulties as a result of the vortical singularity since it is a "weak solution"[5]

of the governing partial differential equations (the radial and circumferential con-

servative variables are continuous through it).

Under certain conditions, the embedded supersonic region can encompass a large

portion of the flow field in the crossflow plane. Results for such conditions are

shown in Figs. 53-55. The cone half-angle for this case is 20° and the free-stream

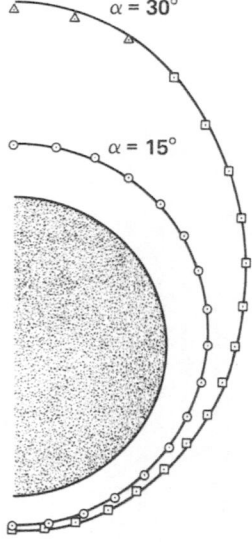

SHOCK-CAPTURING
TECHNIQUE
o BABENKO, et al.
□ COMPUTED
△ EXTRAPOLATED ⎰ BAZZHIN

Fig. 53. Cross-sectional shock shape for cone at angle of attack;
M = 7, α = 15° and 30°, σ = 20°.

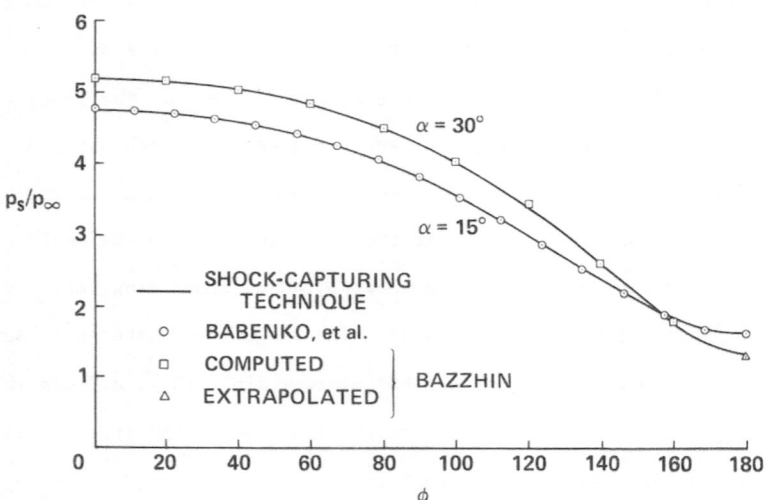

Fig. 54. Density distribution behind shock wave for cone at angle
of attack; M = 7, α = 15° and 30°, σ = 20°.

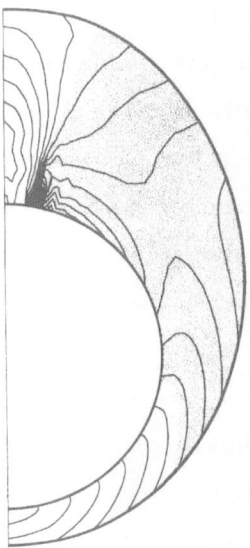

Fig. 55. Crossflow Mach number contour plot for cone at incidence;
          M = 7, α = 30°, σ = 20°.

Mach number is 7. Solutions were obtained for an angle of attack of 15° and are com-

pared with the results of Babenko et al.[59] and for 30°, which were compared with

the incomplete results of Bazzhin.[60] Figure 53 is a plot of the shock shapes for

both angles of attack, and both are in excellent agreement with the results of Baben-

ko and Bazzhin. The density distribution behind the shock wave for the two cases is

shown in Fig. 54, and it also compares well with the other numerical solutions. A

crossflow Mach number contour plot is shown in Fig. 55, and the large shaded portion

indicates the supersonic crossflow region. There is a crossflow shock near the 165°

meridian, and its strength dissipates rapidly with distance from the body.

Cone solutions such as these are easily generated and can be used to supply

starting data for determining the supersonic flow over three-dimensional bodies. When

blunt-nose starting solutions are required, the inverse blunt-body computer code of

Lomax and Inouye[2] is used.

To test the ability of the numerical procedure to describe the multi-shocked

flow field surrounding a shuttle-like configuration, a body was designed that used

segments 2-3 (bottom ellipse) and 6-7 (top ellipse) of Fig. 50. The longitudinal

variation of the required geometrical parameters was obtained from drawings of a delta-wing shuttle orbiter that is now obsolete. The resulting analytical configuration modeled the exact shape in both the planform and profile views but crudely approximated the body cross section in the wing region. The actual models tested in the wind tunnels also varied somewhat from the blueprint designs because of replication problems such as mold shrinkage. Such effects cause additional discrepancy between the analytical and experimental body shapes. The flow conditions for this test were Mach 7.4 and 0° angle of attack. These particular conditions are of interest computationally because available experimental shadowgraphs[61] show two shock-shock intersection regions: one when the canopy shock intersects the bow shock and the other when the wing leading-edge shock intersects the bow shock.

To eliminate any difficulties associated with the thinning entropy layer due to a blunt nose, a 23.07° pointed cone was used to simulate the nose of the vehicle for this test case. The grid size in the radial direction consisted of 18 points initially and was increased to 31 points just before the canopy was encountered. The number of meridional planes was held fixed at 19 for the entire calculation; however, the clustering parameter κ of Eq. (3.4) was varied discretely from 1.0 initially to 0.13 for the last few steps. Its variation was based on the surface-pressure distribution in the region of the wing leading edge as observed on the CRT. This calculation, which consisted of approximately 700 streamwise integration steps and covered more than 90 percent of the body, required about 36 min on an IBM 360/67.

The shock locations in both the planform and profile views are compared in Fig. 56 with the experimental shadowgraphs made by Cleary.[61] When the results obtained by the numerical solution coincide with those of the experiment, only the numerical solution is shown. The agreement is excellent even though the actual nose was approximated by a pointed cone. The canopy shock location disagrees somewhat from that of the experiment. This can be attributed to the fact that the simulated canopy was a little steeper than the actual canopy. When the canopy shock intersected the bow shock, the shock-fitting procedure automatically began to treat the canopy (outermost) shock as a discontinuity and no difficulty was encountered in making the transition. The slip surface or inviscid shear layer that results when two shocks of the same family intersect was observed as a rapid change in a small region of the radial den-

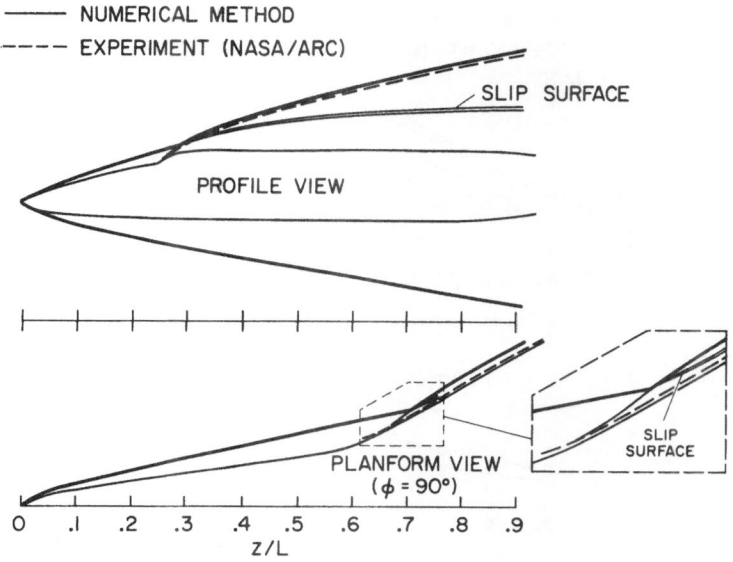

Fig. 56. Shock location for pointed-nose configuration;
M = 7.4, α = 0°.

sity distribution (Fig. 56). It agrees identically with that seen in the shadowgraph. In the planform view, the intersection of the nose and wing leading-edge shocks is shown along with coalesced shock and slip surface. Also shown is the experimental wing shock. There is disagreement between the locations of the numerical and experimental shocks since, in the numerical calculation, the actual wing is crudely approximated on the upper surface by a much thicker wing, which results in a larger standoff distance.

Some interesting results are obtained when a similar configuration is tested—this time with a blunt nose and at angle of attack. The flow conditions are a Mach number of 7.4 and angle of attack of 15.3°. The radial distribution of points was varied from 21 initially to 31 finally. The number of meridional planes was kept constant at 19, while the clustering parameter κ was changed when necessary.

The planform and profile shock shapes are shown in Fig. 57 and compared with the experimental shadowgraphs and with the method of characteristics.[62] The windward portion of the shock agrees quite well with experiment, and the slight variance at its latter stations can be attributed to the disagreement between the numerical and experimental body shapes on the forward portion of the lower surface. On the leeward size,

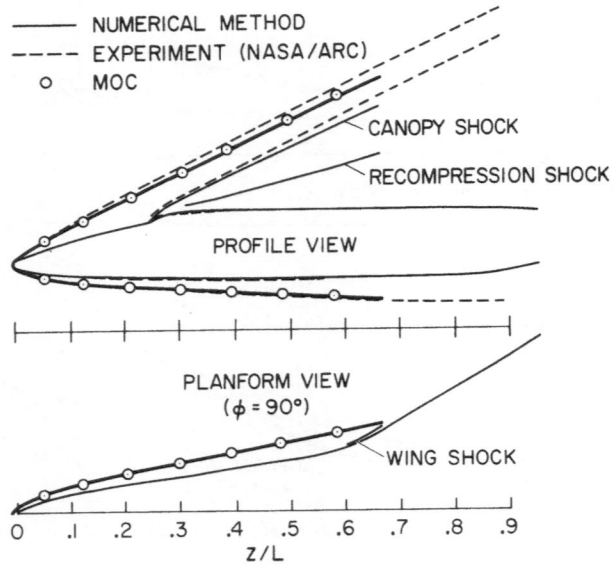

Fig. 57. Shock location for lung-nose configuration;
M = 7.4, α = 15.3°.

the canopy shock was captured, but its location disagreed somewhat from that of the
experiment, again due to the differences in bodies. Because of the overexpansion of
the flow around the canopy, a recompression shock is formed (Fig. 57) that is weaker
than the canopy shock (not observed in this shadowgraph). However, this shock has
been observed in other experiments.[63] The shock generated by the wing leading edge
is shown in the planform view (φ = 90°) of Fig. 57. Its location in this view agrees
well with the experiment since the bottom ellipse closely approximates the lower surface of the shuttle and also controls the shock standoff distance at the leading edge.

Figure 58 shows the pressure distribution in the radial direction for the leeward
plane of symmetry at two longitudinal stations, in which both canopy and recompression
shocks are recognizable. Both shocks are captured within two or three mesh intervals.

Radial pressure and density plots through the merged entropy layers as well as
the wing leading-edge shocks are shown in Fig. 59. The wing leading-edge shock in
this plane is spread over four mesh intervals. Although good definition of the entropy layers is lost, their presence posed no difficulties until their thickness became less than one mesh interval, at which time oscillations began to occur.

The longitudinal surface-pressure distribution for three meridians is shown in

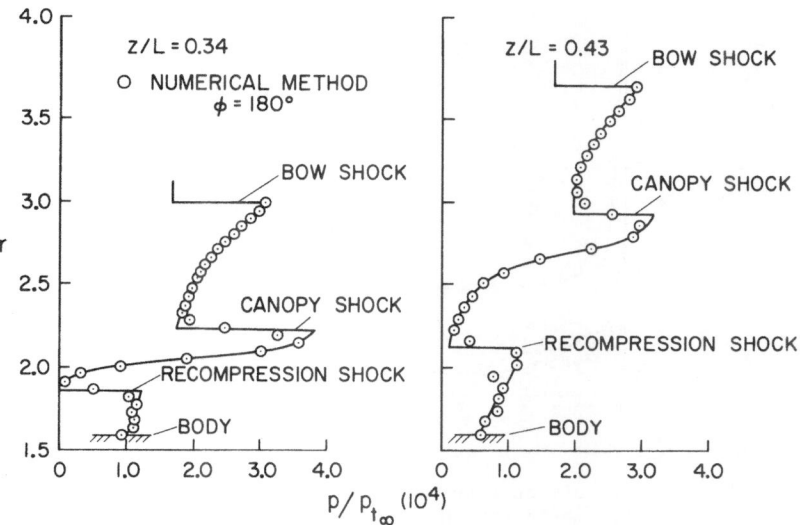

Fig. 58. Propagation of canopy and recompression shocks through shock layer in leeward plane of symmetry; M = 7.4, α = 15.3°.

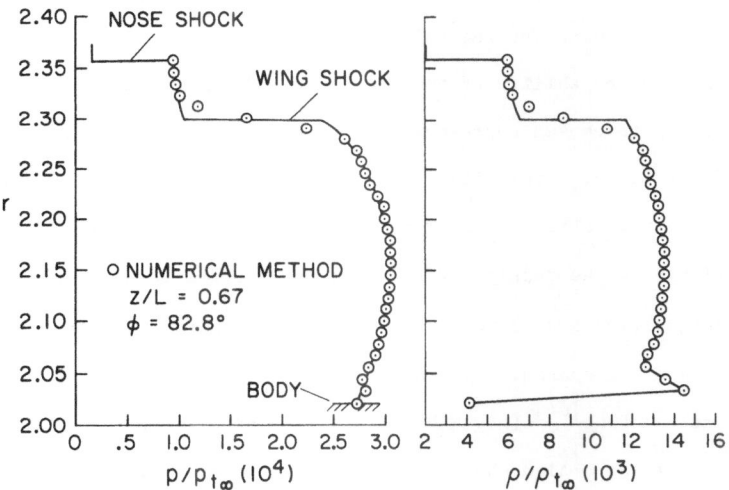

Fig. 59. Radial pressure and density distribution through entropy layer and wing leading-edge shock; M = 7.4, α = 15.3°.

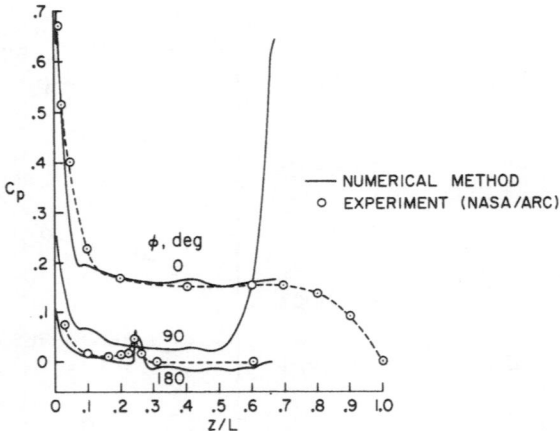

Fig. 60. Longitudinal surface-pressure distribution for the 0°, 90°, and 180° meridians; M = 7.4, α = 15.3°.

Fig. 60 and compared with experimental data obtained from the NASA/ARC 3-1/2-foot hypersonic wind tunnel. As the integration proceeds downstream, the pressure in the 85° plane continually rises, causing a corresponding decrease in the u component of velocity, to the extent that u approaches its local sonic value. This eventually causes the calculations to be terminated since the equations are then no longer hyperbolic. These near-sonic velocities occur only within the entropy layer, and the velocities are well above sonic outside this layer.

To demonstrate the capability of the method to calculate equilibrium air flow, starting conditions were chosen representative of a point on the orbiter's trajectory, i.e., an altitude of 76 km, a velocity of 7.3 km/sec (M = 26.26), and an angle of attack of 15.3°. Such free-stream conditions are difficult to simulate in existing wind-tunnel facilities. The results of calculations for the forward fuselage of a shuttle-like configuration are shown in Fig. 61 where the shock shapes for both perfect and real gases are compared. Also shown are results from a real-gas, method-of-characteristics calculation[64] (given by circles). The agreement between both numerical methods is excellent. Also shown in Fig. 61 is a small segment of the secondary shock originating from the front of the canopy. Here, no differences between the perfect- and real-gas solutions are observed.

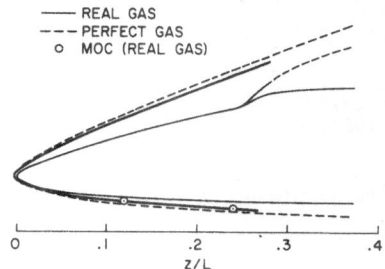

SHOCK SHAPE FOR FUSELAGE OF COMPUTED BODY
M = 26.26   α = 15.3°   ALTITUDE = 76 km

Fig. 61. Comparison of shock location for blunt-nose configurations;
M = 26.26, α = 15.3°, altitude = 76 km.

REFERENCES

1. Rakich, J. V., "Calculation of Hypersonic Flow over Bodies of Revolution at Small Angles of Attack," AIAA J., Vol. 3, pp. 458-464, March 1965.

2. Lomax, H., and Inouye, M., "Numerical Analysis of Flow Properties About Blunt Bodies Moving at Supersonic Speeds in an Equilibrium Gas," NASA TR R-204, 1964.

3. Kutler, P., Reinhardt, W. A., and Warming, R. F., "Multishocked, Three-Dimensional Supersonic Flowfields with Real Gas Effects," AIAA J., Vol. 11, No. 5, pp. 657-664, May 1973.

4. Dekker, T. J., "Constructive Aspects of the Fundamental Theorem of Algebra," B. Dejon and P. Henrici, eds., John Wiley & Sons, Inc., p. 37, 1969.

5. Lax, P. D., "Weak Solutions of Nonlinear Hyperbolic Equations and Their Numerical Computation," Commun. Pure Appl. Math., Vol. 7, pp. 159-193, 1954.

6. Gary, J., "On Certain Finite Difference Schemes for Hyperbolic Systems," Math. Computation, pp. 1-18, 1964.

7. Abbett, M. J., "Boundary Condition Computational Procedures for Inviscid Supersonic Steady Flow Field Calculations," Aerotherm Corp., Mt. View, Calif., Final Rept. 71-41, 1971.

8. Longley, H. J., "Methods of Differencing in Eulerian Hydrodynamics," Los Alamos Scientific Lab., Los Alamos, New Mexico, LASL Rept. LAMS-2379, 1960.

9. Godunov, S. K., "Finite Difference Method for Numerical Computation of Discontinuous Solutions of the Equations of Fluid Dynamics," Math. Skornik, Vol. 47(89), No. 3, p. 271, 1959.

10. Rizzi, A. W., and Inouye, M., "A Time Split Finite Volume Technique for Three-Dimensional Blunt Body Flow," AIAA J., Vol. 11, Nov. 1972.

11. Rizzi, A. W., Klavins, A., and MacCormack, R. W., "A Generalized Numerical Method for Three-Dimensional Supersonic Flow," to be published.

12. Moretti, G., and Abbett, M., "A Time-Dependent Computational Method for Blunt Body Flows," AIAA J., Vol. 4, pp. 2136-2141, 1966.

13. Bohachevsky, I. O., and Mates, R. E., "A Direct Method for Calculation of the Flow about an Axisymmetric Blunt Body at Angle of Attack," AIAA J., Vol. 4, pp. 776-782, 1966.

14. Burstein, S. Z., "Numerical Methods in Multidimensional Shocked Flows," AIAA J., Vol. 2, pp. 2111-2117, 1964.

15. Morretti, G., Grossman, B., and Marconi, F., Jr., "A Complete Numerical Technique for the Calculation of Three-Dimensional Inviscid Supersonic Flows," AIAA Paper 72-192, 1972.

16. Marconi, F., and Salas, M., "Computation of Three-Dimensional Flows about Aircraft Configurations," Computers & Fluids, Vol. 1, No. 2, June 1973.

17. Kutler, P., Lomax, H., and Warming, R. F., "Computation of Space Shuttle Flow Fields Using Noncentered Finite-Difference Schemes," AIAA J., Vol. 11, No. 2, pp. 196-204, Feb. 1973.

18. Blottner, F. G., and Roache, P. J., "Nonuniform Mesh Systems," J. Computational Phys., Vol. 8, pp. 498-499, 1971.

19. Kalnay de Rivas, E., "On the Use of Nonuniform Grids in Finite-Difference Equations," J. Computational Phys., Vol. 10, pp. 202-210, 1972.

20. Chu, C. W., and Powers, S. A., "Determination of Space Shuttle Flow Field by the Three-Dimensional Method of Characteristics," NASA TM X-2508, 1972.

21. Thomas, P. D., Vinokur, M., Bastianon, R., and Conti, R. J., "Numerical Solution for the Three-Dimensional Hypersonic Flow Field of a Blunt Delta Body," AIAA J., Vol. 10, No. 7, July 1972.

22. MacCormack, R. W., "The Effect of Viscosity in Hypervelocity Impact Cratering," AIAA Paper 69-354, pp. 1-7, 1969.

23. Warming, R. F., Kutler, P., and Lomax, H., "Second- and Third-Order Noncentered Difference Schemes for Nonlinear Hyperbolic Equations," AIAA J., Vol. 11, No. 2, pp. 189-196, Feb. 1973.

24. Rusanov, V. V., "On Difference Schemes of Third Order Accuracy for Nonlinear Hyperbolic Systems," J. Computational Phys., Vol. 5, pp. 507-516, 1970.

25. Burstein, S. Z., and Mirin, A. A., "Third Order Difference Methods for Hyperbolic Equations," J. Computational Phys., Vol. 5, pp. 547-571, 1970.

26. Richtmyer, R. D., and Morton, K. W., Difference Methods for Initial-Value Problems, John Wiley & Sons, Inc., New York, 1967.

27. Warming, R. F., and Hyett, B. J., "The Modified Equation Approach to the Stability and Accuracy Analysis of Finite-Difference Methods," J. of Comp. Phys., Vol. 14, No. 2, Feb. 1974.

28. Hopf, E., "The Partial Differential Equation $u_t + uu_x = \mu u_{xx}$," Commun. Pure Appl. Math., Vol. 3, pp. 201-230, 1950.

29. Lax, P. D., and Wendroff, B., "Difference Schemes for Hyperbolic Equations with High Order Accuracy," Commun. Pure Appl. Math., Vol. 17, pp. 381-398, 1964.

30. Hirt, C. W., "Heuristic Stability Theory for Finite-Difference Equations," J. Computational Phys., Vol. 2, No. 4, pp. 339-355, June 1968.

31. Kutler, P., "Application of Selected Finite Difference Techniques to the Solution of Conical Flow Problems," Ph.D. Thesis, Dept. of Aerospace Engineering, Iowa State Univ., 1969.

32. Abbett, M. J., "Boundary Condition Computational Procedures for Inviscid Supersonic Steady Flow Field Calculations," Aerotherm Corp., Mt. View, Calif., Final Rept. 71-41, 1971.

33. Abbett, M. J., "Boundary Condition Computational Procedures for Inviscid Supersonic Flow Fields," Proceedings, AIAA Computational Fluid Dynamics Conference, 1973.

34. Moretti, G., "Importance of Boundary Conditions in the Numerical Treatment of Hyperbolic Equations," High-Speed Computing in Fluid Dynamics, The Physics of Fluids Supplement II, 1969.

35. Roache, P. J., Computational Fluid Dynamics, Hermosa Publishers, Albuquerque, New Mexico, 1972.

36. Bohachevsky, I. O., and Rubin, E. L., "A Direct Method for Computation of Non-equilibrium Flows with Detached Shock Waves," AIAA J., Vol. 4, pp. 600-607, 1966.

37. Kutler, P., and Lomax, H., "The Computation of Supersonic Flow Fields about Wing-Body Combinations by 'Shock-Capturing' Finite Difference Techniques," Second International Conference on Numerical Methods in Fluid Dynamics, Sept. 1970; also Lecture Notes in Physics, Vol. 8, pp. 24-29, 1971.

38. Hayes, W. D., and Probstein, R. F., Hypersonic Flow Theory, 2nd ed., Academic Press, New York, p. 485, 1966.

39. Abbett, M. J., "Sharp Shock Computational Procedures for Inviscid, Supersonic, Steady Flow Field Calculations," Aerotherm Corp., Mt. View, Calif., Final Rept. 72-50, 1972.

40. Thomas, P. D., "On the Computation of Boundary Conditions in Finite Difference Solutions for Multi-Dimensional Inviscid Flow Fields," Lockheed Palo Alto Research Lab., Palo Alto, Calif., LMSC 6-82-71-3, March 1971.

41. Vincenti, W. G., and Kruger, C. H., Jr., Introduction to Physical Gas Dynamics, John Wiley & Sons, Inc., 1965.

42. Kutler, P., and Lomax, H., "Shock-Capturing, Finite-Difference Approach to Supersonic Flows," J. Spacecraft Rockets, Vol. 8, No. 12, pp. 1175-1182, Dec. 1971.

43. Fowell, L. R., "Exact and Approximate Solutions for the Supersonic Delta Wing," J. Aeronaut. Sci., Vol. 23, 1956.

44. South, J. C., and Klunker, E. B., "Methods for Calculating Nonlinear Conical Flows," NASA SP-228, pp. 131-158, 1969.

45. Babaev, D. A., "Numerical Solution of the Problem of Supersonic Flow Past the Lower Surface of a Delta Wing," AIAA J., Vol. 1, No. 9, pp. 2224-2231, Sept. 1963.

46. Beeman, E. R., and Powers, S. A., "A Method for Determining the Complete Flow Field Around Conical Wings at Supersonic/Hypersonic Speeds," AIAA Paper 69-646, 1969.

47. Voskresenskii, G. P., "Numerical Solution of the Problem of a Supersonic Gas Flow Past an Arbitrary Surface of a Delta Wing in the Compression Region," Izv. Akad. Nauk SSSR, Mekh, Zhidk. Gaza, No. 4, pp. 134-142, 1968.

48. Babaev, D. A., "Numerical Solution of the Problem of Flow Round the Upper Surface of a Triangular Wing by a Supersonic Stream," Zh. vychislitel'noi matematiki i matelmaticheskoi fizika, Vol. 2, pp. 278-289, 1962.

49. Oswatitsch, K., and Sun, Y. C., "The Wave Formation and Sonic Boom Due to a Delta Wing," Aeronaut. Quart., Vol. XXIII, May 1972.

50. Roe, P. L., "Wind Tunnel Measurements of Sonic Boom Due to a Plane Lifting Delta Wing," Preliminary results presented to the Euromech 31 Meeting, Aachen, West Germany, May 1972.

51. Bannink, W. J., and Nebbeling, C., "Investigation of the Expansion Side of a Delta Wing at Supersonic Speed," AIAA J., Vol. 11, No. 8, Aug. 1973.

52. Kutler, P., "Numerical Solution for the Inviscid Supersonic Flow in the Corner Formed by Two Intersecting Wedges," AIAA Paper 73-675, 1973.

53. Kutler, P., "Supersonic Flow in the Corner Formed by Two Intersecting Wedges," AIAA J., Vol. 12, No. 5, pp 577-578, May 1974.

54. West, Major J. E., and Korkegi, R. H., "Interaction in the Corner of Intersecting Wedges at a Mach Number of 3 and High Reynolds Numbers," ARL 71-0241, 1971.

55. Charwat, A. F., and Redekeopp, L. G., "Supersonic Interference Flow Along the Corner of Intersecting Wedges," AIAA J., Vol. 5, No. 3, pp. 480-488, March 1967.

56. MacCormack, R. W., and Warming, R. F., "Survey of Computational Methods for Three-Dimensional Supersonic Inviscid Flow with Shocks," AGARD Lecture Series No. 64 on Advances in Numerical Fluid Dynamics, AGARD-LS-64, 1973.

57. Kutler, P., Sakell, L., and Aiello, G., "On the Shock-on Shock Interaction Problem," AIAA Paper No. 74-524, 1974.

58. Merritt, D. L., and Aronson, P. M., "Wind Tunnel Simulation of Head-On Bow Wave-Blast Wave Interactions," NOLTR 67-123, Aug. 1967.

59. Babenko, K. I., Voskrensenskiy, G. P., Lyubimov, A. N., and Rusanov, V. V., "Three-Dimensional Flow of Ideal Gas Past Smooth Bodies," NASA TT F-380, 1966.

60. Bazzhin, A. P., "Some Results of Calculations of Flows Around Conical Bodies at Large Incidence Angles," Lecture Notes in Physics, Vol. 8, 1971.

61. Cleary, J. W., "Hypersonic Shock-Wave Phenomena of a Delta-Wing Space-Shuttle Orbiter," NASA TM X-62,076, 1971.

62. Rakich, J. V., and Kutler, P., "Comparison of Characteristics and Shock Capturing Methods with Application to the Space Shuttle Vehicle," AIAA Paper 72-191, 1972.

63. Cleary, J. W., "Subsonic, Transonic, and Supersonic Stability and Control Characteristics of a Delta Wing Orbiter," NASA TM X-62,066, 1970.

64. Kutler, P., Rakich, J. V., and Mateer, G. G., "Application of Shock Capturing and Characteristics Methods to Shuttle Flow Fields," NASA TM X-2506, Vol. 1, p. 65, 1972.

NUMERICAL AND PHYSICAL EXPERIMENTS
IN VISCOUS SEPARATED FLOWS

Thomas J. Mueller
Department of Aerospace and Mechanical Engineering
University of Notre Dame, Notre Dame, Indiana 46556

1. INTRODUCTION

1.1 General Remarks

Efficient design and/or accurate prediction of performance of fluid dynamic systems requires that the entire flow field be calculable. This requires a simultaneous calculation of the inviscid flow field, the viscous boundary layer along solid surfaces as well as viscous separated or wake flow regions. Numerical calculation of inviscid and attached boundary layer flows using large high speed computers has recently been quite successful because of the simplified mathematical form of the governing equations. Since viscous separated flow regions do not fall within the limitations of the simplified boundary layer equations, one must look to the most complete mathematical model in Newtonian fluid dynamics - the full Navier-Stokes equations. No general solutions of the Navier-Stokes equations are known and it is very unlikely that such a solution will ever be obtained. As a result of the non-linear nature of these equations, in general, and the complex behaviour of separated flows in particular, only approximate methods would appear to be feasible. One approximate method, finite difference numerical analysis, is being used more and more frequently to solve the Navier-Stokes equations for separated flow problems. An excellent description of the formidable difficulties arising in the numerical solution of these equations is given in reference 1. With the advent of faster and larger electronic computers, as well as more efficient numerical techniques, these approximate solutions have and will become more practical and more nearly exact.

It now appears that recent advances in computer technology have steadily reduced the cost of numerical flow field simulation (Ref. 2). In fact, on the basis of cost, numerical experiments are now competitive with physical experiments for many complex fluid dynamics problems. Figure 1 obtained from Ref. 2 illustrates this trend of relative computational cost. It should be remembered that unless the physical experiment is performed full scale and under actual operating conditions, it is often as much of a flow field simulator as is the computer. There is of course no substitute for the real thing.

A recent example of the practical value of finite difference solutions of the Navier-Stokes equations is provided by the complex problem of heat pipe design. For steady operation within the working range of many heat pipes, the Reynolds numbers are quite low so that the flow may be considered laminar. Furthermore, the vapor velocity in the evaporator is low enough so that the density is approximately constant (Ref. 3). Although operable heat pipes and experimental data have been available for several years, the recent work of Bankston and Smith (Ref. 3) presents the quantitative means for laminar heat pipe design. In the introduction to their work, these authors state that "Because of the limitations of previous studies as to Reynolds number or geometry, the rational design of real heat pipes has awaited a solution of the full Navier-Stokes equations for a pipe of finite length and with both evaporation and condensation." Of course, only time will determine the extent to which this statement is true.

Before committees, of either national or international scope, are formed to do away with wind tunnels [1] - existing or planned - an attempt should be made to present a rational perspective. The computer should be viewed as a partner to, not a replacement of, the wind tunnel. In a particular engineering problem one or the other of these simulators may be more advantageous. More advantageous from the point of view of obtaining the desired result either cheaper or faster in real time. Or in some cases one approach may be more flexible and lead to a better understanding of the more general physical phenomena involved. For some extremely complex problems, both physical and numerical experiments may be necessary in order to gain sufficient understanding of important small scale details of the flow.

The basic fluid dynamic problems which have been solved up to the present must be considered to be only a foundation for the future. As always, the foundation must be free from flaws if the future structure is to withstand the test of time or be worth building upon at all. In the spirit of building a foundation, therefore, the great majority of finite difference solutions to the Navier-Stokes equations have been for incompressible, two-dimensional flow with relatively simple geometric boundaries. For an engineer, the decision of whether or not the numerical foundation is adequate must lie in its favorable comparison with the appropriate experimental data. Once this is established for these basic flows, three-dimensional steady and unsteady flows may be approached with confidence. Thus a need exists for careful correlations between finite difference numerical and experimental investigations.

The happy marriage of wind tunnel and computer will undoubtedly result in a greater understanding of the physical processes occurring as well as provide fast, accurate, and therefore practical solutions to the important engineering problems of the present and future. It is to this happy and lasting marriage that the present lecture is directed. May they flourish in each others company and be proud of their offspring.

1.2 Occurrence of Separated Flows

Separated flows have long been known to be detrimental to the performance of hydrodynamic and aerodynamic systems. This type of flow is characterized by relatively slow reverse flows which are distinct from the main flow. They result in an increase in the drag force and usually present complex heat and mass transfer problems. Internal flow separation is common in propulsive nozzles, inlet and wind tunnel diffusers, turbomachinery passageways, and heating and ventilating distribution duct networks to name just a few. External separation often occurs from airfoils or hydrofoils at angle of attack, from surface protuberances such as vortex generators, from struts and rapid changes in external geometry as well as at the termination of the body in the form of a wake. Most of these examples occur at moderate to very high Reynolds numbers (usually turbulent) whether the flow is compressible or incompressible.

---

[1] The term wind tunnel is intended to represent all physical facilities and techniques in this context.

Recently an interest has also developed in moderate to low Reynolds number separated flows occurring in a broad class of environmental and physiological applications. The environmental problems frequently concern the wind or sea loading of stationary structures or the entrainment of pollutants on the lee side of these structures. For problems of this type encountered in water, the flow may be considered to be laminar in many cases. Although the Reynolds number may be only moderate, flows of this type in air are almost always turbulent. Physiological flow separation usually results from aberrations of the circulatory system caused by atherosclerosis (especially near bifurcations), atheroma, heart valve stenosis and aneurysms (particularly common in the aorta). Unnatural separated flows are also introduced into the human circulatory system with such prosthetic cardiovascular devices as occluder heart valves or assist pumps. In these flows where blood is the working fluid serious long term mass transfer problems arise (i.e., thrombi formation) in addition to the immediate performance loss due to increased resistance to the flow. It is indeed difficult to find viscous flows of practical importance where flow separation does not occur.

1.3 Scope of the Present Work

The purpose of this lecture is to compare the results of finite difference solutions to the Navier-Stokes equations with the results of appropriately designed physical experiments. The separated flows reviewed include those incompressible planar flows produced by the blunt base and front step (see Fig. 2a). Other similar separated flows which have received considerable attention in the literature, such as the flow over rectangular cavities and the circular cylinder, will not be presented due to limitations of time and space. A comparison of various finite difference techniques and a comparison of numerical results with experiments for the rectangular cavity may be found in the recent paper by Bozeman and Dalton (Ref. 4). Although there are still more published finite difference works of this type (e.g., the incompressible flow over a backstep, ramp, protuberance, rotating plate as well as compressible backstep, ramp, frontstep, and cavity flows) little or no physical data are reported for these cases. At low to moderate values of Reynolds number, with steady upstream flow, these planar separated flows though laminar may be either steady or unsteady.

The slightly more complicated axisymmetric flow through a fully open disc type prosthetic heart valve shown in Fig. 2b will also be presented. The only other axisymmetric separated flows that have received considerable attention in the literature are the flow past a paraboloid of revolution and a sphere. No physical data appear to be available for the paraboloid of revolution, however, a comparison of physical and numerical experiments for the axisymmetric flow about a sphere may be found in the recent paper of Liu and Lee (Ref. 5). It should be mentioned that for a steady upstream flow, the confining nature of the numerical axisymmetric assumption allows only steady separated flows to be produced.

The finite difference investigations to be presented have used standard, explicit, time-dependent computational methods. The time-dependent approach seems particularly suited to separated flows which often exhibit a time-dependent behaviour, e.g., oscillating wake problems. Although implicit techniques are currently very popular for steady flow solutions, the new parallel process computers (e.g. Illiac IV) appear to favor explicit methods. The important differences in the numerical investigations include the treatment of some boundary conditions, mesh geometry and estimated accuracy of results. These methods have been found, mostly by numerical experimentation, to be consistent, stable and thus convergent to within the desired tolerance over a wide range of Reynolds numbers. Although comments will be made when possible, detailed information concerning these questions of consistency, stability, convergence, and behavioral errors of numerical schemes may be obtained from the recent text of Roache (Ref. 6) and the current literature.

The physical experiments at low Reynolds numbers are actually more difficult than their numerical counterparts. Producing a truly planar or axisymmetric separated flow in the laboratory is always difficult. In addition, one must face the equally difficult problem of measuring velocity and/or pressure in such low Reynolds number flows. Thus it will become evident that the question of accuracy of the physical experiment is often as difficult to answer and/or improve as that of numerical accuracy.

2. THE NAVIER-STOKES EQUATIONS

That a solution to a particular incompressible problem exists is usually either assumed from physical intuition or known from observation. The difficult and yet incomplete mathematical question of the uniqueness of such a solution for the prescribed initial and boundary conditions has apparently been resolved only for two-dimensional incompressible flows (Ref. 7).

The numerical treatment of the Navier-Stokes equations involves deriving the desired form of the equation, choosing a suitable mesh system, developing the finite difference form of the equations and solution procedure, determining the stability of the resultant scheme, and examining the question of accuracy of the solution in view of computational costs as well as the use to be made of the results. The following section presents a brief review of this procedure for the incompressible separated flow problems to be studied later.

2.1 Incompressible Planar Flow

For two-dimensional incompressible flow, the Navier-Stokes equations and continuity equation in a rectangular Eulerian coordinate system may be expressed as

$$\frac{\partial v_x}{\partial t} + v_x \frac{\partial v_x}{\partial x} + v_y \frac{\partial v_x}{\partial y} = -\frac{1}{\rho} \frac{\partial P}{\partial x} + \alpha \left[\frac{\partial^2 v_x}{\partial x^2} + \frac{\partial^2 v_x}{\partial y^2} \right] \tag{1}$$

$$\frac{\partial v_y}{\partial t} + v_x \frac{\partial v_y}{\partial x} + v_y \frac{\partial v_y}{\partial y} = -\frac{1}{\rho} \frac{\partial P}{\partial y} + \alpha \left[\frac{\partial^2 v_y}{\partial x^2} + \frac{\partial^2 v_y}{\partial y^2} \right] \tag{2}$$

$$\frac{\partial v_x}{\partial x} + \frac{\partial v_y}{\partial y} = 0 \tag{3}$$

The vorticity in such a two-dimensional flow can be written

$$\xi = \left(\frac{\partial v_y}{\partial x} - \frac{\partial v_x}{\partial y} \right) \tag{4}$$

Eliminating the pressure terms in equations (1) and (2) by cross-differentiation, applying equation (3) to this result and then using the vorticity relation of equation

(4), the following equation with the vorticity as dependent variable is obtained,

$$\frac{\partial \xi}{\partial t} + v_x \frac{\partial \xi}{\partial x} + v_y \frac{\partial \xi}{\partial y} = \alpha \left[\frac{\partial^2 \xi}{\partial x^2} + \frac{\partial^2 \xi}{\partial y^2} \right] \tag{5}$$

local advection viscous diffusion
term   terms   terms

This form of the Navier-Stokes equations is referred to as the Vorticity Transport equation.

By adding the definition of vorticity (4) to both sides of the continuity equation (3) and introducing the incompressible stream function defined by

$$\frac{\partial \psi}{\partial y} = v_x, \qquad \frac{\partial \psi}{\partial x} = - v_y \tag{6}$$

the following Poisson equation results

$$\frac{\partial^2 \psi}{\partial x^2} + \frac{\partial^2 \psi}{\partial y^2} = - \xi \tag{7}$$

Thus, the Navier-Stokes equations and the continuity equation are reduced to a parabolic Vorticity Transport equation (5) and an elliptic Poisson equation (7) for the stream function. If the slightly modified version of continuity

$$\xi \left(\frac{\partial v_x}{\partial x} + \frac{\partial v_y}{\partial y} \right) = \xi \frac{\partial v_x}{\partial x} + \xi \frac{\partial v_y}{\partial y} = 0 \tag{8}$$

is added to both sides of equation (5), then an alternate and often used form of equation (5) is obtained namely,

$$\frac{\partial \xi}{\partial t} + \frac{\partial (v_x \xi)}{\partial x} + \frac{\partial (v_y \xi)}{\partial y} = \alpha \left[\frac{\partial^2 \xi}{\partial x^2} + \frac{\partial^2 \xi}{\partial y^2} \right] \tag{9}$$

Equation (9) is referred to as the "conservation form" of the Vorticity Transport equation since it may be shown that the transport property ξ is conserved (Ref. 6). Although conservation does not necessarily imply accuracy (Ref. 6), experience indicates that conservation systems generally produce more accurate results (Refs. 8,9).

By rendering the terms of equations (7) and (9) dimensionless, it is possible to extend their range of applicability by using Reynolds principle of similarity. Therefore, if L and v_∞ are the characteristic length and velocity respectively, then

$$x = x^* L \qquad\qquad y = y^* L$$

$$v_x = v_x^* v_\infty \qquad\qquad v_y = v_y^* v_\infty$$

$$\psi = \psi^* v_\infty L \tag{10}$$

$$t = \frac{t^* L}{V_\infty} \qquad\qquad \xi = \frac{\xi^* v_\infty}{L}$$

and the dimensionless conservation Vorticity Transport equation can be written as

$$\frac{\partial \xi^*}{\partial t^*} + \frac{\partial (v_x^* \xi^*)}{\partial x^*} + \frac{\partial (v_y^* \xi^*)}{\partial y^*} = \frac{1}{Re_L} \left[\frac{\partial^2 \xi^*}{\partial x^{*2}} + \frac{\partial^2 \xi^*}{\partial y^{*2}} \right] \tag{11}$$

where the dimensionless vorticity and stream functions are related by the following Poisson equation

$$-\xi^* = \frac{\partial^2 \psi^*}{\partial x^{*2}} + \frac{\partial^2 \psi^*}{\partial y^{*2}} \tag{12}$$

Although it is possible to obtain numerical solutions of the Navier-Stokes equations in the primitive form (i.e., equations 1, 2 and 3), most successful numerical solutions for incompressible flows have used the vorticity-stream function form of equations 11 and 12. The parabolic equation (11) is solved by explicit marching in time from some initial conditions. The steady-state solution (if it exists) is attained asymptotically in time. At each time step, the elliptic equation (12) is solved iteratively by overrelaxation. The velocity components are easily obtained from the $\xi - \psi$ solution by using the definition for the stream function. However, the determination of the pressure requires some care and a reasonable amount of effort (Refs. 6 and 10).

The next logical step in the solution procedure would be to construct the finite difference mesh to fit the particular problem of interest. This will be deferred until the discussion of each of the specific separated flow problems in later sections. Since all of the separated flow problems to be discussed later make use of the same differencing techniques and calculational procedure, a discussion of these common methods and techniques is appropriate.

The basic method used for equation (11) at interior field points is as follows: the velocity components are obtained from the stream function distribution by centered differencing:

$$v_{x_{ij}} = \frac{\psi_{i,j+1} - \psi_{i,j-1}}{2\Delta y} ; \qquad v_{y_{ij}} = -\frac{\psi_{i+1,j} - \psi_{i-1,j}}{2\Delta x} \tag{13}$$

The vorticity is then advanced in time

$$\xi_{ij}^{n+1} = \xi_{ij}^n + \Delta t \left[-\frac{\delta (v_x \xi)^n}{\delta x_{ij}} - \frac{\delta (v_y \xi)^n}{\delta y_{ij}} + \frac{1}{Re_L} \frac{\delta^2 \xi^n}{\delta x_{ij}^2} + \frac{1}{Re_L} \frac{\delta^2 \xi^n}{\delta y_{ij}^2} \right] \tag{14}$$

and the diffusion terms are represented by the centered difference form, as in

$$\frac{\delta^2 \xi}{\delta x_{ij}^2} = \frac{\xi_{i+1,j}^n + \xi_{i-1,j}^n - 2\xi_{ij}^n}{\Delta x^2} \tag{15}$$

The advection or convection terms are represented by the method of upwind differencing. This one-step, explicit, two-time-level method which achieves stability of advection terms involves one-sided, rather than space-centered, differencing. Meteorologists have long known of the stabilizing effects of "upwind" or "weather" differencing. The upwind difference form of the advection terms of equation (11) is

$$
\frac{\delta (v_x \xi)^n}{\delta x_{ij}} = \left\{ \begin{array}{l} \dfrac{(v_x \xi)^n_{i+1,j} - (v_x \xi)^n_i}{\Delta x} \quad \text{for } v_x < 0 \\[3mm] \dfrac{(v_x \xi)^n_{i,j} - (v_x \xi)^n_{i-1,j}}{\Delta x} \text{ for } v_x > 0 \end{array} \right. \tag{16}
$$

Similar forms are used for $\dfrac{\delta (v_y \xi)^n}{\delta y_{ij}}$. In this first order method, information is
advected into a cell only from those cells that are upwind of it[+] (Refs. 10, 11, 12).
Therefore, information is advected from a cell only into those cells which are
downstream of it. This, of course, is physically correct and leads to the following
definition of the Transportive Property. A finite difference formulation of a flow
equation possesses the Transportive Property if the direction of a perturbation in
a transport property is advected only in the direction of the velocity. Upwind differencing methods inspired by this physical reasoning possess this property. In
any method using space centered differences for the advective terms, the effect
of a perturbation in a transport property is advected upstream against the velocity.
A finite difference method which possesses the Transportive Property maintains
the integral kinematic property of the continuum solution in the same way that one
which possesses the Conservation Property maintains the integral Gauss divergence
property of the continuum solution. Of course, space-centered differences are
more accurate than unidirectional upwind differences based upon the formal
Taylor series expansion. However, the transportive property appears to be as
fundamentally important and physically significant as the conservative property
(Ref. 6). Furthermore, it should be remembered that the accuracy of the result
compared to the appropriate physical experiment is the final test of the numerical
procedure and not simply the order of accuracy. Although the upwind differencing
method introduces an artificial viscosity effect which must be carefully examined
when assessing the accuracy of results, it has the further advantage that it is not
stability limited by a cell Reynolds number as are forward-time space-centered
methods. The physical relevance of upwind differencing which seems particularly
appropriate for separated flows with their large backflow regions, has been
discussed by many authors (see Ref. 6 for a list of references).

Numerical stability requires that the time dependent solution cannot proceed
at time increments greater than some critical value. The following expression for
dynamic stability was obtained from a discrete perturbation analysis of a linear
model equation (Refs. 11 and 12).

$$
\Delta t_{crit} \leq \frac{1}{\dfrac{|v_{x_{ij}}|}{\Delta x} + \dfrac{|v_{y_{ij}}|}{\Delta y} + 2\alpha \left[\dfrac{1}{\Delta x^2} + \dfrac{1}{\Delta y^2} \right]} \tag{17}
$$

[+] When v_x reverses sign near a mesh point, a modification to the basic upwind
difference scheme is required in order for strict conservation to hold.

The Δt used for a particular problem is usually some fraction of Δt_{crit}. The stability of this method has been demonstrated over a wide range of Reynolds numbers for various separated flows (Refs. 11,12).

Once the new ξ distribution at time $(n+1)$ is obtained from equation (14), the new ψ distribution is found from equation (12) by iteration using successive over-relaxation. Convergence criteria will be discussed later for each separated flow problem studied.

The accuracy of the upwind differencing method in resolving viscous effects usually deteriorates at high Re_L because it implicitly introduces an effect which resembles an artificial viscosity effect. This numerical or artificial viscosity effect simply reduces the effective Reynolds number $\frac{v}{\alpha}$ to $\frac{v}{\alpha + \alpha_{es}}$. Although this numerical viscosity α_{es} is not unique to upwind differencing it appears to be the predominant source of error in the numerical scheme described above. Recently, Roache (Refs. 13 and 6) has considered the artificial viscosity errors for multi-dimensional viscous flow problems. By considering the application of upwind differencing in a two-dimensional flow and using the appropriate form of equation (14) for constant v_{x_i}, $v_{y_i} > 0$, the following equation was obtained:

$$\frac{\partial \xi}{\partial t} = \frac{\partial (v_x \xi)}{\partial x} - \frac{\partial (v_y \xi)}{\partial y} + (\alpha + \alpha_{ex}) \frac{\partial^2 \xi}{\partial x^2} + (\alpha + \alpha_{ey}) \frac{\partial^2 \xi}{\partial y^2}$$

$$+ \text{ higher terms and differentials} \tag{18}$$

In the transient analysis

$$\alpha_{ex} = \frac{1}{2} v_x \Delta x (1 - C_x) \qquad \alpha_{ey} = \frac{1}{2} v_y \Delta y (1 - C_y) \tag{19}$$

where C_x and C_y are the Courant numbers defined by

$$C_x = \frac{v_x \Delta t}{\Delta t} \qquad C_y = \frac{v_y \Delta t}{\Delta y} \tag{20}$$

In the steady-state analysis

$$\alpha_{ex} = \frac{1}{2} v_x \Delta x \qquad \alpha_{ey} = \frac{1}{2} v_y \Delta y \tag{21}$$

Although these relations for the artificial viscosity only apply in the limit as Δx and Δy approach zero (Ref. 6), they are useful for purposes of discussion. For transient solutions, according to equation (19), the numerical or artificial viscosity effect will be at a minimum for C_x and C_y as close to unity as possible. The question of the values of C_x and C_y is further complicated by the inviscid stability restriction obtained from linear theory that $C_x + C_y \leq 1$ (Ref. 6). In practical problems therefore, it would be impossible, to simultaneously have both C_x and C_y near unity. Consequently some artificial viscous effects will always be present. Equation (21) indicates that for steady-state the directional cell Reynolds numbers must be much less than 2. While this condition is the requirement for formal accuracy, in some physical situations it is not quite this discouraging (Refs. 6 and 11). For example in regions where the boundary layer approximations are valid $\partial^2 \xi / \partial x^2$ is small and the term $(\alpha + \alpha_{ex}) \partial^2 \xi / \partial x^2$ in equation (18) will be small.

Furthermore, v_y is also small so that α_{ey} will be smaller than α. It may also be possible to control this effect to some degree by using a variable mesh where a large number of mesh points are concentrated in regions of steepest gradients. An example of this approach will be presented and discussed later.

In view of equations (19) and (21) it is clear that for Eulerian mesh systems in multidimensional problems, the artificial viscosity varies spatially, e.g., in x- and y-directions for two-dimensional situations. This may occur with a uniform mesh (i.e., $\Delta x = \Delta y$) because v_x and v_y vary throughout the flow field or with a variable mesh because Δx and Δy as well as v_x and v_y vary in the region of interest. For this reason an exact determination of the artificial viscosity error is usually not possible. However, an order of magnitude estimate of this type of error may usually be obtained.

Upwind differencing has often been found to provide the best compromise between accuracy and computer time for separated flow problems (Refs. 11, 12, 4). The choice of this method must be closely related to the desired accuracy and/or the relative difficulty and cost of obtaining similar or more accurate results from other methods. Since directional differencing is not stability limited by a cell Reynolds number, as is the forward-time, space-centered method, it is particularly useful in the study of complex separated flows, where only qualitative accuracy may be required at high Reynolds numbers.

2.2 Incompressible Axisymmetric Flow

The Navier-Stokes equations and the continuity equation in a cylindrical (r, θ, z) Eulerian coordinate system for two dimensional axisymmetric (i.e., $v_\theta \equiv 0$ and $\frac{\partial}{\partial \theta} \equiv 0$) incompressible flow may be expressed as

$$\frac{\partial \xi^*}{\partial t^*} + \frac{\partial (v_r^* \xi^*)}{\partial r^*} + \frac{\partial (v_z^* \xi^*)}{\partial z^*} = \frac{1}{Re_R} \left[\frac{\partial^2 \xi^*}{\partial r^{*2}} + \frac{\partial^2 \xi^*}{\partial z^{*2}} + \frac{1}{r} \frac{\partial \xi^*}{\partial r^*} - \frac{\xi^*}{r^{*2}} \right] \tag{22}$$

This is the conservative Vorticity Transport equation in dimensionless form where the vorticity and stream function are related by the Poisson equation

$$-\xi^* = \frac{1}{r^*} \left[\frac{\partial^2 \psi^*}{\partial z^{*2}} + \frac{\partial^2 \psi^*}{\partial r^{*2}} - \frac{1}{r^*} \frac{\partial \psi^*}{\partial r^*} \right] \tag{23}$$

The comments made in the previous section concerning the solution procedure also apply to the axisymmetric case.

3. PLANAR SEPARATED FLOWS

Although a truly two-dimensional flow is achieved numerically, it is usually very difficult - if not actually impossible in a strict sense - to achieve in physical experiments. Furthermore, duplicating the exact inflow and outflow boundary conditions for a finite computational region is very difficult - if not actually impossible in a strict sense. Thus a large amount of determination and at least some poetic license are necessary if one is to proceed.

For each case studied, a brief description of the physical experiment and an estimate of the accuracy of the measured results will be presented. The procedure for numerical experimentation will then be presented along with the question of accuracy of results. Finally, a comparison of the results of both methods will be discussed.

3.1 Blunt Base

The flow over a blunt-based body represents one of the most convenient and fundamental examples of unsteady separated flow. This oscillating wake is one of the most widely studied problems in fluid dynamics. Particular interest has been generated in the wake of blunt-based airfoils which could provide significant advantages at high speeds, if the drag penalty of the oscillating wake at low speeds could be overcome. A recent numerical and experimental study of this problem is reported in reference 14. The primary goal of this study was to provide numerical solutions which would enhance the understanding of the mechanism by which an oscillating wake was formed. A complete description of the three types of wake oscillations found may be obtained from references 14,15,16. Only the highest Reynolds number case (i.e., the third type of oscillation) will be discussed here since accurate quantitative experimental data are available for comparison with the numerical solution.

3.1.1 Physical Experiments.

The experimentation in the Reynolds number range $Re_h = 1.6 \times 10^4$ was designed to yield: a velocity profile at the same location as the inflow boundary of the finite difference grid; the wake Strouhal number; and the free stream Reynolds number based on the base height $\frac{V_{x\infty} h}{\alpha}$. These experiments were conducted in one of the University of Notre Dame's low-speed, low-turbulence wind tunnels. The model, a 12 in. (30.48 cm) long, 6 in. (15.24 cm) wide and 2 in. (5.08 cm) thick flat plate with an elliptical nose and lucite side plates, was mounted in a two square foot ($0.1858 \, m^2$) test section fitted with a DISA hot wire probe and probe traversing mechanism. The hot wire probe was calibrated with a pitot tube connected to a micromanometer (Ref. 16). The experimental procedure included: traversing the flow field along the inflow station, thereby obtaining the free stream velocity and the velocity profile at the inflow station; and traversing the wake to determine the frequency of oscillation of the wake. The traversals were performed along four lines in the flow field as shown in Fig. 3. The first traverse produced the velocity profile one base height upstream of the base from which the free stream velocity and boundary layer thickness were obtained. The other three traversals were all parallel to the first, but located one-eighth, one, and two base heights, respectively, downstream of the base.

The velocity profile obtained one base height upstream of the base is plotted in Fig. 4 in terms of the ratio of local velocity to free stream velocity versus base heights above the surface. The calibration of the hot-wire probe before and after data were taken indicated that the free stream velocity was measured to within about 2% of the pitot value. The accuracy of this type of velocity measurement is known to decrease as the probe approaches the surface of the model as a result of interference effects as well as the lower velocities encountered. The boundary layer thickness was defined as the distance from the surface at which the local velocity was 99 percent of the free stream value. A fourth order polynomial was then fitted to the data points lying below this height. The resultant curve was considered to be the velocity profile in the boundary layer. The computed velocity profile curve and the data points to which it was fit are shown in Fig. 5. The boundary layer thickness was calculated to be 0.1210 base heights. The Reynolds number was 16,295.

The second section of the experiment was the determination of wake
Strouhal numbers. The frequency of oscillation was determined during the hot
wire traversals. The frequency was taken to be the average frequency recorded
over a ten second measuring period. Hot wire voltages were displayed on the
oscilloscope. Photographs of the oscilloscope trace were taken at the locations
marked on Fig. 3. The letter designation of these locations corresponds to the
letter designations of Fig. 6 which shows the trace photographs.

Figure 6a shows two traces of the hot wire voltage in the boundary layer at
the position where the velocity profile was determined. The small amplitude
oscillatory character of the velocity in this region is immediately apparent. The
oscillations appear to be fed back upstream from the large amplitude oscillations
of the wake. The laminar character of the boundary layer is also confirmed by
this figure. Figure 6b, taken outside the boundary layer, also shows the oscillations to be present in the free stream. Figure 6c shows the oscilloscope trace at
the location of optimum signal clarity. As in the experimental work of Bearman
(Refs. 17 and 18), this position was found to be one base height downstream of the
base and one base height above the centerline. The frequency in this region was
determined to be 23.80 c.p.s. at a free stream velocity of 15.29 f.p.s. (4.66 m.
p.s.), giving a Strouhal number of 0.2594 at a Reynolds number of 16,295.
Frequency measurements in this range are known to be accurate to within 1%.

Moving towards the centerline, Fig. 6d shows that fluctuations in the local
velocity are superimposed on the fluctuations due to the wake oscillation. The
wake oscillation pattern is thus partially obscured. At the centerline, these local
fluctuations are large enough to completely obscure the periodic character of the
wake (Fig. 6e). Displacing the probe an additional base height downstream gives
a single trace similar to Fig. 6e, only having a large amplitude. However, a
multiple trace photograph reveals that the wake oscillation frequency is well defined in this region despite the large local fluctuations, Fig. 6f. Figure 6g shows
the hot wire trace with the probe positioned one-eighth base height behind the base
and on the centerline. The local velocity fluctuations are of small amplitude, and
the wake oscillation is not apparent at this position, probably being of equally
small amplitude.

3.1.2 Numerical Experiments. In the finite difference approximation, to the
solution of the governing equations, approximations to the actual values of the
variables are obtained at a finite number of points within the flow field. When the
grid lines are equally spaced in all areas of the flow field, the spacing is
governed by that necessary to obtain the desired accuracy in the area of largest
gradients, leading to a high density of grid points in areas where there are low
gradients. Fortunately, it is possible, in some cases, to tailor the grid spacing
to the particular problem decreasing the grid point density in the areas of low
gradients and increasing this density in the areas of high gradients. This may be
accomplished without losing the advantages of a rectangular grid.

In the present problem, the largest gradients occur in the boundary layers
on the body and in the region one and one-half base heights either side of the
centerline, downstream of the base. A grid tailored to concentrate points in the
proper locations for the blunt base problem is shown in Fig. 7. The grid used
during computation was finer than that of Fig. 7, consisting of 56 grid spaces in
the stream-wise direction and 112 grid spaces in the cross stream direction. The
grid or mesh aspect ratio, $\frac{\Delta x}{\Delta y}$, varied from 20 to 2. In terms of the base height,

the flow field computed was seven base heights square. The nomenclature for the finite difference mesh is demonstrated in Fig. 8. The nonuniform nature of the mesh or grid structure requires a special formulation of the finite difference approximations to the partial derivatives.

Finite Difference Derivatives

In order to formulate the governing equations in finite difference form, it is necessary to obtain finite difference expressions for the first and second derivatives. Standard Taylor series expansions yield for the first derivative in the x-direction

$$\frac{\partial f_{i,j}}{\partial x} = \frac{\Delta x_{i-1} \; f_{i+j,j}}{\Delta x_i (\Delta x_{i-1} + \Delta x_i)} - \frac{\Delta x_i \; f_{i-1,j}}{\Delta x_{i-1}(\Delta x_{i-1} + \Delta x_i)}$$

$$- \frac{(\Delta x_{i-1} - \Delta x_i) f_{i,j}}{\Delta x_i \Delta x_{i-1}} \tag{24}$$

and for the second derivative in the x-direction

$$\frac{\partial^2 f}{\partial x^2} = \frac{2 f_{i+1,j}}{\Delta x_i (\Delta x_{i-1} + \Delta x_i)} + \frac{2 f_{i-1,j}}{\Delta x_{i-1}(\Delta x_{i-1} + \Delta x_i)} - \frac{2 f_{i,j}}{\Delta x_i \Delta x_{i-1}}$$

$$- \frac{(\Delta x_{i-1} - \Delta x_i)}{3} \frac{\partial^3 f}{\partial x^3} - \frac{\Delta x_i \Delta x_{i-1}}{24} \frac{\partial^4 f}{\partial x^4} \tag{25}$$

If this last expression is truncated before the term containing $\frac{\partial^3 f}{\partial x^3}$, the approximation is not necessarily accurate to the order of magnitude of (Δx^2). However, if the grid is constricted in such a fashion that the order of magnitude of $(\Delta x_{i-1} - \Delta x_i)$ is the same as the order of magnitude of $(\Delta x_{i-1} \Delta x_i)$ truncation prior to this term is permissible, while limiting truncation errors to the order (Δx^2). In this specific case, the expression for $\frac{\partial^2 f}{\partial x^2}$ is given as

$$\frac{\partial^2 f}{\partial x^2} = \frac{2 f_{i+1,j}}{\Delta x_i (\Delta x_{i+1} + \Delta x_i)} + \frac{2 f_{i-1,j}}{\Delta x_{i-1}(\Delta x_{i-1} + \Delta x_i)} - \frac{2 f_{i,j}}{\Delta x_i \Delta x_{i-1}} \tag{26}$$

The derivations of $\frac{\partial f}{\partial y}$ and $\frac{\partial^2 f}{\partial y^2}$ are completely analogous to the above derivations

with Δx replaced by Δy, and the index j varying. As mentioned earlier, it is the absolute accuracy and not simply the order of accuracy which is important. Therefore the grid may be expanded somewhat faster than just described if it is sufficiently fine.

Boundary and Initial Conditions

As in the solution of any time dependent partial differential equation, it is necessary to specify the conditions on the boundaries of the system and the initial state of the system.

Upstream Mesh Boundary

At the inflow or upstream boundary layer, surfaces 1, Fig. 8, the experimentally determined velocity profile, v_x, was integrated to provide the necessary stream function values and v_y was allowed the freedom to adjust with time. The vorticity on this boundary was computed directly from Poisson's equation, using a centered difference in the y-direction and a forward difference in the x-direction.

Upper and Lower Mesh Boundaries

The upper and lower mesh boundaries, surfaces 2, Fig. 8, were assumed to be undisturbed stream lines. The actual condition at an infinite distance from the body was thus brought to within a finite distance of the body. Following the approach of Roache and Mueller (Ref. 10) and Mueller and O'Leary (Ref. 19), the vorticity was specified as equal to the vorticity one grid space within the mesh.

Downstream Mesh Boundary

Three boundary conditions were tried on the outflow boundary, surface 3, Fig. 8. The first was simply setting the vorticity on the boundary equal to the vorticity one grid point upstream. While both Roache and Mueller (Ref. 10) and Mueller and O'Leary (Ref. 19) imply that this condition may be interpreted physically as no production of vorticity between these grid points, it would appear that this is correct only under steady state, converged conditions. If a variation of the vorticity value at the point one grid space upstream of the boundary exists, the interpretation would be quite different. Under such conditions the boundary condition implies the production or dissipation of vorticity in the requisite amount to bring both grid points to the same state at the same time. The condition was abandoned as too restrictive. The second condition tried was a mixed time and space derivative. This boundary condition, first used by Roache and Mueller (Ref. 10), was abandoned as it resulted in the production of large amount of vorticity at the boundary with the subsequent effect of producing a catastrophic instability in the solution. The last approach tried was based on the assumption that vorticity was a linear function of the "x" spatial coordinate over the last three grid points. This resulted in a linear extrapolation of vorticity to the boundary. While still rather restrictive, this condition was felt to allow the flow more freedom at this boundary. The downstream boundary condition on stream function was also handled by a linear extrapolation.

Solid Surfaces

As the stream function on the body surface was chosen to be identically zero there was never any need to re-compute this value. Following the work of Roache and Mueller (Ref. 10), the vorticity on surface 4, Fig. 8, was computed using a one sided difference. The stream function and vorticity on the base of the body, surface 5, Fig. 8, were computed in a similar manner to those on the upper surface of the body.

Corners

The value of the stream function at the corners, like all other places on the body, was set to zero. Again using the approach of Roache and Mueller (Ref. 10), the corner points were assigned two values of vorticity. If, while computing a point above or below the corner, the value of vorticity at the corner was required, this value was computed as if the corner point lay on the upper or lower surface, respectively. If, however, the value at the corner was required while computing a point behind the base, the corner vorticity was computed as if it lay on the base of the body.

Initial Conditions

Three types of initial conditions were used successfully. The first defined the stream lines as straight lines, parallel to the surface upstream of the base, which were turned through an angle at the plane of the base and projected to the downstream boundary. The second defined the stream line to be parallel to the center-line both upstream and downstream of the base except within one grid space downstream of the plane of the base. In this region the stream lines underwent a sharp vertical deflection. This condition was felt to best approximate the impulsive starting condition. In both of these initial conditions a slight asymmetry was introduced by deflecting the zero stream line one grid point vertically from the centerline. The third condition was the use of a previously attained solution as the starting point for a new case.

3.1.3 Method of Solution. The upwind differencing technique described earlier was applied to the solution of the planar Vorticity Transport equation (11). The effect of upwind differencing was to bias the solution so that vorticity could be advected only in the direction of flow. The technique, however, presented no such bias to the diffusion of vorticity. It was felt that the use of this technique approximated more closely the physical situation.

Once new values of vorticity were obtained at all interior grid points, it was necessary to compute the new values of the stream function, which satisfied Poisson's equation and were compatible with the vorticity field values. Poisson's equation was solved (Ref. 16) by an iteration technique called successive over-relaxation by Young and extrapolated Liebmann by Frankel (Ref. 20). The iteration was continued until the net change between two successive iterates, at each point, was less than 5×10^{-6}.

Once the solution to Poisson's equation was obtained, the boundary conditions on vorticity were applied. A new time step was then computed, the appropriate data recorded and the sequence restarted at the calculation of vorticity at the new time level. The repetition of this sequence continued until the flow pattern repeated itself at a definite frequency. The oscillatory solution was then considered to have been reached. This took from 50 to 70 hours of UNIVAC 1107 time for three cycles.

Interpretation of Plotted Fields

Nearly all results from the numerical phase of this work are presented in the form of plots of constant stream function value, stream lines. This method of presenting the results was chosen as it enhanced the investigators' understanding of the flow field. Only the portion of the grid, which was required to observe the region of interest, was plotted. The plots therefore, do not necessarily represent the entire calculated field.

A linear interpolation between grid points was used to locate the intersections of stream lines with grid lines. Therefore, it should be noted that the locations and contours of the plotted stream lines are only as accurate as a linear interpolation along grid lines. However, the high grid point density, particularly in the region

of high gradients, resulted in a reasonably smooth and accurate plot. The increments in stream function between adjacent plotted stream lines were not uniform
throughout the plotted field. Rather, they were chosen to best demonstrate the
character of the flow and minimize plotting time.

A bifurcation of the zero stream line, where it separates from the body, is evident
in nearly all plots. This bifurcation is not indicative of two separation points.
Rather, it is a characteristic of the plotting routine and the limits of the solution.
The finite difference solution cannot locate the position of separation, it can only
specify the two grid points on the body between which separation occurs. In a
manner consistent with this limit of the solution, the plotting routine draws a line
to both grid points. The point of separation, then lies between the two intersections
of the legs of the bifurcation with the body.

3.1.4 Discussion of Results. Using the results of the experiment, the inflow
velocity profile with boundary layer thickness of 0.121 h and Reynolds number of
$Re_h = 16,295$ were inserted into the numerical procedure and the impulsive initial
conditions were used. A circulation region formed behind the upper portion of the
base and subsequently entrained stream lines which had passed as far as a half
base height below the lower surface of the body. This circulation region fed back
upstream on the upper surface of the body. Eventually, two circulation regions
were formed by the division of the single original region at the corner. As the
circulation in both regions was in the same sense, an interface between them was
unsupportable, leading to the production of a third circulation region between the
two original ones. This region entrained the stream lines from above the base
carrying them entirely across the base to the lower corner. The region originally
behind the base moved downstream and the circulation region on the upper surface
decreased in intensity. Eventually, the region on the upper surface had completely
dissipated, the loop of the stream lines from above the body to below and back to
above had moved away from the base and a small region of circulation had formed
on the upper portion of the base. This recirculation region grew and entrained
stream lines that had passed below the base. This sequence is shown in Fig. 9. The
oscillatory solution for this case is shown in Fig. 10.

The experimental criterion for transitioning of the shear layer presented by
Roshko and Lau (Ref. 21) indicates that the shear layers should transition prior to
the formation of the detached circulation regions. The obscuring of the periodic
oscillation on the oscilloscope trace is most likely due to the onset of turbulence.
It appears, however, that the finite difference solution continues to closely approximate the gross flow patterns of the wake, even though it does not take this
transitioning into account. For this type of oscillation, at $Re_h = 16,295$, the
numerically determined Strouhal number for the full period was 0.222 compared
to the experimental value of 0.259. These two values differ by about 14%.
Physically, one could attribute some of this difference to the facts that the experiment shows signs of turbulence and that the flow produced was not strictly two-
dimensional due to the finite size of the model. The predominant source of error
in the numerical scheme is the numerical or artificial viscosity effect. In addition, for oscillatory flows phase errors caused by dispersion could also become
serious (Ref. 8). Both of these types of errors will be discussed.

Although an exact determination of the artificial viscosity effect cannot be
given for this multidimensional viscous flow, arguments - mostly from computational experience - will be given. The simplest argument is based upon the
estimate that the artificial viscosity error in the boundary layer is about $\mu \frac{\Delta y}{\delta}$

or simply proportional to $\frac{\Delta y}{\delta}$ (Ref. 11). In the present case, Δy varied considerably because of the variable mesh. The region of largest gradients in the boundary layer was approximately contained between the minimum value, $(\Delta y)_{min}$ = 0.00625 h, and the value Δy = 0.0214 h. Since the experimentally determined boundary layer thickness was 0.121 h, one could estimate that the local artificial viscosity error would be between

$$\frac{(\Delta y)_{min}}{\delta} \times 100 = 5.17\% \qquad \text{and} \qquad \frac{\Delta y}{\delta} \times 100 = 21.4\%.$$ The region of largest

gradients in the wake was about three fourths of a base height either side of the centerline. However, there is no convenient parameter (i.e., like δ) available with which to estimate the error in the wake. Because the boundary layer assumptions are not valid in the wake, one would expect a larger artificial viscosity effect in this free shear layer. In view of Equation (19), although $\Delta x > \Delta y$, v_x in the shear layer where viscous effects are important is usually only a fraction of the free stream value. This is the reason given by Roache (Ref. 6) for the fact that viscous finite difference solutions with non zero $(\alpha_e)_{steady\ flow}$ are often more accurate than might be expected from evaluating $(\alpha_e)_{steady\ flow}$ based on free stream conditions. Based on the work of Runchal, Spalding and Wolfshtein (Ref. 22), it appears that artificial viscosity may be examined from the point of view of stream line angle. They found that the artificial viscosity could be approximated by $\alpha_e = 0.36\ v_x\ \Delta x\ \sin(2\theta)$, where θ was the stream line angle. In reference 22, Runchal, Spalding and Wolfshtein used an upwind difference method for the driven cavity problem. At Reynolds number of 100 with their non-uniform mesh of 13 x 13 they found that their results compared favorably with computations made with a uniform mesh of 51 x 51. Their reasons for this surprising result is that their mesh not only concentrated more points in regions of steepest gradients, but that the mesh was generally aligned with the stream line pattern. In regions where this was not true (i.e., in the corners θ is large and approaches 45°) the velocity was small so that α_e was probably small. These same arguments also appear to apply to the oscillating wake solution presented above.

By studying a damped oscillation, Cheng(Ref. 8) arrived at the following conclusion concerning phase errors caused by dispersion. A spatial resolution of 30 meshes per wave length and 300 time steps per cycle can provide engineering accuracy for the first wave length for a second order accurate scheme. In the blunt base oscillating wake solution, there were more than 30 meshes per wave length and between 6000 and 7000 time steps per cycle. It therefore seems reasonable, in accordance with Cheng's result, to conclude that phase errors were not serious in this case.

Finally, the recent work of Bozeman and Dalton (Ref. 4) for the cavity problem shows clearly that directional differencing with the conservation form of the equation produced better solutions than using the non-conservative equation at Reynolds number of 1000. Furthermore, with centered differencing at this Reynolds number, neither the conservative nor the non-conservative equations converged to the steady state solution.

3.1.5 Concluding Remarks. If you are interested in Strouhal numbers, velocity profiles or lift and drag values, with accuracy of 1 or 2%, physical experiments will probably be the fastest and cheapest approach, especially if the necessary wind tunnel and instrumentation are available. If, however, you are interested in studying the detailed mechanism of formation of the oscillating wake and/or the oscillatory pressure and shear stress components throughout the entire flow field, numerical experiments can be obtained if a large amount of computer time is available.

3.2 Symmetric Channel Contraction - Frontstep

The complexity of internal flows is a result of the simultaneous presence and interaction of different fluid phenomena. One such example is that of the flow in an abruptly converging duct as shown in Fig. 2a. Here there is superposition of two opposing factors, the pressure drop associated with the acceleration of the main flow in the duct and the locally increasing pressure near the duct walls due to blockage. The latter effect results in separation of the flow from the walls. To date there has been little numerical or experimental research directed toward the low Reynolds number frontstep problem (Ref. 23). A more complete description of this problem may be obtained from reference 23. Only the zero bleed case of reference 23 will be discussed here. Since accurate experimental data have been difficult to find in the literature for this low Re case, a qualitative experiment was performed in support of this discussion.

3.2.1 Physical Experiments. A two-dimensional symmetrical channel with an approximate contraction of 2 to 1 was set on the water table. The channel extended 50 cm from the rounded entrance to the contraction and 30 cm from the contraction to the downstream end. The two sections of the channel were approximately 5 cm and 2.5 cm apart with 1.25 cm frontsteps. The centerline velocity was determined by measuring the time necessary for a very small particle of wood to move 15 cm on the water surface in the upstream channel section. The average of three runs was used, the approximate separation point was found visually by alternately injecting a small amount of blue dye far upstream at the wall, and in the separation region. A summary of these qualitative data is given below in the order taken.

Upstream Water Depth (cm)	$v_{x_{\mathbb{C}}}$ (cm/sec)	Re_D	Re_h	$\dfrac{x_s}{h}$	Comments
2.5	5.00	1400	700	0.71	3-D effects
2.5	2.73	770	385	0.50	" "
2.5	6.53	1830	915	0.86	" "
4.5	3.62	1162	581	0.71	" "
5.0	2.34	656	328	0.57	Plexiglas lid i.e. no free surface.

Note that all but the last case were true water table experiments containing a free surface. The probable error in these experiments is at least $\pm$ 10%. These data are plotted in Fig. 11. No noticeable upstream influence of the contraction was observed about two step heights upstream.

3.2.2. Numerical Experiments. The location of the mesh points was found by overlaying the geometry of the upper-half channel, as shown in Fig. 12, with a rectangular mesh which was uniform throughout the flow field. The aspect ratio, $\dfrac{\Delta x}{\Delta y}$, of the mesh was two. The number of grid lines, corner locations and mesh nomenclature are shown in Fig. 12. The upstream boundary was located four step heights from the contraction and the downstream boundary three step heights from the contraction.

Boundary and Initial Conditions

Somewhat different boundary conditions were used for this problem than for the blunt base problem. The initial conditions, however, were almost identical.

Upstream Boundary

The inflow or upstream boundary is surface (2) in Fig. 12. Initially, Poiseuille flow was specified at the inflow boundary (i.e., $v_x = v_{x_{C_L}} (1-(\frac{y^2}{D^2}))$ and $v_y = 0$).

During the numerical solution, however, these conditions were allowed to vary from time step to time step permitting the inflow v_x and v_y to be sensitive to the solution as it developed downstream. This represents an extension of the concept proposed by Roache (Ref. 12) for the treatment of the upstream boundary. Roache allowed only v_y the freedom to adjust with time. This proved insuitable for the internal flow studied here because the blockage effect of the front step produced a larger upstream influence than that produced by the back step studied by Roache. In the present study, at the conclusion of a time step, new values of ψ at the inflow boundary points were obtained by a 2nd order extrapolation of ψ from the next three points downstream. A compatible value of vorticity was then generated by applying the Poisson equation at the inflow boundary under the assumption that ψ_{xx} remains constant over the first Δx.

Plane of Symmetry

Because the flow field studied was specified to be symmetric, the center surface (1) of Fig. 12, represents the plane of symmetry. The stream function along surface (1) was constant and was arbitrarily set equal to zero and the vorticity was set equal to zero.

Solid Wall Boundaries

The solid walls, surfaces (3), (4), (5) and (6) in Fig. 12 were no-slip boundaries. The stream function was constant along each of these boundaries and all equal to ψ along surfaces (3) if there was no bleed flow. Vorticity was computed using a one-sided difference (Ref. 12).

Downstream Boundary

The stream function on this boundary surface (7) of Fig. 12, was obtained by a linear extrapolation from adjacent interior points. The vorticity at surface (7) was set equal to the vorticity at the adjacent interior points. No computational instabilities were encountered using these downstream boundary conditions.

Step Corner

The stream function at the step corner, since it is part of the solid walls, was a constant. Using the approach of Roache and Mueller (Ref. 10), the corner point was assigned two values of vorticity. When computing the velocity of a point upstream of the corner, this value was computed as if the corner lay on the step surface. However, if the value at the corner was required while computing a point below the corner, this value was computed as if the corner lay on the down-stream wall, i.e., surface (6). Another investigation supporting the use of this double-valued treatment of the vorticity in flows with sharp corners was provided recently by Stevenson (Ref. 24).

Initial Conditions

It has been shown that the initial condition need not have physical significance (e.g., Ref. 19). In the present study, the transient solution serves only as a numerical technique in approaching the asymptotic case and need not represent a physical development of the flow field. Naturally, it is advantageous to pick the initial condition as close to the steady state solution as is conveniently possible to reduce the ensuing computation time. In the present study, two types of initial conditions were used. One, termed the "cold start" initial condition similar to the first type discussed in the blunt base problem, was used for the initial solution of a series of solutions, all having the same mesh structure and inflow condition. The second was used for succeeding cases, in which only the Reynolds number or the value of u_B varied from a previous case. The converged solution of that previous case was used as an initial condition for the new case.

3.2.3 Method of Solution. The method of solution including the upwind differencing technique was described earlier. In evaluating the relative convergence of the numerical solutions, several criteria were employed. At the conclusion of each step, the following four quantities were computed:

$$\sum_{i,j} \left| \xi_{i,j}^{n+1} \right| \tag{27}$$

$$\sum_{i,j} \left| \psi_{i,j}^{n+1} \right| \tag{28}$$

$$\sum_{i,j} \left| \xi_{i,j}^{n+1} \right| - \sum_{i,j} \left| \xi_{i,j}^{n} \right| \tag{29}$$

$$\sum_{i,j} \left| \psi_{i,j}^{n+1} \right| - \sum_{i,j} \left| \psi_{i,j}^{n} \right| \tag{30}$$

The first two expressions simply represent the total magnitude of vorticity and stream function throughout the finite difference field. They indicate the production or dissipation of vorticity and the corresponding development of the separated flow pattern with increasing time steps. Expressions (29) and (30) represent the total production or dissipation rates of ξ and ψ in the field.

In judging the relative convergence of the numerical scheme, the solution was continued until expressions (27) and (28) remained constant to five significant figures over at least 50 time steps. Usually expression (29) had reached a value less than 5.0×10^{-4} and expression (30) less than 3.0×10^{-4} by that time. In the solution of Poisson's equation by successive over-relaxation, a relative convergence criteria was used to terminate the iterative solution of the stream function. The iterative solution was continued until the change between two successive iterates, at each point, was less than some fixed fraction of the maximum value, usually 10^{-6}. Criteria such as this represent a subjective judgment based on an examination of the computer output and the total computer time available.

3.2.4 Discussion of Results. Numerical solutions were obtained over a range of Reynolds numbers based on the channel half-width of from 500 to 1500. The corresponding Reynolds number range based on the frontstep height was 250 to 750. Only the no bleed results will be discussed here (see Ref. 23 for the bleed results). Contour plots of stream function and vorticity for six Reynolds numbers from 500 to 1500 are presented in Fig. 13. Stream function is reflected about the channel centerline and plotted in the lower half of the figure with the vorticity directly above. There are several physical characteristics apparent in these contour plots. Although the stream lines present a general view of the flow pattern that occurs, more information can often be found in the vorticity plots. The most obvious feature is the large wall vorticity gradient near the sharp corner as well as the downstream wall. This is of course due to the high shearing stress in these regions. Note the volume flow beneath the lower two stream lines in the stream function plot. All of this fluid squeezes past the sharp corner and is confined to a very thin viscous region near the downstream wall. Also, note the dotted curved line, representing $\xi = 0$, which extends between the upstream wall and the step face in the stagnant corner region. Where the wall vorticity goes through zero, the wall shear stress is also identically zero. Thus, this vorticity contour line locates both the separation and reattachment points of the dividing stream line. With increasing Reynolds number, the corner eddy increases in size and intensity and the vorticity gradients become more severe near the walls.

Figure 14 depicts typical velocity profiles for two different Reynolds numbers, Re_D of 500 and 1500. Profiles at six different stations are plotted for each Reynolds number, at x equal to 4, 1, and 1/2 step heights upstream and 1/2, 1, and 3 step heights downstream of the corner. The profiles show the anticipated development from the Poiseuille profile at $\frac{x}{h} = -4$ to an inflected profile, characteristic of a point near separation, at $\frac{x}{h} = -\frac{1}{2}$. Downstream, the flow rapidly approaches the Poiseuille profile again. At $Re_D = 500$, the actual separation point occurs at approximately $\frac{x}{h} = -0.540$; at $Re_D = 1500$, it occurs at approximately $\frac{x}{h} = -0.791$.

Figure 11 summarizes results for the no bleed cases. This figure is a plot of the location of the separation point, in non-dimensional step heights upstream of the step, versus Reynolds number. The distance x_s was determined by locating $\xi_w = 0$ along the upstream wall in the vorticity solution. Figure 11 indicates that the effect of Reynolds number is to produce an upstream movement of x_s for an increase in R. This observation is in agreement with existing experimental results for laminar flow separation in general and with the qualitative experiments in particular. Wall shear stress distribution were also obtained from the numerical solution and are presented and discussed in reference 23.

The present numerical study was conducted on a CDC 6600 computer. A typical solution, consisting of 1514 lattice points, required approximately 15 min. of central processor time. Stability analysis of the upwind differencing method (Ref. 12) results in a Δt restriction that depends upon the maximum v_x and v_y components in the field. During the transient phase of the solution, the critical time step was re-evaluated after each time step, utilizing the current maximum velocities, so as to always use the largest Δt permissible. Typically Δt was on the order of 0.016 and about 800 time steps were required to satisfy the convergence criteria. Successive over-relaxation of Poisson's equation for ψ required only a single iteration to satisfy the relative convergence criteria after the initial 50 or so time steps.

In any computer simulation of a time dependent flow field, a choice is made on the fineness of the lattice grid, size of the computational time step, and convergence criteria. The choices are not arbitrary but influenced by computer core limitations and the excessive costs of lengthy computation. The question of artificial or effective viscosity inherent with the approximate numerical methods used here will be discussed in view of the results obtained. As in all approximate techniques this represents an error, the severity of which must be evaluated in light of the cost and effort one would expend to reduce or eliminate it.

Roache has shown in reference 6 that a multi-dimensional analysis of the upwind differencing method leads to the conclusion that a steady state solution is subject to an artificial viscosity which is dependent upon the size of Δt as well as Δx. This dependence of artificial viscosity on Δt was not observed in the present study so that one must resort to numerical experimentation in this matter.

An examination of artificial viscosity was performed on the converged solution of $Re_D = 500$, $u_B = 0$. Figure 11 implies that this case would be most sensitive to changes in viscosity (i.e., Reynolds number) if a change in x_s/h was used as a judging criteria. The converged solution represented a non-dimensional time, $t = 13.26$. To that point Δt had been set at 90% of the critical time step for stability (Δt_c). At this point Δt was reduced to 20% of Δt_c and the solution continued until $t = 25.46$. Consistent with previous experience with this numerical methods, no detectable change was observed in the location of separation or reattachment points of the corner eddy. No significant changes occurred in the other flow parameters. Hence, one concludes that the steady solution is not subject to an artificial viscosity that depends on Δt. A second test was conducted to investigate the artificial viscosity dependence on Δx. A case of $Re_D = 500$, $u_B = 0$ was computed with Δx and Δy set at $\frac{1}{2}$ their previous values, resulting in a lattice consisting of 5,842 points. This solution was computed to $t = 12.75$ x_s/h and y_R/h changed by 4.2% and 2.3% respectively. From a practical engineering point of view, these changes would usually be considered insignificant. The increase in accuracy could not justify the expenditure of approximately 1.75 hours of CDC 6600 central processor time required to compute this solution.

In this case it is not surprising that the artificial viscosity effect introduced by the directional differencing is small since, except for the small region near the step, the boundary layer approximations are valid. Therefore, $\frac{\partial^2 \xi}{\partial x^2}$ is small and the term $(\alpha + \alpha_{ex}) \frac{\partial^2 \xi}{\partial x^2}$ in equation (18) will be small. Also, v_y is small so that α_{ey} will be small and not appreciably affect the $(\alpha + \alpha_{ey}) \frac{\partial^2 \xi}{\partial y^2}$ term. Another way of arriving at the same conclusion is to consider the argument presented by Runchal, Spalding and Wolfshtein (Ref. 22) based on the stream lines angle measured with respect to the grid. The stream line angles are very small except near the step (see Fig. 13). In the separated portion of this region the velocities are low so that even when θ is large, the overall effect on $\alpha_{ex} = 0.36\, v_x \Delta x$ $\sin(2\theta)$ is probably small. This overall situation is aided, of course, by the low values of Reynolds number used.

3.2.5 <u>Concluding Remarks</u>. Low Reynolds number separated flow of this type may be studied quickly and relatively cheaply by numerical methods. Numerical experiments offer greater flexibility than do physical experiments since the effects of upstream velocity profile shape, positive and negative base bleed,etc., may be easily obtained. Furthermore, values of shear stress and pressure may be readily determined throughout the flow field. To obtain accurate data of this type from physical experiments as these low Reynolds numbers would be much more difficult and costly.

4. AXISYMMETRIC SEPARATED FLOW

4.1 Disc-Type Artificial Heart Valve

Artificial heart valves produce serious short and long term problems associated with the separated flows they produce. Separated flows,characterized by relatively slow reverse flows which are distinct from the main flow, may occur upstream and downstream of the sewing ring and behind the occluder as shown in Fig.2b. The slow reverse flow traps lipids, platelets, and other debris making this type of region thrombogenic. Furthermore, it is thought that all artificial valves are sufficiently hemolytic to enhance thrombus formation through the accumulation of the resulting ghosts,fragments and other debris. Although artificial heart valves have been available for well over ten years, there is relatively little quantitative data concerning the extent of thrombogenic separated flow regions produced by these valves since most experiments have been of the qualitative flow visualization type. Because flows of this type occur at relatively low values of Reynolds number,they appear to be very interesting from the computational viewpoint. The numerical and experimental investigation described in this section is for the flow through an axisymmetric fully open disc-type heart valve (Refs. 25,26,27).

4.1.1 <u>Physical Experiments</u>. The experiments were performed in a closed loop steady flow apparatus consisting of reservoir tank, a bellmouth entrance section to 1.00 inch (2.54 cm) inside diameter transparent acrylic tube. Five anti-turbulence screens were attached to the inlet of the bellmouth and a disc type heart valve in the fully open position was located 18 tube diameters from the tube entrance. The acrylic tube continued downstream of the heart valve for about 16 tube diameters where an elbow directed the liquid to the pump. The flow rate and therefore the Reynolds number (Re_D) was varied by means of the pump bypass system.

These experimental studies were directed toward studying the largest separated region produced, i.e., the region behind the disc as shown in the sketch inset in Fig. 15. Data were obtained for Reynolds numbers based on the tube diameter and average velocity upstream of the sewing ring (Re_D) of from 50 to almost 10,000. The Reynolds number in vivo in the aortic region varies from zero to a maximum of about $Re_D = 6300$. The length and shape of this separated region was determined using two methods of flow visualization.

An electrochemical technique referred to as the Thymol Blue (thymolsulphonephthalein) method was found to be useful only at very low Reynolds numbers, i.e., from 50 to 100. This method takes advantage of a proton transfer reaction near a current carrying wire in a Thymol Blue solution which turns the

the solution dark blue. This dark blue fluid is neutrally buoyant and is carried downstream by the flowing yellow fluid. Photographs of the flow patterns for values of Re_D were obtained and the length of the separated region behind the disc was taken from the photographs.

For the dye injection experiments, a special sliding tube injector was constructed. This dye injector, attached to the center of the disc, consisted of a slotted guide hypotube, 0.050 inch (0.0197 cm) outside diameter, internally fitted with a movable dye injection hypotube. The injection hypotube contained an orifice which could therefore be translated along the geometric center of the test section. A very small amount of neutrally buoyant dye was injected at a constant rate using an infusion-withdrawal pump. The reattachment location or downstream end of the separated region behind the disc was determined by moving the dye injection orifice until a change in flow direction was observed.

These experiments and additional experiments using the hydrogen bubble method indicated that the separated region behind the disc was steady, axisymmetric and laminar up to $Re_D = 200$. From $Re_D = 200$ a small swirl component appears in the flow while for $400 < Re_D \leq 2000$ the separated region becomes unstable though laminar. This instability is the precursor of a complete breakdown into turbulent flow. At Reynolds numbers greater than 2000, the flow was very difficult to observe because of the rapid diffusion of dye caused by the high mean flow velocities and random small scale motion. For the largest Reynolds numbers the fluid motion appeared to be turbulent.

4.1.2 Numerical Experiments. The location of the grid or mesh points was determined by overlaying the geometry to be studied (shown in Fig. 16) with a square mesh (i.e. $\Delta r = \Delta z$) which was uniform throughout the flow field. Figure 16 along with Table 1 describes the mesh nomenclature and the various mesh sizes used.

Boundary and Initial Conditions

Since the boundary and initial conditions are almost identical to those used in the previously described cases, they will be presented as briefly as possible.

Upstream Boundary

The inflow or upstream boundary, surface 1 of Figure 16, was specified as a parabolic v_z-velocity profile for all the cases to be discussed in this article. Although v_z was specified in this manner, v_r was allowed the freedom to adjust with time.

Centerline

The centerline is labeled surface 2 in Fig. 16. Assuming axial symmetry, the vorticity and stream function at the centerline were set equal to zero and the velocity in the r-direction was also set equal to zero.

Solid Boundaries

The solid boundaries are labeled (3) in Fig. 16. The value for stream function along the top boundary, including the sewing ring, were given a constant value determined from the integration of the inflow velocity profile. The value of stream function for the disc was the same as for the centerline, zero. The velocity in both r and z directions were zero. Vorticity on the solid boundaries was computed using the definition of vorticity.

Corners

The values of vorticity for the corners of the solid boundaries were computed using one sided differences as described in the frontstep and blunt base cases.

Downstream Boundary

The outflow or downstream boundary is labeled 4 in Fig. 16. The boundary condition for stream function and vorticity used by Fanning and Mueller (Ref. 15) and described in the blunt base problem was used for the outflow. Thus a linear extrapolation in the z direction over the last three points was made for the values of stream function and vorticity. This allows the flow to continue in the general direction dictated by the two interior points adjacent to the outflow boundary.

Initial Conditions

Initial conditions similar to those described in the frontstep and blunt base cases were used.

4.1.3 Method of Solution. The method of solution including upwind differencing for advection terms and convergence criteria, was essentially the same as for the frontstep problem described earlier, except of course equations (22) and (23) were used.

4.1.4 Discussion of Results. Figure 16 along with Table 1 describe the geometries and mesh sizes used. The first mesh used was the coarse mesh. A converged solution at $Re_D = 100$ was reached rapidly, however, the nature of this solution was questionable since no wake was apparent behind the disc. Experiments were performed using mesh sizes two and three times smaller than the coarse mesh. Unfortunately, the Univac 1107 did not have sufficient storage for the finest mesh. In order to accommodate the finest mesh it was necessary to shorten the geometry by one disc length on both upstream and downstream ends. The results for the three runs at $Re_D = 100$ on the length of the separated region downstream of the disc were very instructive. The Reynolds number Re_D is based upon the average inflow velocity and the tube diameter. The medium mesh gave results which appear reasonable, i.e., wake behind the disc. The length of this wake or separated region behind the disc was even larger for the fine mesh. The flow also separated closer to the corner of the disc for the fine mesh than for the medium mesh. While still finer meshes might give more accurate results, these calculations indicated that the improvement would be quite small and probably not compensate for the increased computer time required. Further numerical experiments were performed to study the effects of inflow velocity profile shape and inflow location and are reported in reference 26. Recently, the components of the shear stress have been extracted from these solutions in order to locate regions with very high or very low values. Results of this type have considerable significance from the medical point of view (Ref. 28). Numerical calculations using the fine mesh produced the same trend as the physical experiments up to $Re_D \approx 500$ as shown in Fig. 15. The corrected numerical results shown in Fig. 15 are significantly different from those presented earlier in references 25, 26 and 27. The latest results were obtained after finding and correcting a local conservation error in the computer program. This agreement might be further improved if the actual velocity profile approaching the heart valve could be used in the numerical calculations.

Although much more time and effort was involved in obtaining the experimental data shown in Fig. 15, than in the frontstep case, it was only possible to improve the measurement of the Reynolds number (i.e., probable error of the order of 3%). The accuracy of determining the length of the wake behind the disc was never better than $\pm$ 10%. In the high Reynolds number region, where the wake was unstable this value was often exceeded as the interpretation of the wake length became more difficult and thus more subjective.

It appears that in this case the artificial viscosity effect will be rather large as either the Reynolds number increases or as the mesh size increases from the fine mesh used. One would not expect the boundary layer assumptions to be valid except in small regions near the upstream and downstream ends of the geometry, i.e., from the sewing ring until downstream of the wake $\frac{\partial^2 \xi}{\partial z^2}$ and v_r are both large. From the streamline angle point of view (Ref. 22), significant areas of this flow have large values of θ with respect to the grid together with large values of velocity so that α_{er} and α_{ez} are probably large. In particular, the regions between the sewing ring and the disc and near the downstream end of the wake have streamline angles around 45°.

4.2.5 <u>Concluding Remarks</u>. While it appears that for this simple geometry, the numerical experiments offer some advantages over the physical experiments at low Reynolds numbers, great care must be exercised in the choice of the method and mesh configuration used. This seems to be the type of problem where both physical and numerical experiments should be performed simultaneously as a check on each other.

5. SUMMARY

From this study of three quite different viscous separated flow problems using the same numerical techniques, it is apparent that experimentation is as important in the numerical sense as it is in the physical sense. The time-dependent conservative equations with upwind differencing for advection terms produced stable solutions over a wide range of Reynolds numbers. The actual accuracy of the numerical solutions varied with the flow problem and mesh geometry. A variable mesh, concentrating mesh points in regions of steepest gradients, appears to be a convenient method of reducing the, ever present, artificial viscosity effect. Numerical solutions with engineering accuracy can be obtained if great care is taken to control this artificial viscosity effect. Furthermore, they allow the study of basic flow phenomena which would be more difficult to produce and study in physical experiments. These methods, however, are still in the developmental stages and are not available in foolproof form for use by design engineers.

There should be no doubt that, at present and for the foreseeable future, the computer and wind tunnel are and will be dependent upon each other for the practical solution of complex separated flow problems.

REFERENCES

1. WIRZ, H. J. and SMOLDEREN, J. J.: Numerical integration of Navier-Stokes equations.
   AGARD Lecture Series 64: Advances in numerical fluid dynamics, 5-9 March 1973, pp 3-1 - 3-13.

2. RAKICH, J. V.: Introduction to the Proceedings of the Symposium on Computational Fluid Dynamics.
   Proceedings of the AIAA Computational Fluid Dynamics Conference, Palm Springs, Cal. July 19-20, 1973.

3. BANKSTON, C. A. and SMITH, H. J.: Vapor flow in cylindrical heat pipes.
   ASME Transact. Series C, J. of Heat Transfer, Vol. 95, No. 3, August 1973, pp 371-376.

4. BOZEMAN, J. D. and DALTON, C.: Numerical study of viscous flow in a cavity.
   J. of Computational Physics, Vol. 12, No. 3, July 1973, pp 348-363.

5. LIU, C. L. and LEE, S. C.: Transient state analysis of separated flow around a sphere.
   Computers and Fluids, Vol. 1, No. 3, Sept. 1973; pp 235-250.

6. ROACHE, P. J.: Computational fluid dynamics.
   Hermosa Publishers, P. O. Box 8172, Albuquerque, New Mexico 87108, 1972.

7. LIONS, J. L. et PRODI, G.: Un theoreme d'existence et unicite dans les equations de Navier-Stokes en dimension 2.
   C. R. Academie Sci. Paris, 248, 1959, pp 3519-3521.

8. CHENG, S. I.: Accuracy of difference formulation of Navier-Stokes equations.
   The Physics of Fluids, Supplement II, High speed computing in fluid dynamics; Vol. 12, No. 12, Part II, December 1969, pp II-34 - II-41.

9. ALLEN, J. D.: Numerical solution of the compressible Navier-Stokes equations for the laminar near wake in supersonic flows.
   Princeton University, Ph. D. Dissertation, Princeton, New Jersey, 1968.

10. ROACHE, P. J. and MÜELLER, T. J.: Numerical solutions of laminar separated flows.
    AIAA J. Vol. 8, No. 3, March 1970, pp 530-538.

11. THOMAN, D. C. and SZEWCZYK, A. A.: Time dependent viscous flow over a circular cylinder.
    The Physics of Fluids, Supplement II, High speed computing in fluid dynamics; Vol. 12, No. 12, Part II, December 1969, pp II-76- II-86.

12. ROACHE, P. J.: Numerical solutions of compressible and incompressible laminar separated flows.
    Ph. D. Dissertation, Dept. of Aero-Space Engineering, University of Notre Dame, Notre Dame, Indiana, Nov. 1967.

13. ROACHE, P.J.: On artificial viscosity.
    J. Computational Physics, Vol.10, October 1972, pp 169-184.

14. FANNING, A.E. and MUELLER, T.J.: Numerical and experimental investigation of the oscillating wake of a blunt based body.
    AIAA J., Vol.11, No.11, Nóv.1973, pp 1486-1491.

15. FANNING, A.E. and MUELLER, T.J.: A numerical and experimental investigation of the oscillating flow in the wake of a blunt based body.
    AIAA Paper 71-603, Palo Alto, Cal., 1971.

16. FANNING, A.E.: A numerical and experimental investigation of the oscillating flow in the wake of a blunt based body.
    M.S.Thesis, U. of Notre Dame, Notre Dame, Indiana, August 1970.

17. BEARMAN, P.W.: Investigation of the flow behind a two-dimensional model with a blunt trailing edge and fitted with splitter plates.
    J.Fluid Mechanics, Vol.21, Part 2, 1965, pp.241-255.

18. BEARMAN, P.W.: The effect of base bleed on the flow behind a two dimensional model with a blunt trailing edge.
    The Aeron.Qua., Vol. XVIII, August 1967, pp 207-224.

19. MUELLER, T.J. and O'LEARY, R.A.: Physical and numerical experiments in laminar incompressible separating and reattaching flows.
    AIAA Paper 70-763, presented at the AIAA 3rd Fluid and Plasma Dynamics Conference, Los Angeles, Cal. June 29-July 1,1970.

20. SMITH, G.D.: Numerical solution of partial differential equations.
    Oxford University Press, 1965.

21. ROSHKO, A. and LAU, J.C.: Some observations on transition and reattachment of a free shear layer in incompressible flow.
    Proc.1965 Heat Transfer and Fluid Mechanics Institute, Los Angeles, Cal., June 21-23, 1965.

22. RUNCHAL, A.K., SPALDING, D.B., WOLFSHTEIN, M.: Numerical solution of the elliptic equations for transport of vorticity, heat and matter in two-dimensional flow.
    The Physics of Fluids, Supplement II, High speed computing in fluid dynamics, Vol.12, No.12, Part II, December 1969, pp II-21 - II-28.

23. CAMPBELL, D.R. and MUELLER, T.J.: Effects of mass bleed on an internal separated flow.
    Proc. Symp. on Application of Computers to Fluid Dynamics Analysis and Design at Polytechnic Institute of Brooklyn, New York, Jan.3-4, 1973. (Submitted to Computers and Fluids for publication).

24. STEVENSON, J.F.: Flow in a tube with a circumferential wall cavity.
    ASME Transact., Series E, J. of Applied Mechanics, Vol.40, No.2, June 1973, pp 355-361.

25. UNDERWOOD, F.N. and MUELLER, T.J.: Numerical studies of the steady, axisymmetric flow through a disc-type prosthetic heart valve. Proc. 25th ACEMB, Bal Harbour, Florida, October 1972, p. 273.

26. UNDERWOOD, F.N.: Numerical and experimental studies of the steady, axisymmetric flow through a disc-type prosthetic heart valve. M.S. Thesis, U. of Notre Dame, Notre Dame, Indiana, August 1972.

27. MUELLER, T.J., et al.: On the separated flow produced by a fully open disc-type prosthetic heart valve. ASME 1973, Biomechanics Symposium Proceedings, AMD-Vol.2, Atlanta, Georgia, June 1973, pp 97-98.

28. MUELLER, T.J., LLOYD, J.R. and UNDERWOOD, F.N.: Unpublished data. University of Notre Dame, Notre Dame, Indiana, 1973.

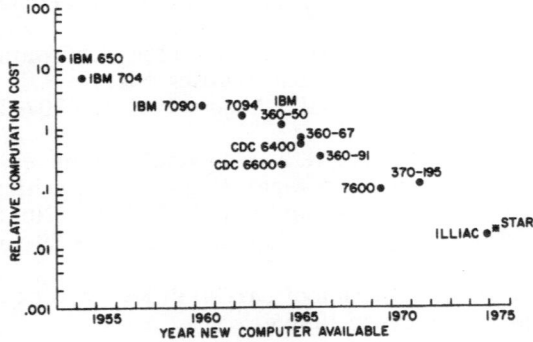

Fig. 1 - Trend of computational cost for computer simulation of a given flow (from Ref. 2)

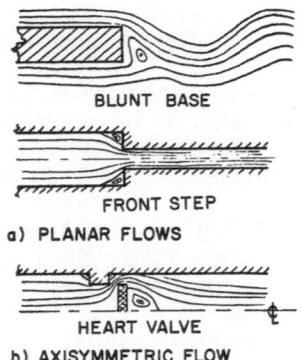

Fig. 2 - Separated flows

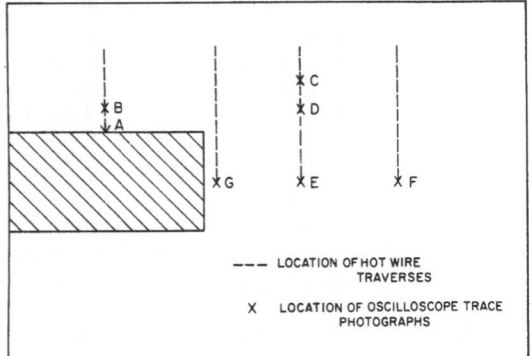

Fig. 3 - Locations of anemometer traverses and oscillograph photographs

403

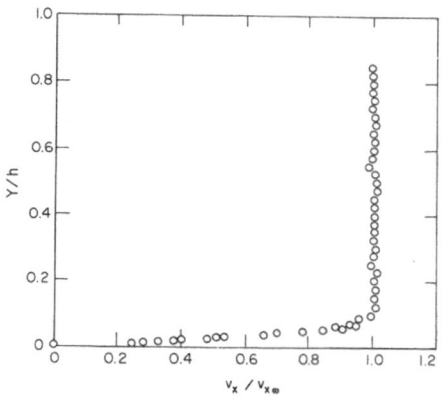

Fig. 4 - Inflow velocity profile, $Re_h = 16,295$

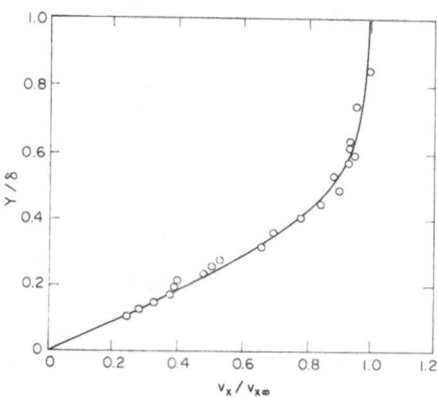

Fig. 5 - Boundary layer velocity profile, $Re_h = 16,295$, $\delta = 0.121\,h$

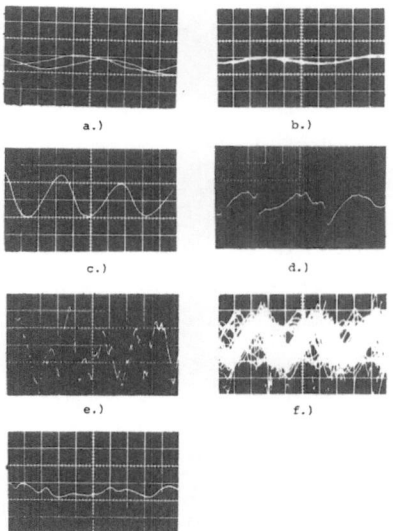

Fig. 6 - Hot wire voltage-oscilloscope photographs

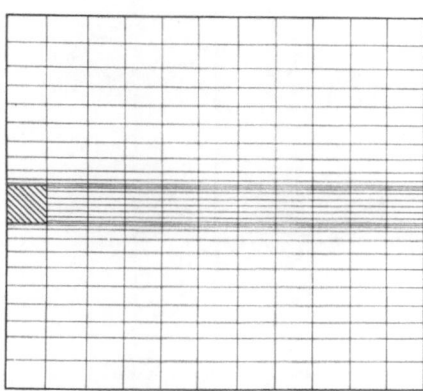

Fig. 7 - Variable rectangular grid

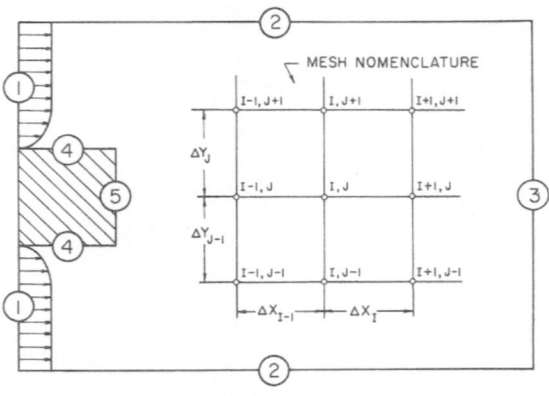

MESH NOMENCLATURE

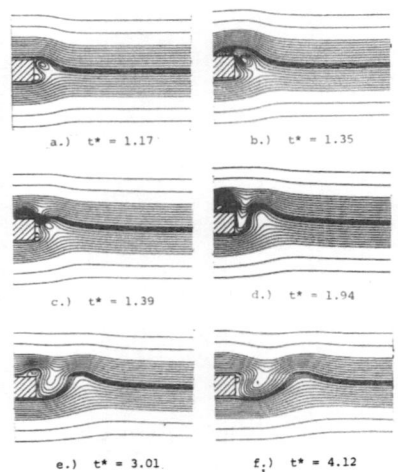

a.) t* = 1.17   b.) t* = 1.35

c.) t* = 1.39   d.) t* = 1.94

e.) t* = 3.01   f.) t* = 4.12

Fig. 8 - Grid and boundary nomenclature

Fig. 9 - Numerical development
Re$_h$ = 16,295,
δ = 0.121 h

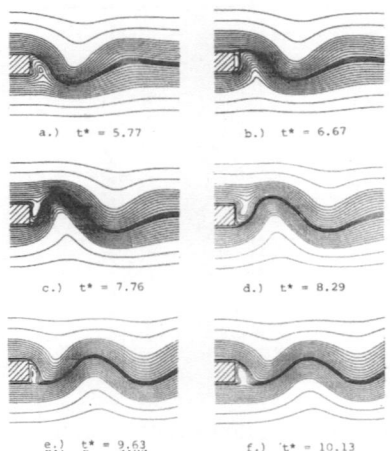

a.) t* = 5.77   b.) t* = 6.67

c.) t* = 7.76   d.) t* = 8.29

e.) t* = 9.63   f.) t* = 10.13

Fig. 10 - Numerical solution,
Re$_h$ = 16,295,
δ = 0.121 h

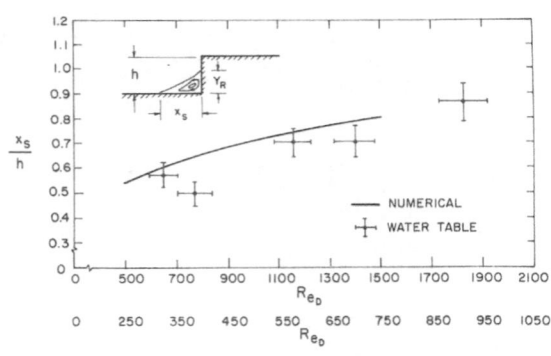

Fig. 11 - Separation point versus Reynolds
number for zero bleed (i.e., u_B=0)

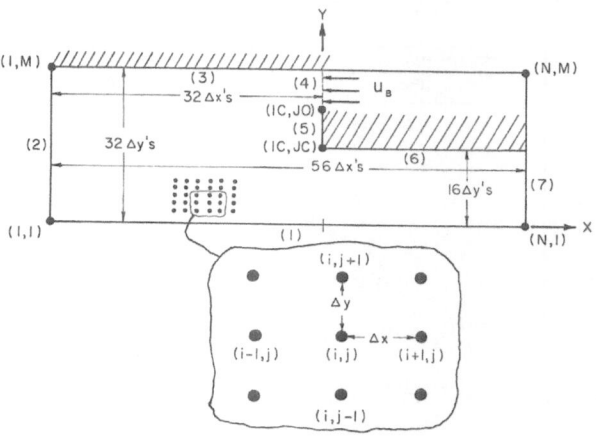

Fig. 12 - Front step geometry, boundary
labeling convention, mesh nomenclature

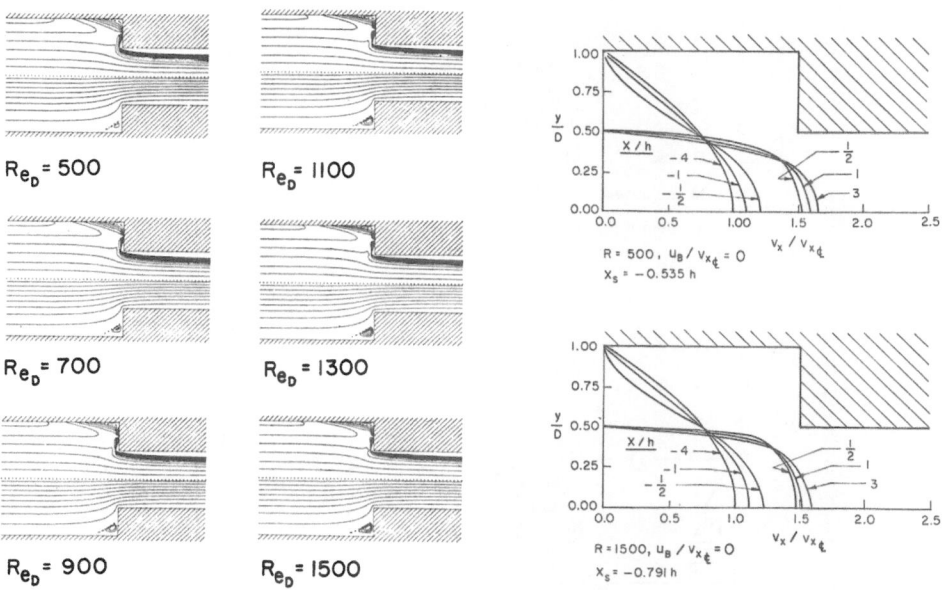

Fig. 13 - Contour vorticity-stream
function plots for zero bleed

Fig. 14 - Velocity profiles up and
downstream of the step
at Re_D = 500 and 1500,
zero bleed

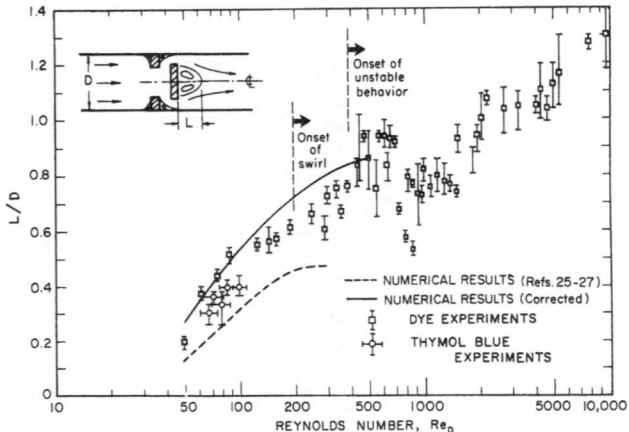

Fig. 15 - Length of separated region behind
disc versus Reynolds number

Table 1

Mesh Coordinates

Mesh	B		C		D		E		F	
	I	J	I	J	I	J	I	J	I	J
Coarse	21	9	26	9	30	11	32	11	62	15
Medium Test	41	17	51	17	59	21	63	21	123	29
Medium Reg.	61	17	71	17	79	21	83	21	165	29
Fine	61	17	76	25	88	31	94	31	184	43

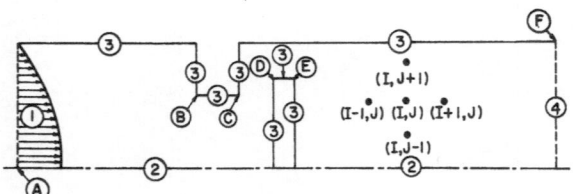

Fig. 16 - Geometry, mesh nomenclature, and
boundary labeling convention

LIST OF FIGURES

Figure No.

NOMENCLATURE

C	Courant number
D	diameter or channel half width
f	arbitrary function
h	base height or step height
L	length of separated region behind disc or characteristic length
P	pressure
Re_D	Reynolds number based upon D
Re_h	Reynolds number based upon h
Re_L	Reynolds number based upon L
r	radial direction
t	time
v	velocity
x, y	cartesian coordinates
z	axial direction
α	kinematic viscosity
α_e	effective or artificial viscosity
Δ	increment
δ	increment or boundary layer thickness
θ	streamline angle or cylindrical coordinate
μ	absolute viscosity
ξ	vorticity
ρ	density
τ	shear stress
ψ	stream function

NOMENCLATURE (continued)

Subscripts

B	base bleed
$\mathcal{CL}$	centerline
r	radial
ij	condition at ij mesh point
crit	critical value
max	maximum value
s	steady state value or separation
x	Cartesian coordinate
y	Cartesian coordinate
z	axial
∞	free stream

Superscripts

n	time level
*	dimensionless

STABILITY OF EXPLICIT TIME DEPENDENT TREATMENT

OF HYPERBOLIC BOUNDARY PROBLEMS

J. SMOLDEREN

von Karman Institute for Fluid Dynamics
Rhode St Genèse, Belgium

1. INTRODUCTION

It is well known that the explicit time-marching finite difference
reatment of hyperbolic equations may lead to severe stability problems.
von Neumann (Ref. 1) has given a necessary and often sufficient condition for stability in the case of linear equations with constant coefficients and uniform mesh. The criterion has been widely and successfully used in more general cases, including linearized versions of non
linear problems, the only limitation in its use being the complexity
of the algebraic manipulations required to work out the criterion for
partial differential systems of high order or involving more than two
independent variables.

Unfortunately, the von Neumann analysis does not generally apply to
problems involving boundary conditions. Many examples show that instability may occur even if the von Neumann criterion is satisfied. The
condition is still necessary but far from sufficient.

A general theory, due to Godunov and Ryabenkii (Ref. 2) leads to a
more restrictive criterion which is usually found to be sufficient to
ensure stability in the presence of boundary conditions.

A simplified presentation of this theory and the resulting criterion will be outlined and illustrated by a simple example.

A more thorough mathematical treatment will be found in a recent
paper by Kreiss (Ref. 3).

2. EXPONENTIAL SOLUTIONS OF FINITE DIFFERENCE EQUATIONS

Finite difference discretization of a partial differential equation
or system with constant coefficients leads to a linear difference equation or system, the coefficients being also constant, if a uniform
mesh is used. Exponential solutions exhibiting separation of variables
therefore also exist for the difference equation.

Consider, for instance, the simplest case of two independent variables t and x, t being the time-like marching variable (in a steady
state supersonic problem, the time-like variable will be the streamwise

coordinate). Let the regular mesh system be defined by

$$t_k = t_0 + k\Delta t \qquad (k = 0,1,2,\ldots)$$

$$x_j = x_0 + j\Delta x \qquad (j = 0,1,2,\ldots)$$

and let u_j^k be the approximate values of the unknown function(s) at the mesh point (t_k, x_j), resulting from the difference equations. These equations being linear with constant coefficients, it is easy to construct particular solutions with separated variables of the form

$$u_j^k = \text{const. } \rho^k \lambda^j \qquad (1)$$

which are analogous to the exponential and Fourier modes for the corresponding differential equations.

Substituting the above expression(s) in the difference equation(s) one obtains after cancelling the common factor, a relation connecting ρ, the temporal amplification factor and λ, the spatial amplification factor :

$$F(\rho,\lambda) = 0 \qquad (2)$$

This represents the characteristic relation for the difference scheme. In the case of a system of equations, this relation is obtained by expressing that the determinant of the linear algebraic system for the constant coefficients introduced by (1) is zero.

The von Neumann criterion expresses that the temporal amplification factor ρ (generally a complex number) must have a modulus bounded by $1 + O(\Delta t)$ for all values of λ which represent spatial Fourier modes, i.e., for

$$\lambda = \exp \ (i\pi n/N) \qquad (n = 1,2,\ldots,N) \qquad (3)$$

(N being the number of spatial meshes used).

If boundary conditions are imposed at either one or both boundaries, x_0 and $x_0 + N\Delta x$, of the spatial domain considered, then these Fourier modes will generally not be acceptable solutions of the problem because of incompatibility with the boundary conditions.

However, during an initial part of the progression in time, the influence of the boundary conditions will not be felt inside the numerical region of dependence of the initial data. The solution in this region will blow up if the von Neumann criterion is not satisfied so that the criterion still represents a necessary condition for stability.

Let us assume that the boundary conditions are also linear with constant coefficients. Considering the perturbations to a given solution, one may reduce the study to the case of homogeneous boundary conditions of the form

$$\sum_{j,\ell} a_{j,\ell} u_j^{k-\ell} = 0 \tag{4}$$

where the summation involved a fixed number of indices, defining a few neighbouring points to the boundary and a few earlier time steps.

It is obviously possible to combine solutions of the form (1) corresponding to the same value of ρ to satisfy a suitable set of boundary conditions of the type (4). Indeed, for ρ given, the characteristic relation (2) represents an algebraic equation for λ say of the rth degree with r roots : $\lambda_1, \ldots \lambda_r$. (In the case of equal roots there will exist polynomial-exponential solutions of the form

$$u_j^k = \text{const } P(j) \, \rho^k \, \lambda^j$$

where P is a polynomial in j).

A linear combination of the form

$$u_j^k = \rho^k \sum_{\nu=1\ldots r} A_\nu \, \lambda_\nu^j \tag{5}$$

will be solution of the boundary problem if the boundary conditions (4) allow unique determination of the coefficients A_ν. It can easily be shown that the number of boundary conditions to be specified in order to define the time marching procedure is indeed equal to r, the number of λ roots of the characteristic equation (2). This number may actually be higher than the order in x of the partial differential equation or system to be treated, if one uses a discretization of order of accuracy higher than the first. In such case, additional boundary conditions, which are not at all specified by the physical situation to be treated, must be invented, with suitable regard for accuracy require-

ments. This need for a choice involving a degree of arbitrariness, represents a serious difficulty as it may strongly influence the stability of the calculations.

3. STABILITY OF MODES COMPATIBLE WITH THE BOUNDARY CONDITIONS

Substitution of the expression (5) of the combined mode in the r
boundary conditions of type (4) will yield a linear algebraic system
of r equations for the r unknown coefficients A_1, ..., A_r. A non trivial solution will exit only if the determinant of the system is zero.
This yields a new algebraic condition connecting ρ, λ_1, ..., λ_r.

Expressing that λ_1, ..., λ_r are the r roots of the characteristic
equation (2) will lead, after considerable algebraic manipulations,
to a new algebraic equation for the temporal amplification ρ, the roots
of which may be considered to be eigenvalues for our linear homogeneous boundary problem.

A new necessary condition for stability may then be established, by
requiring that the roots of the final equation in ρ are all of modulus
smaller than $1 + O(\Delta t)$.

The inextricable complexity of this approach will be understood if
it is realized that conditions may be prescribed at both boundaries
of the spatial region, so that factors of the type λ^N will occur in
the equations defining the coefficients A and in the corresponding
determinant, N being the number of spatial meshes. We must therefore
expect that the final equation for ρ will be of a degree which increases indefinitely with N, and the same will be true for the number
of its roots.

This apparently hopeless situation can fortunately be resolved by
showing that the number of unstable modes is necessarily bounded for
all values of N.

4. UNSTABLE BOUNDARY MODES

The number of exponential modes compatible with suitable boundary
conditions have been shown to be increasing with the number of spatial
meshes, as is the case for the Fourier modes (eq. 3). However, the new
family of modes has very different properties, in general, and the

coefficients of an expansion in terms of these modes may not be bounded for increasing N.

However, we are interested only in stability conditions here, and may therefore limit ourselves to the study of the eventual unstable modes corresponding to $|\rho| > 1 + \varepsilon$ (ε positive, fixed).

It may be shown that if the finite difference scheme satisfies the von Neumann criterion, then there occurs, for $N \to \infty$, a decoupling of the influence of the boundary conditions applied at different boundaries of the spatial domain, for the unstable modes. Each boundary may therefore be considered separately and this introduces powers of λ with low exponents independent of N. The resulting equation in ρ for the unstable modes will therefore also be of a relatively low degree and independent of N.

Of course, the number of stable modes will always be unbounded, for increasing N, because the total number of modes is unbounded.

It may actually be shown that if s boundary conditions are imposed at the lower boundary ($j = 0$) and $r - s$ at the upper boundary ($j = N$), then the characteristic relation (2) will have exactly s roots λ with modulus smaller than unity and ($r-s$) roots with modulus larger than unity, for all ρ such that $|\rho| > 1$ (unstable modes).

This important lemma leads to the following property. At the limit $N \to \infty$, the determinant of the algebraic system for the coefficients A defining the mode (5) may be factored and the condition on the λ reads

$$P_s(\lambda;\rho) \; P_{r-s}(\lambda;\rho) = 0$$

where P_s, P_{r-s} are polynomials of degree s and r-s respectively. The coefficients of P_s depend only in the lower boundary conditions and those of P_{r-s} on the upper boundary conditions. This is the decoupling property which may be expressed as follows. Possible unstable modes will be of the form

$$u_j^k = \rho^k \; \Sigma \; A_\nu \; \lambda_\nu^j \qquad\qquad \text{with } |\lambda_\nu| < 1 \text{ for } 1 \leq \nu \leq s \qquad (6)$$

$$u_j^k = \rho^k \; \Sigma \; A_\nu \; \lambda_\nu^j \qquad\qquad \text{with } |\lambda_\nu| > 1 \text{ for } s+1 \leq \nu \leq r \qquad (7)$$

The modes (6) are called lower boundary modes and must satisfy the s
boundary conditions imposed at the lower boundary. The upper boundary
modes (7) must satisfy the upper boundary conditions. Such a decoupling
will not occur for stable modes.

These considerations, which dramatically reduce the algebraic manipulations, may also be generalized to multi-dimensional spaces and
even curved boundaries may then be treated, using a local approach.

It should be mentioned that the establishment of a stability criterion, expressing that the moduli of ρ for all boundary modes are
smaller than $1 + O(\Delta t)$, is still very tedious in non trivial cases and
considerably more difficult to work out than the von Neumann criterion.

Kreiss has shown that the conditions so obtained are not always
sufficient (Ref. 3) but the example he used to show the existence of
non exponential, polynomial modes, must be considered as rather mar-
ginal (Leap-fog scheme).

5. EXAMPLE : SUFFICIENT STABILITY CRITERION FOR AN ADDITIONAL BOUNDARY CONDITION

Using the concept of boundary modes, it is sometimes possible to
obtain a sufficient condition for the absence of unstable exponential
boundary modes, without even specifying the scheme and its characteristic relation (eq. 2). We illustrate this using the example of a single
hyperbolic equation of the first order for the unknown function $u(x,t)$.
Assume that a higher order scheme is used so that an additional condition must be introduced at say $j = 0$. This assumes that the characteristics are running downward in the (t,x) plane.

The new value u_0^{k+1} must then be defined to proceed with the marching
scheme. Let us select a linear condition expressing u_0^{k+1} in terms of
the earlier values at the neighbouring points $j = 0$ and $j = 1$:

$$u_0^{k+1} = a\, u_0^k + b\, u_1^k$$

First order accuracy at the boundary (which is compatible with
second order accuracy of the scheme) will require

$$a + b = 1$$

According to the theory described in section 4, there will be only one possible unstable boundary mode of the form

$$u_j^k = A\rho^k \lambda^j \qquad\qquad |\rho| > 1; \; |\lambda| < 1$$

This leads to the following homogeneous equation for A :

$$\rho A = aA + (1-a) A\lambda$$

Therefore, the mode will exist only if

$$\rho = a + (1-a)\lambda \qquad\qquad\qquad (8)$$

This mode will represent an unstable lower boundary mode only if $|\rho| > 1$, $|\lambda| < 1$. If these inequalities cannot be simultaneously satisfied, taking account of (8), then no unstable exponential mode will exist (note that this will represent a sufficient but not always necessary condition, because λ and ρ must also satisfy the characteristic relation of the scheme, which may not be possible for $|\rho| > 1$).

Obviously, the image in the complex ρ plane of the unit circle of the λ plane, under the linear transformation (8), is a circle with radius $|1-a|$ centered at $\rho = a$. No unstable exponential mode will exist if this circle is entirely inside the unit circle $|\rho| = 1$. This will be the case if

$$-1 < 2a-1 < 1, \qquad -1 < a < 1$$

hence, if

$$0 < a < 1 \qquad\qquad\qquad\qquad (9)$$

(9) therefore represents a sufficient condition for stability as far as exponential modes are concerned.

The following interpretation of condition (9) is interesting: u^{k+1} is taken equal to the value of u corresponding to the preceding time step k, at some point located between $j = 0$ and $j = 1$ (weighted average). But we have assumed that the characteristics of the hyperbolic equation run towards decreasing values of x for increasing t (otherwise an essential boundary should have been imposed at $j = 0$ to define the solution). Therefore, the additional condition (8), taking account of the criterion (9), turns out to be in qualitative agreement with the "characteristic rule" advocated by Moretti (Ref. 4), on the basis of physical arguments. One should, however, be careful about the use of physical arguments in the discussion of a purely numerical problem.

REFERENCES

1. O'BRIEN, G.G., HYMAN, M.A. and KAPLAN, S.: A study of the numerical solution of partial differential equations. J. Math.and Phys., Vol. 29, 1950, pp 223-251.

2. GODUNOV, S.K. and RYABENKII, V.S.: Special stability criteria for boundary condition problems for non self-adjoint finite difference equations. Uspekhi Mat.Mauk. Vol. 18, 1969, p. 3.

3. KREISS, H.O.: Boundary conditions for difference approximations of hyperbolic differential equations. in Advances in numerical fluid dynamics, AGARD Lecture Series No 64, 1973.

4. MORETTI, G.: The importance of boundary conditions in the numerical treatment of hyperbolic equations. Brooklyn Polytechnic Institute, PIBAL Report No 68-34, 1968.

IMPROVING OF THE NUMERICAL SOLUTIONS
BY USING ANALOGUE SUBROUTINES

Listing of the graphs
========================

Table of contents
==================

IMPROVING OF THE NUMERICAL SOLUTIONS
BY USING ANALOGUE SUBROUTINES

G.C. VANSTEENKISTE
Professor of Engineering
University of Ghent
Coupure Links 533
B-9000 Ghent
Belgium

INTRODUCTION

There exist many approaches to the reduction of the time required to obtain numerical solutions using general-purpose digital computers. A number of these involve the utilization of special-purpose peripheral devices which function as subroutines, to be called by the digital computer program as required. Most frequently these peripheral devices are digital in nature, containing the memory, arithmetic, and control required to perform computations which would be much more time-consuming if performed by the central processor unit (CPU). A less widely used technique involves the use of analog devices in the peripheral unit. In this case, the digital computer is coupled to an analog unit by means of interface hardware.

Such a system is effective in increasing the rate at which the digital computer can solve certain classes of problems because the analog device is essentially a fully parallel computer, the solution time being independent of the problem complexity (number of arithmetic operations). The use of such a peripheral, therefore, involves essentially a trade-off between the time required for the data to pass through the interface in the two directions and the time that would be required to perform the calculations in the CPU.

The present text is directed to a specific application : the treatment of fluid flow problems. A variety of basic techniques are available for the numerical solution of transient fluid flow problems, such as Monte Carlo methods, Methods of Characteristics, Finite Element Methods and Finite Difference Methods. The last ones however are used almost exclusively for the treatment of complex systems (fig. 1).

HYBRID SYSTEMS

The possibility of operating many analog computing elements in parallel permits real-time simulation of large dynamical systems. As a matter of fact, computing-element bandwidths make it possible to simulate many systems on a speed-up ("fast") time scale. Unlike the situation in digital machines, this computing speed is difficult to trade off for increased accuracy.

The term true hybrid is applied to those combined computing systems containing analog as well as digital hardware in appreciable amounts. The ways in which analog and digital computer installations can be used together can be classified into two broad categories : *Unilateral operation,* in which information flows across the interface between the analog and the digital sections in only one direction ; and *bilateral operation,* in which the flow across this interface is in both directions. Both methods require conversion equipment at the interface. Figures 2a and 2b illustrate two types of unilateral systems. Note that only one type of converter, either analog-digital or digital-analog is required at each interface. In such systems the analog or the digital computer can be regarded as playing the part of a complex and elegant input or output device. Fig. 3 illustrates a bilateral hybrid system. Such systems are characterized by a closed loop formed by the digital computer, the digital-analog conversion devices, the analog computer, and the analog-digital converters. In addition to those major units bilateral hybrid computer systems also include a number of other important devices. *Multiplexers* and *demultiplexers* are employed to permit the converters, which translate continuous DC voltages into digital code and vice versa, to be time-shared. *Hold devices* are required to maintain the continuously varying analog signals at constant values for a time sufficiently long to permit conversion, and to maintain the output voltage of digital-analog converters at constant levels while conversion is taking place. *Buffers* are required to adjust voltage levels and pulse shapes in a manner compatible with the digital and analog computers in use. Finally, *timing and control circuitry* is required to synchronize the processes in the various units comprising the hybrid loop. Each of these units and subunits manifests input-output relationships which deviate from the ideal or specified behavior. In designing hybrid computer systems and in evaluating their performance it is therefore necessary to determine the effect of the non-ideal behavior of each sub-unit upon the overall system dynamics. An important major application of hybrid computer systems involves the interconnection of the analog-digital loop with major items of hardware. Because of their wide-spread application in telemeter and communications systems, unilateral hybrids no doubt outnumber bilateral systems by a wide margin.

It is possible to identify certain general computational requirements which suggest the hybrid computer approach. In this connection, it should be noted that hybrid techniques are frequently employed to overcome certain shortcomings in present-day analog or digital computers. As these limitations are reduced or eliminated, it is possible that some of the applications for hybrid computing systems will disappear (table I).

The increasing speed and decreasing cost of general-purpose digital computers as well as the advent of time-shared on-line digital systems may well narrow the range of applications for hybrid equipment. On the other hand, the introduction of integrated circuitry makes possible the realization of linkage equipment in increasingly simple, reliable and economical form thus making hybrid techniques increasingly attractive in areas in which pure analog or pure digital computers may be adequate. The following are, the chief motivations for interconnecting digital and analog computers :

- *To combine the speed of an analog computer with the accuracy of a digital computer.*
- *To permit the use of system hardware in a digital simulation.*
- *To increase the flexibility of an analog simulation by using digital memory and control.*
- *To increase the speed of a digital computation by utilizing analog subroutines.*
- *To permit the processing of incoming data which are partially discrete and partially continuous.*

Not only analog hardware but also human operators make necessary the development of an analog-digital interface in such sophisticated computer systems. Once the necessity for an interface has been established, it becomes possible to examine each computational task to determine whether it can better be performed on the analog or the digital side of the hybrid computer system.

Combined simulation is motivated by the hope of combining analog-computer speed in some parts of the simulation with digital accuracy and function-generating power in other parts of a simulation. Digital storage and analog output of several complicated functions of two to four variables is a frequent application of combined-simulation techniques.

Applications of this type combine not only the *best* features of analog and digital computation, but also their *worst,* viz., analog-computer inaccuracy and the sampled-data bandwidth limitations of the digital computer. Digital operations imply *sampled-data operations, digital processing delays,* and *quantization,* all of which produce essential errors in the representation of continuous dynamical-system variables. These errors must be evaluated and mitigated through careful choice of the portion of the simulation carried out on the digital computer. While quantization errors rarely cause trouble, digital sampled-data operations and processing delays limit the digital part of a combined simulation to slowly varying variables.

Even so, subtle loop-instability effects are encountered ; it is best to check each problem on a reduced time scale to see whether sampling rates are fast enough for stability.

The most effective utilization of hybrid computers has to be expected for those problems which allow the full use of the pass-band of new analog elements ; in these cases, the competitivity of hybrid computers vs. purely digital ones may be higher than for analog computers in the past.
Often this point is not sufficiently taken into account, and the evaluation of hybrid computers is done for applications concerning the solution of sets of slow ordinary differential equations, leading to poor comparison with full digital simulation.

Let us assume that today's typical digital computer used for simulation purposes has a word length of 24 bit, at least. Hence, the local round-off error, which is equivalent to the 'static' error in analog computers, is less than 10^{-7}. The dynamic error in a digital simulation program is caused by the truncation error of numerical integration and depends on the actual integration formula, used in the program, and on the step size. Because a digital simulation program has to be executed in a quasi-parallel way, the step size of any integration is equal to what is called the "frame-time", i.e. the total execution time of all operations within one step of integration.

As given in reference [1] fig. 4 shows a comparison of the error vs. frequency for the above digital computer and a modern analog computer (with 0,01 % static accuracy and a 300 KHz, 3dB pass-band for amplifiers). The comparison is made for a typical dynamic model, of a size corresponding to a 100 amplifiers analog computer (with a typical implementation of integrators, summers, multipliers and function generators) ; the execution on the digital computer of one time step for this model, with leasttime programming (in machine language), requires about 4000-5000 memory cycles, depending on the integration formula used (frame-time of 8-10ms, for a memory cycle of 2 µs.). Fig. 4 shows the result of an error analysis for three different integration formulas (Euler-, trapezoidal- and Adams-Bash-forth-integration) assuming a frame time of 10 ms. As the diagram shows, the "bandwidth-to-accuracy ratio" even of the Adams-Bashforth-formula (which is of relatively high order, resulting in small dynamic errors) is 1 - 2 decades smaller than that of the analog computer. The situation is much worse in the case of one-step formulas such as the trapezoidal rule or even the Euler integration. Additionally, one has to consider, that for large-scale problems a frame time of 10 ms can be considered as being the lower boundary, even if the computer is very fast.

The difficulties of pure digital simulation are not entirely characterized by yet unsatisfying computation speed of appropriate computers.

Even more serious problems arise by the potential instability of numerical integration. If truncation errors would be the only problem, this could easily be overcome by using multistep methods of adequate high order (disregarding for the moment the starting problems). What makes the point crucial is that the risk of causing instabilities in the solution of differential equations by discrete variable methods increases with the order of a multistep formula.

For a detailed analysis of hybrid systems, the reader is referred to reference [1].

ANALOG COMPUTER ORIENTED APPROACH

In analog computer oriented hybrid systems, relatively elegant and expensive analog computers perform the bulk of the computations, usually involving the integration of systems of ordinary differential equations, while the digital partner is employed in a subsidiary capacity for control, function generation, memory, etc.

In the continuous-time-discrete-space (CTDS) hybrid method, continuous integration with respect to time is performed by analog integrators. The space domain to be simulated is broken up into two or more sections, and analog elements are provided only for a single section. The analog system is then used repeatedly to provide solutions for each of the sections of the space domain. If now an analog system is used to represent only a portion of the overall field, the "boundary" conditions at the two ends of this system are generally not known but are rather a part of the solution being sought. It is therefore necessary to employ iterative techniques to "match" all the time-shared sections.

Fig. 5 shows a simple three-integrator system. A digital computer together with the necessary linkage equipment is employed to simulate adjacent sections in the positive and negative x-directions by applying to the analog network the transient potentials which would, as indicated in fig. 5, be generated by integrators immediately to the left and to the right of the section shown, and to record the potentials generated by the analog system. The latter transients become the "boundary" potentials to be applied to the analog system when it is used subsequently to simulate adjacent sections in the space domain. The analog system is used successively to represent the sections of the space domain starting with the left-most section and ending with the right-most section. Initially, the "boundary values" for the interior sections are arbitrarily assumed. After the first iterative pass through the system, the digital computer will have stored in its memory the first "solution values" generated by the analog system in each section. These, then, serve as the boundary excitations for the succeeding iterative cycle in which the analog system again sequentially simulates each of the sections in the space domain. These iterative cycles are repeated until convergence to the correct solution has been obtained.

The need for this high number of iterations as well as the errors introduced in the hybrid loop ; greatly limit the applicability of this approach. It has merit when large numbers of nonlinear one-dimensional diffusion equations must be treated.

In the continuous-space-discrete-time (CSDT) hybrid method, the problem time variable is discretized ; so that solutions are generated at successive time levels. In approximating the time derivative, two possibilities exist : forward difference and backward difference approximations. Equations employing forward differences, are solved explicitly, a process which is relatively simple computationally but which has the inherent possibility of computational instability. If the ratio of the time increment to the space increment is too large, round-off errors made in the course of the solution will gradually build up until they overshadow the solution, thus making it worthless. In order to obtain satisfactory solutions by this method, it is necessary to make the time increment relatively small, that is to take many time-consuming steps. Using backward differences, the one-dimensional diffusion equation results in :

$$\frac{d^2 u_i}{dx^2} - \frac{1}{\Delta t} u_i = u_{i-1}(x) - S_i(x)$$

with boundary conditions

$$u_i(0) = u_0(i\Delta t) \ , \ \frac{du_i}{dx}(1) = 0$$

and with

$$S_i(x) = S(x, \ i\Delta t)$$

A digital computer offers an obvious means to provide the functions $S_i(x)$ and $u_{i-1}(x)$. The $S_i(x)$ functions are stored originally as part of the problem statement whereas $u_{i-1}(x)$ has been stored as a result of the previous solution at the i-1 interval (fig. 6).

Unfortunately, there are a number of severe problems which limit drastically the usefulness of this scheme. First of all, at each time interval we must solve a boundary condition problem. Starting at x = 0 and integrating towards x = 1 we must guess initially $\frac{du_i}{dx}(0)$ in order to have the derivative vanish at x = 1.

This implies an iterative process in itself to converge on the initial condition which generates the correct solution satisfying the final condition. The real problem however, is that the above equation is unstable, and the smaller we make Δt, the more severe is the instability.

This will make it difficult to repeat solutions and will put a definite lower limit on Δt.

On the other hand the backward difference representation for $\partial u / \partial t$ is rather inaccurate, having a large residual error given by

$$\frac{\Delta t}{2} \left. \frac{\partial^2 u}{\partial t^2} \right|_i$$

This implies that Δt should be made quite small for reasonable accuracy, which is quite incompatible with the stability problem described in the previous paragraph. Thus the continuous-space, discrete-time method for hybrid solution of partial differential equations would appear to be inferior to the conventional discrete-space continuous-time method.

DIGITAL COMPUTER ORIENTED APPROACH

Since the accuracy of analog operations is limited by the quality of the electrical components, the use of analog subroutines may entail a degradation of solution accuracy. However, as is the case in the method described below, the use of the analog subroutine in an error-correction mode entirely obviates this disadvantage. The approach is applicable to elliptical, parabolic, hyperbolic as well as biharmonic partial differential equations. Problems are programmed for convention digital computation, and the digital computer flow chart is analyzed to determine if some particularly time-consuming blocks could be better handled by analog rather than digital techniques, that is, by an analog subroutine. An analog unit is then especially constructed for this computational task. Because of its special-purpose nature, such a unit is far less expensive than a general-purpose analog computer and entails few of the usual programming or maintenance problems. A computer system, embodying this concept has been developed and has been operating successfully at the University of California, Los Angeles, for over five years. The system was designed specifically to assist in the solution of a class of nonlinear partial differential equations, and a wide variety of engineering problems have been successfully treated (reference [2]. A similar hybrid system is now under development at the University of Ghent and will be used for partial differential equation studies in collaboration with the Von Karman Institute.

When transient field problems are treated on a digital computer, all partial derivatives must be approximated by finite difference expressions. As seen above, in approximating the time derivative, for parabolic partial differential equations, backward differences are employed. The computational procedure is implicit with no danger of computational instability but requires now the solution of a large number of simultaneous algebraic equations at each time level which involves time-consuming matrix inversion of large sparse matrices. A variety of techniques for solving these simultaneous difference equations have been introduced over the years ; the so-called successive over-relaxation method and the alternating direction implicit method are by far the most widely used. It is here that the hybrid computer makes its principal contribution. The analog subroutine approach, as utilized in the so-called discrete-space-discrete-time (DSDT) computer constitutes indeed an alternative, designed to save computer time at the expense of additional hardware. A principal advantage of the DSDT hybrid method, is furthermore its interactive capability permitting the user to combine his heuristic insight into the physics of the problem with formal mathematical and algorithmic methods. The difficulty of studying systems whose dynamics are characterized by partial differential equations in more than one space dimension lies not only in the complexity of the problem but also in the fact that it is virtually impossible to express all the constraints in an algorithmic form. This bottleneck has severely limited the utility of the conventional computer approach to complex inverse problems. The DSDT hybrid computer method can make an important contribution by permitting powerful and useful computational algorithms to be programmed on the digital computer, while allowing the specialist to introduce his judgments and insights as they are required in the computational process, without first translating them into an abstract programming language. Another advantage of the approach, when compared to competing digital and analog methods, is its efficiency in handling nonlinear problems. A disadvantage lies in the fact that the number of finite difference grid points which can be treated efficiently with a given system is limited by the available hardware.

Consider the general mathematical problem (see ref. [2])

$$L(\phi) = d \tag{1}$$

where ϕ and d are either vectors, functions, or matrices, and L is an operator. Such a problem is frequently treated by iterative techniques. For the rth iterative cycle,

$$M(\delta^{(r)}) = d - L(\phi^{(r)}) \tag{2a}$$
$$\phi^{(r+1)} = \phi^{(r)} + \theta\delta^{(r)} \tag{2b}$$

where M is an invertable linear operator, θ is a scalar relaxation factor, and δ is a correction term. Provided that the convergence criteria are satisfied, $\phi^{(r)}$ will converge to ϕ, the solution of (1). The rate of convergence is influenced by the choice of M and θ. If L is a linear operator, the solution ϕ is obtained after a single iteration by choosing, M equal to L and $\theta = 1$. Using analog subroutines, the computations represented by (2) are carried out as follows :

- An error vector $\Delta^{(r)} = d - L(\phi^{(r)})$ is computed digitally.
- $\Delta^{(r)}$ is scaled and converted into a set of analog signals.
- The equation $M(\delta^{(r)}) = \Delta^{(r)}$ is solved for $\delta^{(r)}$ by an analog technique.
- The analog signals $\delta^{(r)}$ are converted into digital form and unscaled.
- $\phi^{(r+1)}$ is computed digitally using (2b).

This sequence is repeated until

$$||\Delta^{(r)}|| < \varepsilon \tag{3}$$

where $|| \;\; ||$ is a suitably chosen norm, and ε is a scalar parameter controlling solution accuracy.

The basic reason for calculating $\delta^{(r)}$ by analog means is to reduce the time required to solve (2). The analog subroutine presents a computing speed advantage relative to digital methods for two reasons. The parallel- computing nature of analog circuitry permits the term $\delta^{(r)}$ to be calculated more rapidly ; more important, however, as demonstrated below is that the number of iterations (r) required to attain convergence of (2) is usually far smaller.

It is important to recognize that errors or inaccuracies in the analog circuitry do not directly affect the solution accuracy provided only that convergence is achieved. This is the case because the analog circuitry is employed only to determine the error term $\delta^{(r)}$. By suitably scaling the analog signals so as to utilize the maximum analog dynamic range during each iteration, the analog errors are made to constitute only a small percentage of $\delta^{(r)}$, regardless how small this correction term becomes as convergence is approached. It is this feature of the method that makes it possible to employ relatively simple and inexpensive analog components.

The structure of the analog unit is determined by the choice of the linear operator M. This operator should be selected so as to produce as rapid a rate of convergence as possible. For optimum speed advantage, each problem to be solved would require a different analog unit. As is shown below, however, it is possible to design analog subroutines which can accommodate large classes of problems without change of structure, and at the same time effect impressive increases in computing efficiency.

APPLICATION TO THE DSDT APPROACH

An important class of problems of the form of (1) is the linear algebraic system

$$L\phi = d \tag{4}$$

where ϕ and d are N-vectors, and L is a N x N matrix. If all the eigenvalues of L have positive real parts, (4) can be solved by the iterative scheme

$$\Delta^{(r)} = d - L\phi^{(r)} \tag{5a}$$

$$\delta^{(r)} = M^{-1}\Delta^{(r)} \tag{5b}$$

$$\phi^{(r+1)} = \phi^{(r)} + \theta\delta^{(r)} \tag{5c}$$

where M is a N x N matrix while $\Delta^{(r)}$ and $\delta^{(r)}$ are N-vectors.

Fig. 7 is a block diagram of the hybrid computer implementation of (5), where the analog network constitutes a realization of the matrix M. Since all mathematical operations, which must be performed by the analog network in order to generate the solution vector, are algebraic in nature, no integrators, capacitors, or other memory or delay elements are required in the analog network. If, moreover, M is a Stieltjes matrix (a symmetric positive-definite matrix all of whose off-diagonal elements are zero or negative), no electrical components other than passive resistors are required to realize M. The analog subroutine functions in the following manner :

- During each iterative cycle (r) the vector $\Delta^{(r)}$ is read out of the digital computer in a serial fashion. Each component of this vector is transformed into analog form (i.e., a scaled analog voltage proportional to its magnitude) by means of the digital/analog converter (DAC).

- The analog voltages generated sequentially by the DAC are distributed to an array of sample-hold units. Each of these units accepts an analog voltage (present for a brief period of time at the output of the demultiplexer) and maintains this constant voltage, until its input voltage is updated during the following iterative cycle. Each sample-hold unit, therefore, acts as a memory of one element of the vector $\Delta^{(r)}$. At the end of the digital computer readout, all these elements exist simultaneously at the outputs of the array of sample-hold units.

- The output of each sample-hold unit forms a separate input to the analog network. This network is essentially a parallel-operating algebraic equation solver. As soon as the entire vector $\Delta^{(r)}$ is available at the outputs of the sample-hold units, this network "relaxes" to the correct solution of (5b). Since the analog network contains no reactances, the setting time required for the "relaxation" is less than 1μs. The vector $\delta^{(r)}$, therefore, is immediately available at the output node points of the analog network.

- By means of the multiplexer, the output node points of the network are sampled sequentially.

- Each component of the vector $\delta^{(r)}$ is translated in turn into digital form by means of the analog/digital converter (ADC).

- The elements of the vector $\delta^{(r)}$ are read into the digital computer.

- Equation (5c) is solved on the digital computer to complete the rth iterative cycle.

Any errors introduced by imperfections in the hybrid loop affect only the error vector $\delta^{(r)}$, and are, therefore, second-order effects. If, for practical reasons, M cannot be made closely equal to L, then it is important that an optimal or near-optimal value be chosen for θ. If both L and M are symmetric and positive definite, it can be shown that the eigenvalues of $M^{-1}L$ are all real and positive. Let λ_{max} and λ_{min} denote the largest and smallest of these eigenvalues, respectively. The value of θ which maximizes the rate of convergence is then given by

$$\theta_{opt} = \frac{2}{\lambda_{min} + \lambda_{max}}$$

Although the exact calculation of λ_{max} and λ_{min} can be time-consuming, simple formulas for estimating them are often available.

This method is applicable to all systems of the type of (4) in which L is a stability matrix. Of particular interest, however, are classes of problems in which L is a sparse matrix of a special type. In the solution of partial differential equations by finite-difference techniques, the system of difference equations leads to a matrix L having only 3,5, or 7 nonzero elements in each row, depending upon whether the problem is formulated in one, two, or three space dimensions. N is then equal to the number of finite-difference grid-points in the space domain. The development of efficient techniques for the solution of such a system of algebraic equations is important because N is often very large and because the system of equations must usually (in the case of nonlinear problems) be solved several times at each of a large number of time levels.

To illustrate the efficacy of the analog subroutine approach to such a problem consider the equation

$$\frac{\partial^2 \phi}{\partial x^2} + \frac{\partial^2 \phi}{\partial y^2} = \frac{\partial \phi}{\partial t} \tag{7}$$

with the boundary conditions : $\phi = 1$ everywhere on the boundary for all t ; and the initial conditions : $\phi(x, y, 0) = 1 - \sin \pi x \sin \pi y$. Using a backward-difference approximation, the finite-difference approximation for (7) is

$$\frac{-\phi_{i-1,j,k} - \phi_{i+1,j,k} + 4\phi_{i,j,k} - \phi_{i,j+1,k} - \phi_{i,j-1,k}}{h^2}$$

$$= -\frac{\phi_{i,j,k} - \phi_{i,j,k-1}}{\Delta t} \tag{8}$$

where

$$\phi_{i,j,k} \equiv \phi(i\Delta x, j\Delta y, k\Delta t)$$

Letting $\Delta x = \Delta y = h = 0.10$, and $\Delta t = 0.01$, (8) becomes

$$- \phi_{i-1,j,k} - \phi_{i+1,j,k} + 5\phi_{i,j,k} - \phi_{i,j-1,k} - \phi_{i,j+1,k} = \phi_{i,j,k-1}$$

This set of finite-difference equations may be expressed in vector-matrix form as

$$L\phi_k = d_k = \hat{d}_k + b$$

where

$$
L = \begin{vmatrix}
L^\alpha & -I & 0 & 0 & 0 & 0 & 0 & 0 & 0 \\
-I & L^\alpha & -I & 0 & 0 & 0 & 0 & 0 & 0 \\
0 & -I & L^\alpha & -I & 0 & 0 & 0 & 0 & 0 \\
0 & 0 & -I & L^\alpha & -I & 0 & 0 & 0 & 0 \\
0 & 0 & 0 & -I & L^\alpha & -I & 0 & 0 & 0 \\
0 & 0 & 0 & 0 & -I & L^\alpha & -I & 0 & 0 \\
0 & 0 & 0 & 0 & 0 & -I & L^\alpha & -I & 0 \\
0 & 0 & 0 & 0 & 0 & 0 & -I & L^\alpha & -I \\
0 & 0 & 0 & 0 & 0 & 0 & 0 & -I & L^\alpha
\end{vmatrix}
; b = \begin{matrix}
b^{\alpha 1} \\
b^{\alpha 2} \\
b^{\alpha 2} \\
b^{\alpha 2} \\
b^{\alpha 2} \\
b^{\alpha 2} \\
b^{\alpha 2} \\
b^{\alpha 2} \\
b^{\alpha 1}
\end{matrix}
$$

$$\hat{d}_k = \phi_{k-1}$$

and where

$$
L^\alpha = \begin{vmatrix}
5 & -1 & 0 & 0 & 0 & 0 & 0 & 0 & 0 \\
-1 & 5 & -1 & 0 & 0 & 0 & 0 & 0 & 0 \\
0 & -1 & 5 & -1 & 0 & 0 & 0 & 0 & 0 \\
0 & 0 & -1 & 5 & -1 & 0 & 0 & 0 & 0 \\
0 & 0 & 0 & -1 & 5 & -1 & 0 & 0 & 0 \\
0 & 0 & 0 & 0 & -1 & 5 & -1 & 0 & 0 \\
0 & 0 & 0 & 0 & 0 & -1 & 5 & -1 & 0 \\
0 & 0 & 0 & 0 & 0 & 0 & -1 & 5 & -1 \\
0 & 0 & 0 & 0 & 0 & 0 & 0 & -1 & 5
\end{vmatrix}
$$

$$
b^{\alpha 1} = \begin{vmatrix}
2 \\
1 \\
1 \\
1 \\
1 \\
1 \\
1 \\
1 \\
2
\end{vmatrix}
\qquad
b^{\alpha 2} = \begin{vmatrix}
1 \\
0 \\
0 \\
0 \\
0 \\
0 \\
0 \\
0 \\
1
\end{vmatrix}
$$

and I is the 9 by 9 identity matrix.

The analog network has nine internal node points in the x and the y directions. The solution was carried out for 50 time levels with 6 decimal-place precision for each solution point. A total of 110 iterative cycles were required to traverse the entire time domain. The same problem was then solved by the successive overrelaxation method using the optimum overrelaxation factor. In this case, 334 iterative cycles were needed. The solutions by the two methods agreed at all points to better than five decimal places, demonstrating that the analog subroutine method, although using analog components limited by the quality of the electrical components, does not introduce any appreciable degradation in solution accuracy while providing an appreciable reduction in the number of computer runs. The analog subroutine is used in an error-correction mode which entirely obviates the disadvantage of component quality.

In the solution of problems of the type of (7), the so-called alternating direction implicit (ADI) method is sometimes used. In this technique, the iterations at each time level are usually minimized through the utilization of the tridiagonal algorithm. Where applicable, this method has been found to be somewhat faster than the analog subroutine approach. On the other hand the ADI method is often found to be impractical where the field geometry is irregular, where the problem is formulated in three space-dimensions, or where irregular finite-difference grids are employed. The analog subroutine approach is not subject to any of these limitations.

CONCLUSION

Experience with a wide variety of engineering field problems (references [3] and [4]) has demonstrated that the approach described in this text is highly effective in reducing the digital computer time required for the solution of partial differential equations. This gain in speed results primarily from the reduction in the number of iterations. The analog subroutine approach is especially well adapted to the treatment of systems of algebraic equations since in that case the analog network has a particularly simple configuration. There exist, however, many other problem areas for which the general approach described here may prove to be beneficial.

REFERENCES

[1] VANSTEENKISTE G.C., *Analog and Hybrid Computation*, course notes Von Karman Institute for Fluid Dynamics, Brussels (1970)

[2] KARPLUS W.J. and DRACUP J.A., *Technical Completion Report*, UCLA - ENG - 7142

[3] SHIH CHAU and DRACUP J.A., *Simulation of Evaporation from Constant Source with Finite Areas*, Water Resources Research, vol.5, n°1, Feb. 1969, pp.281-290

[4] STOIKE D. and KARPLUS W.J., *Heat transfer in pyrolytic graphite in a re-entry environment*, Journal Spacecraft Rockets, vol.5, Dec. 1968, pp. 1491-1493.

TABLE I

ANALOG COMPUTER
————————————————————

DIGITAL COMPUTER
————————————————————

- DEPENDENT VARIABLES TREATED IN CON-
  TINUOUS FORM

- HANDLING OF ALL DATA IN QUANTIZED OR
  DISCRETIZED FORM

- ACCURACY LIMITED BY THE QUALITY OF
  THE COMPUTER COMPONENTS

- ACCURACY DETERMINED BY THE NUMBER OF
  BITS CONTAINED IN MEMORY REGISTERS
  AND UPON THE SPECIFIC NUMERICAL TECH-
  NIQUE SELECTED FOR A SPECIFIC PROBLEM

- PARALLEL OPERATION, ALL ELEMENTS
  OPERATING SIMULTANEOUSLY

- SERIAL OPERATION, TIME-SHARING OF ALL
  OPERATIONAL AND MEMORY UNITS

- HIGH SPEED OR REAL-TIME OPERATION,
  COMPUTING SPEEDS LIMITED BY BAND-
  WIDTH OF THE ELEMENTS AND NOT BY
  COMPLEXITY OF THE PROBLEM

- SOLUTION TIMES RELATIVELY LONG, DETER-
  MINED BY THE COMPLEXITY OF A PROBLEM;
  ABILITY TO REDUCE ERRORS BY INCREAS-
  ING LENGTH OF TIME REQUIRED TO OBTAIN
  THE SOLUTION ON THE COMPUTER

- ABILITY TO PERFORM EFFICIENTLY MUL-
  TIPLICATION, INTEGRATION, ADDITION,
  NONLINEAR FUNCTION GENERATIONS, LI-
  MITED ABILITY TO MAKE LOGICAL DECI-
  SIONS, STORE NUMERICAL DATA, PROVIDE
  TIME-DELAYS

- ABILITY TO PERFORM A LIMITED NUMBER
  OF ARITHMETIC OPERATIONS : ADDITION
  AND MULTIPLICATION; INTEGRATION AND
  DIFFERENTIATION REQUIRE APPROXIMATE
  TECHNIQUES; FACILITY FOR MEMORIZING
  DATA INDEFINITELY AND TO PERFORM LO-
  GICAL OPERATIONS AND DECISIONS

- PROGRAMMING TECHNIQUES BASED ON SUB-
  STITUTION OF ANALOG COMPUTING ELEMENTS
  FOR PHYSICAL SYSTEM ELEMENTS

- PROGRAMMING TECHNIQUES BEAR LITTLE
  DIRECT RELATIONSHIP TO THE PROBLEM
  UNDER STUDY, BUT ARE FACILITATED BY
  COMPILERS AND ROUTINES

- FACILITY FOR INCLUDING ANALOG HARD-
  WARE FROM A SYSTEM UNDER STUDY IN THE
  COMPUTER SIMULATION

- FLOATING-POINT OPERATIONS ELIMINATE
  SCALE FACTOR PROBLEMS

- PROVISIONS TO PERMIT THE ENGINEER TO
  EXPERIMENT BY ADJUSTING POTENTIOMETER
  SETTINGS ON THE COMPUTER; THEREBY
  GAINING DIRECT INSIGHT INTO SYSTEM
  OPERATION

- FACILITY FOR ALTERING AND CONTROLLING
  THE TOPOLOGY OF THE DATA FLOW WITHIN
  THE MACHINE ON THE BASIS OF CALCULA-
  TIONS

- SUITED FOR : PHYSICAL SIMULATION
              REAL TIME SIMULATION
              ENGINEERING EXPERIMEN-
              TATION
              INTEGRATION OF DIFFE-
              RENTIAL EQUATIONS

- SUITED FOR : STORAGE OF DATA
              LOGICAL FUNCTIONS AND
              DECISIONS
              HIGH SPEED ARITHMETIC
              SOLUTION OF ALGEBRAIC
              EQUATIONS

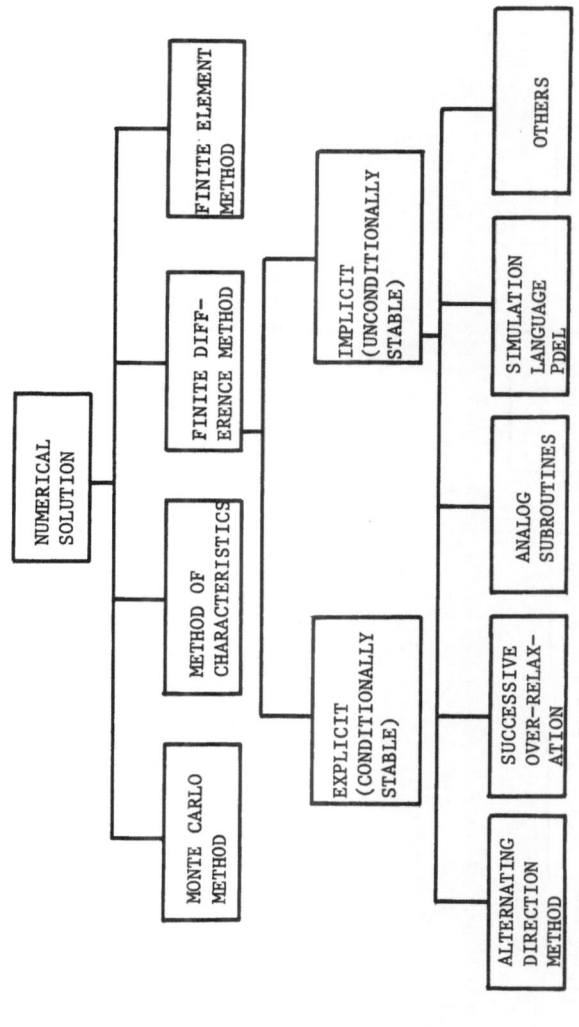

Fig. 1. Computation spectrum

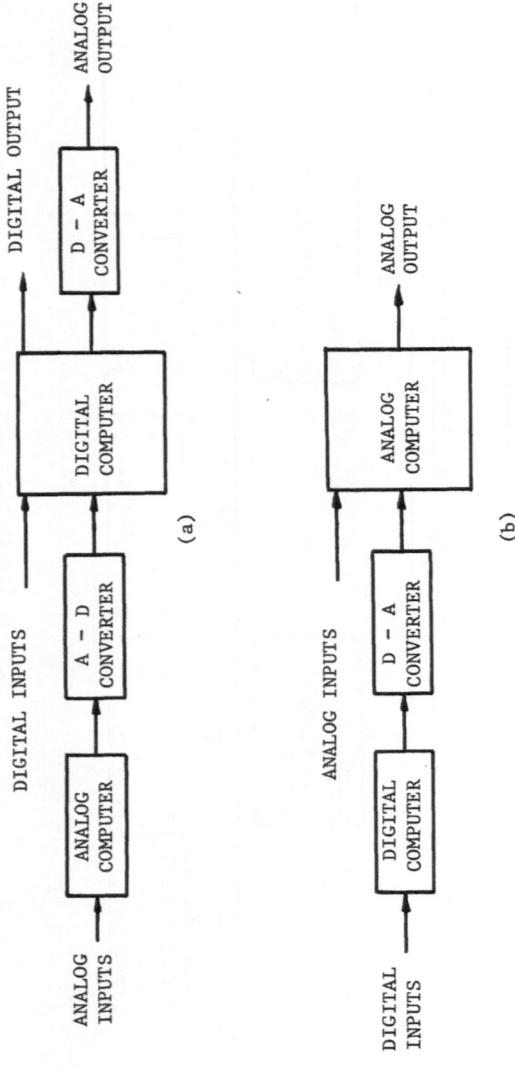

Fig. 2. Unilateral hybrid analog/digital systems

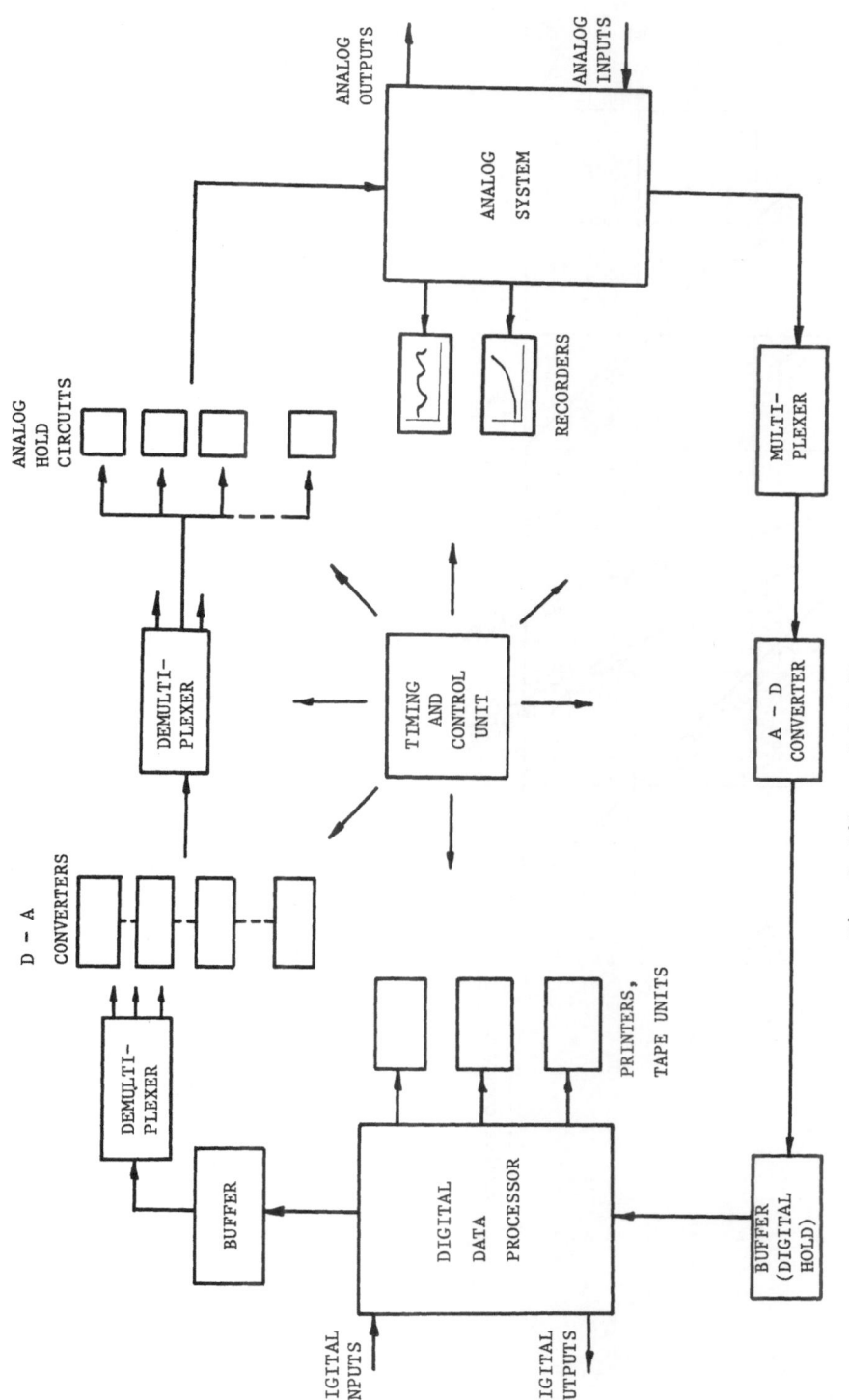

Fig. 3. Bilateral hybrid system

438

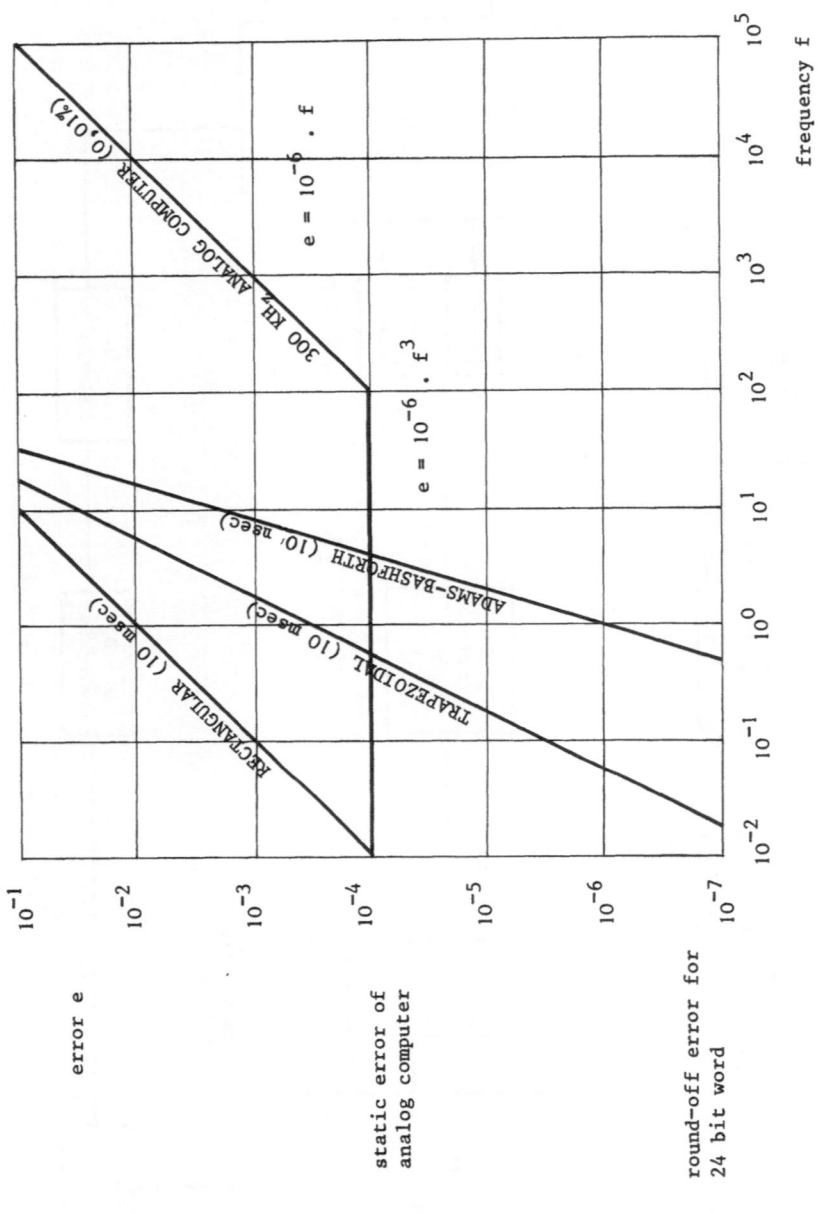

Fig. 4. Error vs. frequency diagram comparison of analog and digital implementation (from W. Giloi)

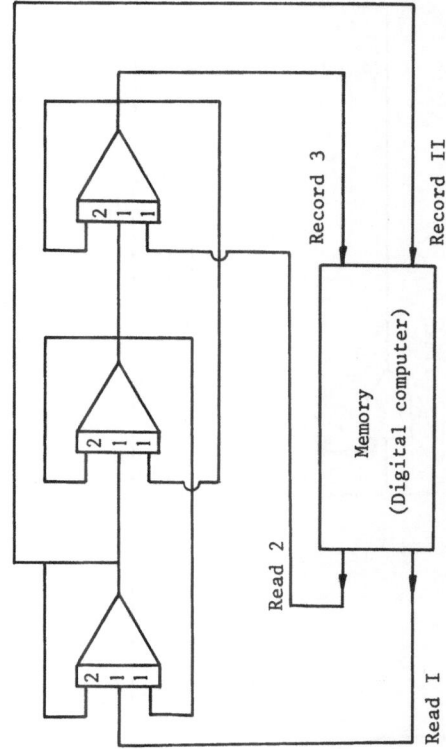

Fig. 5. Discrete – space continuous – time method

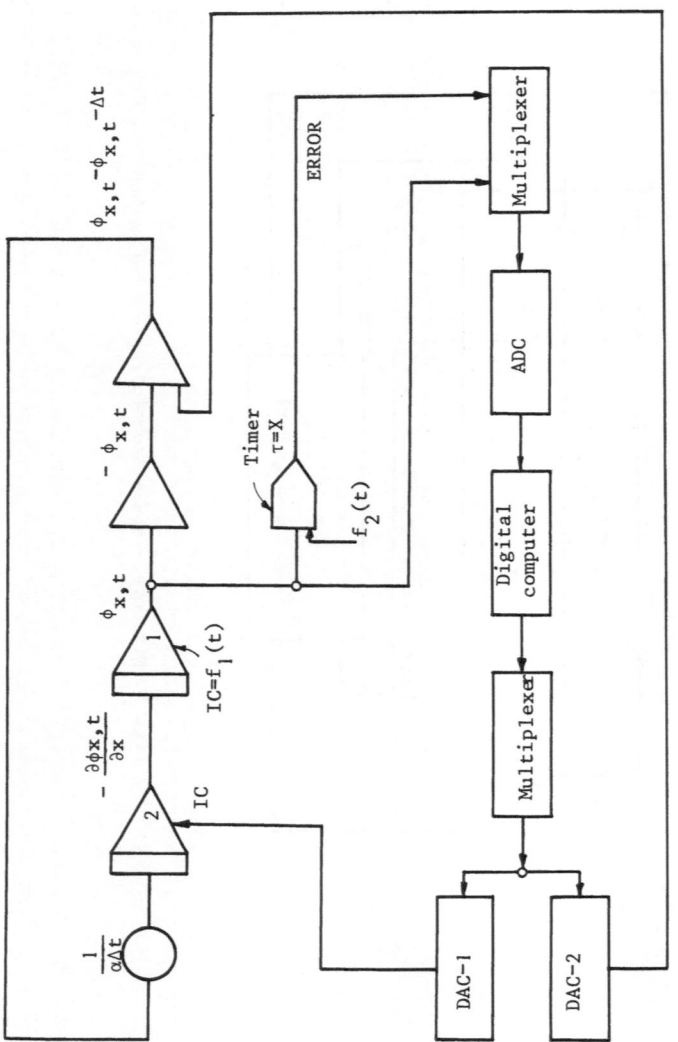

Fig. 6. Continuous – space discrete – time method

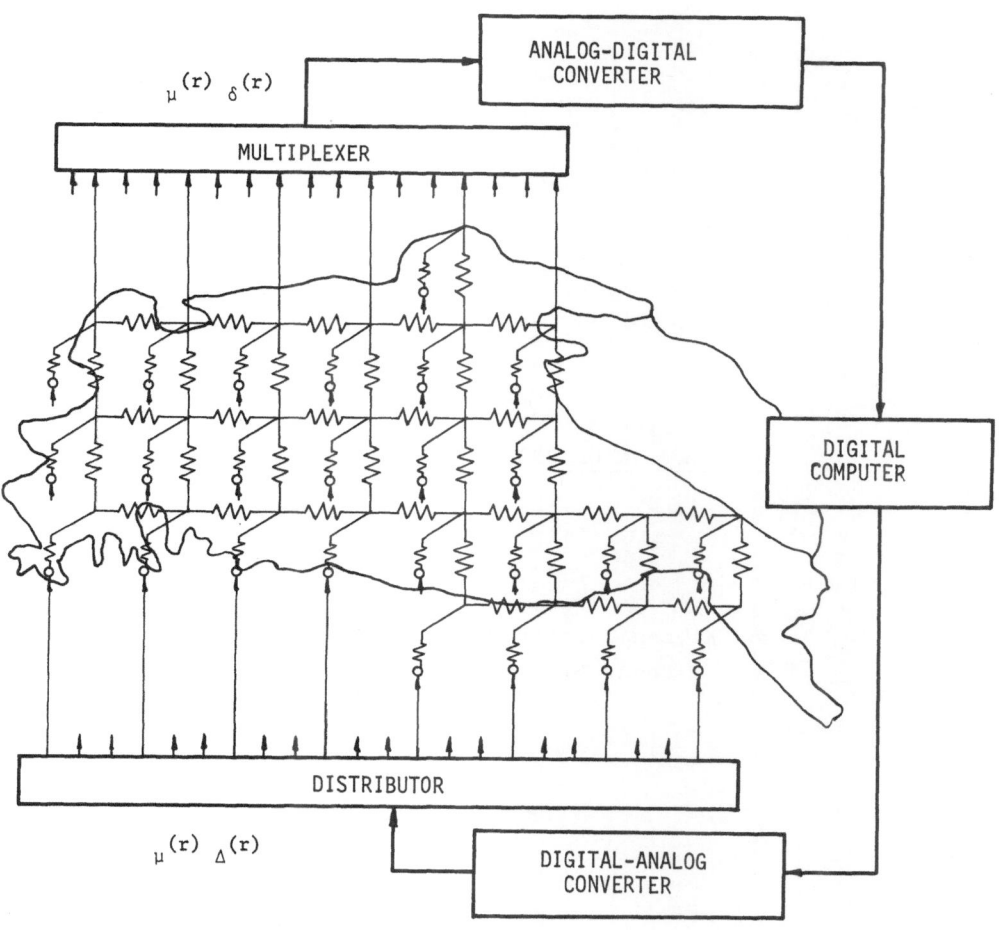

$_\mu(r)$ $_\delta(r)$

$_\mu(r)$ $_\Delta(r)$

Fig. 7. DSDT Hybrid System

COMPUTATION OF UNSTEADY BOUNDARY LAYERS

H.J. WIRZ

von Karman Institute for Fluid Dynamics
Rhode St Genese, Belguim

and

DFVLR, Germany

ABSTRACT

Starting with a brief discussion of some basic unsteady boundary layers (Rayleigh, Stokes), the fundamental equations describing three dimensional unsteady, laminar and turbulent compressible boundary layers are presented.

In order to facilitate the computation of unsteady boundary layers, various boundary layer transformations have been employed; their applicability is investigated.

Two different approaches to solve approximately (integral and finite difference method) the governing equations are discussed in some detail. Special attention is given to finite difference methods.

Finally, some available results are reported, indicating that the increasing capabilities of digital computers have made it possible to obtain solutions of some of these complex flow patterns.

1. INTRODUCTION

The subject of unsteady boundary layers, that is the addition of unsteadiness to viscous flows near boundaries, has again attracted considerable attention in recent years. The proceedings of the IUTAM symposium on Unsteady Boundary Layers (Ref. 1), 1971 trace out an impressive and vivid picture of the various activities in this particular field.

Unsteady boundary layers play an important role for the understanding of particular flow problems in many technical applications such as turbomachines, helicopter rotors, shock tubes, propellers, unsteady flight problems, duct and nozzle flows, combustion chambers, charging and discharging motions, etc.

The increasing capabilities of digital computers have made it possible to obtain solutions of some of these complex flow patterns in a reasonable amount of time. These results, however, are still only of

limited usefulness to design engineers due to the many difficulties
which remain to be solved.

Although several comprehensive summaries of the state of the
theory of unsteady boundary layers have been prepared by Schlichting
and Gersten (Ref. 2),, Rott (Ref. 3), Stuart (Ref. 4) and Stewartson
(Ref. 5), the present state and the development of computational methods
to solve approximately the basic equations is usually treated rather
briefly.

At this point it is worthwhile to take note of some basic unsteady boundary layers which may serve as instructive examples for more
general boundary layers.

Although the concept of boundary layer theory was introduced
first by Prandtl 1904, as well known, the very first and simplest unsteady boundary layers, the Stokes and the Rayleigh layer, being linear,
stem for an earlier period, the mid 19th century (Stokes 1851).

Consider a two-dimensional semi-infinite region of incompressible
fluid in the domain y > 0, bounded by a rigid impermeable wall y = 0.
If the flow is assumed to be independent of the coordinate x along the
wall, then the Navier-Stokes and the continuity equations reduce to

$$\frac{\partial u}{\partial t} = -\frac{1}{\rho}\frac{\partial p}{\partial x} + \frac{\partial^2 u}{\partial y^2} \;; \qquad \frac{\partial p}{\partial y} = 0 \qquad\qquad (1)$$

where $u(y,t)$ is the velocity in the x-direction, to is the time, p is
the pressure (required to be linear in x), ρ is the density and ν is
the kinematic viscosity. These equations represent essentially an analogue to the heat conduction equation with heat addition. In the case
of zero pressure gradient, equation 1 reduces to the simpler diffusion
equation

$$\frac{\partial u}{\partial t} = \frac{\partial^2 u}{\partial y^2} \qquad\qquad (2)$$

The first basic layer, the Stokes layer, is periodic in t and satisfies
the boundary conditions (zero pressure gradient)

$$y = 0 \qquad\qquad u = u_w\cos\omega t \qquad\qquad (3)$$
$$y \to \infty \qquad\qquad u = 0$$

and the solution is

$$u = u_w \, e^{-\eta} \cos(\omega t - \eta) \qquad (4)$$

where

$$\eta = y \left(\frac{\omega}{2\nu} \right)^{1/2} \qquad (5)$$

ω being the frequency. Another form of the Stokes layer arises from the boundary conditions:

$$y = 0 \; : \; u = 0$$
$$\qquad \qquad (6)$$
$$y \to \infty \; : \; u = U_\infty \cos \omega t$$

with

$$-\frac{1}{\rho} \frac{\partial p}{\partial x} = -\omega U_\infty \sin \omega t$$

and the solution is

$$u = U_\infty \cos \omega t - U_\infty e^{-\eta} \cos(\omega t - \eta). \qquad (7)$$

As can be seen from these solutions there is no outflow from the boundary layers. Secondly, the thickness of the Stokes layer is proportional to $\left(\frac{\nu}{\omega} \right)^{1/2}$, and thirdly, the solutions (4) and (7) exhibit an important phase shift. The skin friction leads the velocity at the edge of the layer by $\frac{\pi}{4}$.

The Rayleigh layer, which represents the flow generated by a sudden movement of the plane $y = 0$, arises from the boundary conditions

$$t = 0 \qquad u(y,o) = 0; \qquad y > 0$$
$$y = 0 \qquad u = U_0 \qquad \qquad \Bigg\} \qquad t > 0$$
$$y \to \infty \qquad u = 0$$
$$\qquad \qquad (8)$$

and the solution is

$$u = U_0 \; \text{erc} \; c \left(\frac{y}{2(\nu t)^{1/2}} \right) \qquad (9)$$

The analogue form of (9) satisfies the following boundary conditions

$$t = 0 \qquad u(y,t) = U ; \qquad y > 0$$

$$y = 0 \qquad u = 0$$

$$y \to \infty \qquad u = U_0$$

$$t > 0 \tag{10}$$

and the solution is

$$u = U_0 \; \text{erf} \; \frac{y}{2(\nu t)^{1/2}} \tag{11}$$

Again, these solutions do not exhibit an outflow since the velocity at infinity is not allowed to depend on x, whilst the thickness of this layer is proportional to $(\nu t)^{1/2}$.

A third basic unsteady viscous layer is the critical layer (Stuart, Ref. 6) which occurs when a vorticity wave propagates in a shear flow u(y). This flow is governed by the Heisenberg-Tollmien equation and its characteristic feature is the promotion of phase differences between the velocity components, yielding concomitant Reynolds stresses.

As long as stability problems are not considered, the Stokes and the Rayleigh layer serve as the simplest examples for many unsteady boundary layers.

Although it is convenient (Stuart, Ref. 7) to subdivide the subject into the following sections :
(i) the response of steady boundary layers to imposed fluctuations;
(ii) the generation of steady streaming by a purely oscillatory motion;
(iii) impulsive and other non-oscillatory boundary layers;
(iv) effects of rotation;
(v) stability of unsteady flows,
we here prefer to discuss a number of problems arising from the development of computational methods which numerically solve the basic parttial differential equations. Thus we may classify this work as follows
(i) governing fluid flow equations;
(ii) boundary layer transformations;
(iii) computational methods;
(iv) discussion of results.

2. GOVERNING FLUID FLOW EQUATIONS

For three-dimensional unsteady compressible laminar boundary layer flow, these equations may be written in terms of a cartesian coordinate system (inertia system) as

Continuity :

$$\frac{\partial \rho}{\partial t} + \frac{\partial}{\partial x} (\rho u) + \frac{\partial}{\partial y} (\rho v) + \frac{\partial}{\partial z} (\rho w) = 0 \tag{12}$$

Momentum :

$$\rho \frac{Du}{Dt} = - \frac{\partial p}{\partial x} + \frac{\partial}{\partial y} (\mu \frac{\partial u}{\partial y}) \; ; \tag{13}$$

$$\rho \frac{Dw}{Dt} = - \frac{\partial p}{\partial z} + \frac{\partial}{\partial y} (\mu \frac{\partial w}{\partial y}) \; ; \qquad \frac{\partial p}{\partial y} = 0 \tag{14}$$

Energy :

$$c_p \rho \frac{DT}{Dt} - \frac{D_e p_e}{Dt} = \frac{\partial}{\partial y} (\lambda \frac{\partial T}{\partial y}) + \mu \left((\frac{\partial u}{\partial y})^2 + (\frac{\partial w}{\partial y})^2 \right) \tag{15}$$

where u, v, w are the boundary layer velocities in the x, y, z directions respectively; t is the time coordinate, ρ is the density, ν is the viscosity, T is the temperature, p is the pressure, λ is the heat conduction coefficient, c_p is the specific heat at constant pressure. Finally, we have the operators

$$\frac{D}{Dt} = \frac{\partial}{\partial t} + u \frac{\partial}{\partial x} + v \frac{\partial}{\partial y} + w \frac{\partial}{\partial z} \tag{16}$$

and

$$\frac{D_e}{Dt} = \lim_{\substack{u \to u_e \\ w \to w_e \\ v \to 0}} \frac{D}{Dt} \tag{17}$$

where the subscript e denotes the flow at the edge of the boundary layer. The outer flow velocities and the temperature are connected with the pressure through

$$\rho_e \frac{D_e u_e}{Dt} = - \frac{\partial p}{\partial x} \; ; \qquad \rho_e \frac{D_e w_e}{Dt} = - \frac{\partial p}{\partial z} \tag{18}$$

$$\frac{\partial p}{\partial t} = \rho_e \frac{D_e}{Dt} \left(c_p T_e + \frac{(u_e^2 + w_e^2)}{2} \right) \tag{19}$$

In order to complete these equations a thermal and caloric equation of state together with a specification of the fluid properties (μ,λ) must be given. Usually the following laws are applied (perfect gas)

$$p = \rho RT \ , \quad c_p = \text{const.} \tag{20}$$

and $\mu = \mu(T)$ together with $\lambda = \dfrac{\mu c_p}{Pr}$ (21)

where R is the gas constant and Pr is the Prandtl number, which is assumed to be constant.

The incompressible case of these equations is derived by setting ρ constant. Additionally, in the energy equations (15), the second term on the left side does not appear, and the dissipation term in (15) is usually neglected.

The very simplest case for two-dimensional and incompressible flows has been treated to some extent, so that it is worthwhile to write down these equations explicitly.

Continuity :

$$\frac{\partial u}{\partial x} + \frac{\partial v}{\partial y} = 0 \tag{22}$$

Momentum :

$$\frac{\partial u}{\partial t} + u \frac{\partial u}{\partial x} + v \frac{\partial u}{\partial y} = - \frac{1}{\rho} \frac{\partial p}{\partial x} + \nu \frac{\partial^2 u}{\partial y^2} \tag{23}$$

where ν is the kinematic viscosity. The pressure gradient is given by

$$- \frac{1}{\rho} \frac{\partial p}{\partial x} = \frac{\partial u_e}{\partial t} + u_e \frac{\partial u_e}{\partial x} \tag{24}$$

The set of partial differential equations given so far has to be completed by appropriate boundary and initial conditions. The formulation of boundary conditions (zero velocity difference between fluid and wall at the wall together with either a prescribed temperature or heat flux at the wall, and at the edge of the layers, the free stream conditions for velocities and temperature) usually do not involve any problems. However, it must be realized that in most of the practical problems the outer flow field is not known, especially for three dimensional unsteady problems.

Additionally, initial conditions for the t, x, z coordinates must be prescribed. Their explicit formulation is usually a difficult and complicated problem.

We will discuss some aspects of this problem in more detail in connection with the presentation of solved problems.

So far we have mentioned only laminar flows. In most of the practical problems, however, turbulent flows are of even greater importance than laminar flows.

In the simplest case of an incompressible two-dimensional flow the usual derivation of such equations leads to the following system

$$\frac{\partial u}{\partial x} + \frac{\partial v}{\partial y} = 0 \tag{25}$$

$$\frac{\partial u}{\partial t} + u \frac{\partial u}{\partial x} + v \frac{\partial u}{\partial y} = -\frac{1}{\rho} \frac{\partial p}{\partial x} + \frac{1}{\rho} \frac{\partial \tau}{\partial y} \tag{26}$$

where the shear stress is given by

$$\tau = \mu \frac{\partial u}{\partial y} - \rho \overline{u'v'} \tag{27}$$

the latter term being the Reynolds shear stress.

Cebeci and Keller (Ref. 8) use the eddy viscosity concept (composite layers) to relate the Reynolds shear stress to the mean flow velocity, and discuss differences and similarities between an spatially one dimensional unsteady flow and a steady two dimensional flow.

Another concept to close the above system of equations is achieved by making use of the turbulent kinetic energy equation. This approach has been used by Patel and Nash (Ref. 9) in the manner suggested by Townsend (Ref. 10) and Bradshaw, Ferris and Altwell (Ref. 11).

Based on three essential assumptions, the following equation for the Reynolds shear stress τ (the shear stress involving the molecular viscosity is regarded as negligible) has been treated by Patel and Nash to compute some unsteady turbulent boundary layers:

$$\frac{\partial \tau}{\partial t} + u \frac{\partial \tau}{\partial x} + v \frac{\partial \tau}{\partial y} - 2a_1 \tau \frac{\partial u}{\partial y} + 2a_1 \frac{\tau_{max}}{\rho U_e} \frac{\partial}{\partial y} (a_2 \tau)$$

$$+ \frac{a_1}{L} \tau^{3/2} \rho^{1/2} = 0 \tag{28}$$

For the present purpose, here it is sufficient to note that the coef-
figient a_1 is assumed to be a universal constant (equal to 0.15), while
a_2 and L are universal empirical functions of $\frac{y}{\delta}$, δ being the boundary
layer thickness. A sketch of these functions is shown in figure 1.

Two different cases have been treated numerically. The first is
a flat plate turbulent boundary layer subjected to periodic oscillations
in the free stream. Only two sets of measurements, those due to Karlson
(Ref. 12) and Miller (Ref. 13), seem to be available for the purpose of
comparison. Additionally, only time averaged (over a complete cycle)
mean velocity profiles are reported, so that a detailed comparison is
not possible.

The second case is a boundary layer subjected to an adverse free
stream velocity gradient which remains constant in the x-direction and
is periodic in time. No experimental results are available.

The essential difficulty in predicting such flows arises from
the so-called 'closure problem' and not primarily from computational
aspects. This problem has not been resolved satisfactorily up to now,
even for steady turbulent boundary layer flows.

3. BOUNDARY LAYER TRANSFORMATIONS

The set of equations describing compressible three dimensional
unsteady laminar boundary layer flows (12-21) have been written in physical cartesian coordinates, which are of course usually the simplest
ones. In most practical problems, however, further transformations are
often needed. The necessity for such transformation arises mainly from
three sources.

The boundary layer flow over a real body requires, in general,
the formulation of the boundary layer equations in curvilinear coordinates (see for example reference 14). The choice of these coordinates
(usually orthogonal ones) may depend on special physical conditions, on
the availability of the free stream data, but also on the (numerical)
determination of the metric coefficients. Streamlines of the outer flow
at a fixed time and their orthogonal trajectories together with the
outer normal of the body may form a suitable system of curvilinear coordinates. Such a system is reported by Piquet and Zeytounian (Ref. 15).

Many unsteady flow phenomena may be better described in moving or rotating coordinates, which are then no longer inertia systems. Additional 'forces' as for example Coriolis and centrifugal ones, will appear in a rotating coordinate system. The system of equations for unsteady, incompressible boundary layer flow about a rotating airfoil has been recently studied by McCroskey and Yaggy (Ref. 16) in order to numerically investigate a few problems of helicopter rotor boundary layers.

Many of the difficulties associated with the computation of unsteady boundary layers may be removed by special coordinate transformations. We shall discuss these aspects in some detail.

In order to avoid the explicit calculation of the density in the basic set of equations, the following transformation may be employed. Additionally, the continuity equation (12) is integrated identically in defining two functions ,

$$\xi = x; \qquad \eta = \int_0^y \frac{\rho}{\rho_\infty} \, dy \; ; \qquad \zeta = z \; ; \qquad \tau = t \tag{29}$$

$$u = \frac{\rho_\infty}{\rho} \frac{\partial \psi}{\partial y} \; ; \qquad w = \frac{\rho_\infty}{\rho} \frac{\partial \chi}{\partial y} \; ; \qquad v = -\frac{\rho_\infty}{\rho} \left(\frac{\partial \psi}{\partial x} + \frac{\partial \chi}{\partial z} + \frac{\partial \eta}{\partial t} \right) \tag{30}$$

where ρ_∞ is a constant reference density. Introducing a velocity v^+ through

$$v^+ = -(\psi_\xi + \chi_\zeta) \tag{31}$$

the operator $\frac{D}{Dt}$ remains nearly unchanged

$$\frac{D}{Dt} = \frac{D}{D\tau} \equiv \frac{\partial}{\partial \tau} + u \frac{\partial}{\partial \xi} + v^+ \frac{\partial}{\partial \eta} + w \frac{\partial}{\partial \zeta} \tag{32}$$

The resulting set of equations thus may be written as
Continuity :

$$u_\xi + v_\eta^+ + w_\zeta = 0 \tag{33}$$

Momentum :

$$\frac{Du}{D\tau} = -\frac{1}{\rho_e} \frac{p}{\partial \xi} \frac{T}{T_e} + \frac{\partial}{\partial \eta} \left(\frac{\rho \mu}{\rho_\infty \rho_\infty} \frac{\partial u}{\partial \eta} \right) \tag{34}$$

$$\frac{Dw}{D\tau} = -\frac{1}{\rho_e} \frac{p}{\partial \zeta} \frac{T}{T_e} + \frac{\partial}{\partial \eta} \left(\frac{\rho \mu}{\rho_\infty \rho_\infty} \frac{\partial u}{\partial \eta} \right) \tag{35}$$

Energy :

$$c_p \frac{DT}{D\tau} - \frac{1}{\rho_e} \frac{D_e p_e}{D\tau} \frac{T}{T_e} = \frac{\partial}{\partial \eta} (\frac{\lambda \rho}{\rho_\infty \rho_\infty} \frac{\partial T}{\partial \eta}) + \frac{\mu \rho}{\rho_\infty \rho_\infty} ((\frac{\partial u}{\partial \eta})^2 + (\frac{\partial w}{\partial \eta})^2) \tag{36}$$

This transformation is originally due to Dorodnitzyn (Ref. 17), and may be further extended.

   In order to remove some of the difficulties associated with the leading edge singularities, the following transformation is often employed. Consider a two dimensional unsteady boundary layer, written in a non dimensional form :

$$u_x + v_y = 0 \tag{37}$$

$$u_t + uu_x + vu_y = \frac{\partial U_e}{\partial t} + u_e \frac{\partial U_e}{\partial x} + u_{yy}$$

subjected to the following transformations :

$$\xi = x ; \qquad \eta = (\frac{U_e}{2x})^{1/2} y ; \qquad \tau = t \tag{38}$$

$$f' \equiv \frac{\partial f}{\partial \eta} = \frac{u}{U_e} ; \qquad \bar{v} = v + \frac{1}{2} (\beta_\xi - 1) f' + \frac{1}{2} \beta_\tau \eta \tag{39}$$

The momentum equation then takes the form

$$\frac{\xi}{U_e} \frac{\partial f'}{\partial \tau} + f' \frac{\partial f'}{\partial \xi} + v \frac{\partial f'}{\partial \eta} = \beta_\xi (1 - f'^2) + \beta_\tau (1 - f') + \frac{1}{2} \frac{\partial^2 f'}{\partial \eta^2} \tag{40}$$

where β_ξ and β_τ are

$$\beta_\xi = \frac{\xi}{U_e} \frac{\partial U_e}{\partial \xi} ; \qquad \beta_\tau = \frac{\xi}{U_e^2} \frac{\partial U_e}{\partial \tau} \tag{41}$$

The advantages of such a trandformation, which may be applied also to the three dimensional case, (see Dwyer, Ref. 18), are :
(i) possible leading edge singularities are removed;
(ii) an equation(s) to determine the initial conditions along $\xi = 0$
     may be obtained by taking the limit $\xi \to 0$;
(iii) the boundary layer thickness is very nearly constant in terms of
      the transformed coordinate ($\eta \simeq 6$);
(iv) the derivatives of the independent variables are stretched so that
     high accuracy may be obtained with relatively large step sizes.

Another transformation (Crocco, Ref. 19) which removes the usual difficulty with the location of the outer edge of the boundary layer (the coordinate y has to be taken finite), is defined through

$$\xi = x \; ; \quad \eta = \frac{u}{U_e} \; ; \quad \tau = t \tag{42}$$

$$\phi = \frac{1}{U_e} \frac{\partial u}{\partial y} \tag{43}$$

The boundary layer equation (37) in the new form is :

$$\phi^2 \frac{\partial^2 \phi}{\partial \eta^2} + A \frac{\partial \phi}{\partial \eta} - B\phi - \frac{\partial \phi}{\partial \tau} - \eta U_e \frac{\partial \phi}{\partial \xi} = 0 \tag{44}$$

while the usual boundary conditions, now transformed, yield

$$\eta = 0 \qquad\qquad \phi \frac{\partial \phi}{\partial \eta} + C = 0$$

$$\eta = 1 \qquad\qquad \phi = 0.$$

The coefficients A, B, C are given functions of the independent variables

$$A = (\eta^2 - 1) \frac{\partial U_e}{\partial x} + (\eta - 1) \frac{1}{U_e} \frac{\partial U_e}{\partial \tau}$$

$$B = \eta \frac{\partial U_e}{\partial x} + \frac{1}{U_e} \frac{\partial U_e}{\partial \tau} \tag{46}$$

$$C = \frac{\partial U_e}{\partial x} + \frac{1}{U_e} \frac{\partial U_e}{\partial \tau}$$

This type of transformation has been extensively used in the work of Piquet (Ref. 20). The advantages are obvious : (i) the integration domain in the η direction is bounded ($0 \leq \eta \leq 1$), (ii) the main equation (44) is linear except for the first term, (iii) the extension of this transformation to systems of two dimensional unsteady boundary layers is possible.

On the other hand, this transformation has also disadvantages, of which the problem of uniqueness of the solutions in cases where the velocity profile exhibits an overshoot, is the severest one. The necessary and sufficient condition is $\phi \neq 0$ in $0 < \eta < 1$. Since it is well known that unsteady boundary layers are especially liable to exhibit the effect of an overshoot, care has to be taken.

Finally, a group of transformations needs to be considered which reduces the partial differential equations for unsteady boundary layers to differential equations with fewer independent variables than the original ones. The solutions of these equations (usually called similarity solutions) are extremely helpful for many reasons.

In cases where one succeeds in reducing the partial differential equations to nonlinear ordinary ones, their solutions are normally easy to obtain. Besides their physical relevance, they may be used to check computational procedures designed to solve the complete partial differential equations. Sometimes these solutions may be used in the construction of approximate methods of the von Karman-Pohlhausen type.

For incompressible two dimensional unsteady boundary layers, similarity solutions with one independent varaible have been reported by Schuh (Ref. 20) and Geis (Ref. 21). The free stream is of the form $U_e = \frac{mx}{t}$ or $U_e = ct^n$, while Yang (Ref. 22) treated the case $u_e = x/(a+bt)$, where a, b are constants. A class of similarity solutions for unsteady spatially one dimensional compressible boundary layers is considered by Yang and Huag (Ref. 23), the free stream velocity being again of the form $u_e = ct^n$. Gabbert (Ref. 24) and Wirz (Ref. 25) finally derived a set of similarity solutions for spatially two dimensional compressible flows with free stream velocities of the form $u_e = \frac{mx}{t}$; $u_e = mx$; $u_e = \frac{m}{t}$.

Another important aspect of these similarity (only one independent variable) and semi-similarity (more than one independent variable) solutions is associated with the problem of obtaining initial conditions to start a general computational method designed to solve the complete partial differential equations. It is often possible to derive approximate initial conditions in making use of special similarity or semi-similarity solutions. Semi-similarity solutions have been reported by Tani (Ref. 26), who considered the case $u_e = U_0 - \frac{x}{T-t}$, U_0 , T being constants. A more general class of this type of solution is discussed by Hassan (Ref. 27) (see also Hayasi, Ref. 28).

4. COMPUTATIONAL METHODS

The problem of theoretically predicting unsteady boundary layer flows with sufficiently general boundary and initial conditions usually requires, in the simplest case, the calculation of approximate solutions of at least one partial differential equation with three independent variables. The situation is even more complicated if three dimensional unsteady boundary layers are considered, leading to systems of differential equations with four independent variables.

Two basic approaches have been successfully developed, which we will discussed here in some detail.

The first one, usually referred to as the integral method, is a natural extension of the well known von Karman-Pohlhausen procedure to unsteady boundary layers. Mathematically speaking, these methods belong to a wider class of approximate methods based upon weighted residuals.

Schuh (Ref. 29) presented in 1953 the first integral method of this type to calculate unsteady two dimensional incompressible boundary layers on bodies with arbitrary shape and arbitrary varying free stream velocities.

Although this type of approximate methods is still employed, more recently, the finite difference approach has attracted more attention because of its wider and easier applicability to complicated problems.

The basic idea of the integral method applied to unsteady boundary layers may best be explained by considering the following set of equations describing the development of momentum and thermal laminar boundary layers in an incompressible laminar fluid.
Continuity

$$\frac{\partial u}{\partial x} + \frac{\partial v}{\partial y} = 0 \tag{47}$$

Momentum

$$\frac{\partial u}{\partial t} + u \frac{\partial u}{\partial x} + v \frac{\partial u}{\partial y} = \frac{\partial U_e}{\partial t} + U_e \frac{\partial U_e}{\partial x} + \frac{\partial^2 u}{\partial y^2} \tag{48}$$

Energy

$$\frac{\partial T}{\partial t} + u \frac{\partial T}{\partial x} + v \frac{\partial T}{\partial y} = \frac{1}{\sigma} \frac{\partial^2 T}{\partial y^2} \tag{49}$$

together with the boundary conditions :

$$y = 0, \quad u = v = 0, \quad T = T_w$$
$$y \to \infty, \quad u = U_e(x,t), \quad T = T_e, \tag{50}$$

T_w and T_e being constants.

These equations, written here in a nondimensional form, where u,v are the velocities and T is the temperature and σ is the Prandtl number, have been treated to some extend by Yang (Ref. 30) and Miller (Ref. 31). The work of Schuh (Ref. 29) is based on the momentum equation only, plus an additional equation for the balance of mechanical energy.

If the velocity component v is eliminated from the momentum and energy equation by using the continuity equation and each term in equation 48 and 49 is integrated with respect to y from 0 to ∞, the so called integral momentum and energy equations are obtained :

$$\frac{\partial}{\partial t} (U_e \delta_1) + U_e^2 \frac{\partial \delta_2}{\partial x} + U_e \frac{\partial U_e}{\partial x} (\delta_1 + 2\delta_2) = \left(\frac{\partial u}{\partial y}\right)_0 \tag{51}$$

$$\frac{\partial}{\partial t} (\theta_1) + U_e \frac{\partial \theta_2}{\partial x} + \frac{\partial U_e}{\partial x} \theta_2 = -\frac{1}{\sigma} \left(\frac{\partial \theta}{\partial y}\right)_0 , \tag{52}$$

where

$$\delta_1 = \int_0^\infty \left(1 - \frac{u}{U_e}\right) dy, \qquad \delta_2 = \int_0^\infty \frac{u}{U_e}\left(1 - \frac{u}{U_e}\right) dy \tag{53}$$

are the displacement and the momentum thickness, and θ_1 and θ_2 are defined with $\theta = (T - T_e) \Big/ (T_w - T_e)$ by

$$\theta_1 = \int_0^\infty \theta \, dy \qquad \theta_2 = \int_0^\infty \frac{u}{U_e} \theta \, dy \tag{54}$$

If each term in the original momentum equation is multiplied by u before integrating as before, the so called integral (mechanical) energy equation is obtained

$$\frac{U_e}{2} \frac{\partial \delta_1}{\partial t} + \frac{1}{2} \frac{\partial}{\partial t} (U_e^2 \delta_2) + \frac{1}{2} \frac{\partial}{\partial x} (U_e^3 \delta_3) = \int_0^\infty (\frac{\partial u}{\partial y})^2 \, dy \qquad (55)$$

where

$$\delta_3 = \int_0^\infty \frac{u}{U_e} \left(1 - (\frac{u}{U_e})^2 \right) \, dy \qquad (56)$$

is the "energy-" thickness of the boundary layer.

The generalization of obtaining additional integral equations is obvious. Such an approach has been explored in the work of Koob and Abott (Ref. 32).

The system of integral equations (51) and (52) contains, in principle, six unknowns, whereas only two equations are given so that a closure of the problem is only achieved if additional equations (algebraic or differential equations) for the remaining quantities can be formulated. This is the key problem of all approximate methods based on integral equations.

The integral methods have been frequently used and are still in common use to calculate steady boundary layers, where the basic equations reduce to ordinary differential equations. However, this advantage of the method is lost, if unsteady problems or steady three dimensional boundary layers are considered. In all these cases still partial differential equations must be solved numerically by finite difference methods, although one might say that the reduction of one independent variable is always a clear advantage with respect to the computer storage to be needed.

In order to close the basic integral equations, the following ideas have been employed :
Assuming that all unsteady velocity and temperature profiles for a specific problem, satisfying the boundary conditions, belong to a single parameter family of curves, the quantitites δ_1, δ_2, θ_1, θ_2 and $(\frac{\partial u}{\partial y})_0$ and $\frac{1}{\sigma} (\frac{\partial \theta}{\partial y})_0$ are functions of this parameter only, so that all unknowns may be expressed in terms of the thicknessess δ_2 and θ_2, for instance. These relations, often termed as "universal functions" are usually rather complcated but may be calculated in advance. The essential parameter is

usually derived from a so called compatibility condition, which may be
the partial differential equation evaluated at the wall. Schuh (Ref. 29),
Yang (Ref. 30) and others use the following parameter

$$\lambda \equiv - \frac{\delta_2^2}{U_e} \left(\frac{\partial^2 u}{\partial y^2}\right)_0 = \delta_2^2 \left(\frac{\partial U_e}{\partial x} + \frac{1}{U_e} \frac{\partial U_e}{\partial t}\right) \tag{57}$$

which is the natural extension of the parameter frequently used in steady
two dimensional boundary layers.

The accuracy of the results depends strongly on the choice of
the single parameter family of profiles for velocity and temperature,
Fourth order polynomials, Hartree-profiles and similarity solutions of
the unsteady equations have been applied, yielding slightly different
results, if the region of separation where remarkable differences occur,
is excluded.

In order to improve the accuracy, profiles with more than one
parameter need to be considered. Although these profiles may be found
rather easily, a greater number of integral equations (partial differential equations in x and t) must be solved numerically. In addition,
all the "universal functions" are now two parametric functions, which
need to be calculated and stored. Thus, we see that the involved numerical work is strongly increasing being comparable with the pure finite
difference methods. Because of the lack of any convergence proof, even
with higher order methods, the accuracy of the solutions cannot be
estimated.

A clear advantage, however, needs finally to be mentioned. The
formulation of the initial conditions for unsteady boundary layers does
not involved the dependency on the y-coordinate any more and is therefore often much more easy to obtain.

In trying to assess the relative merits of the integral methods,
which have been successfully applied to a number of unsteady problems,
the limitations (lack of generality and accuracy) indicate that only
special unsteady boundary layers may be treated with easy success.

The second method is entirely characterized by the use of finite
differences to transform the partial differential equations into a set
of algebraic equations, which then are solved simultaneously on a com-

puter. The exploration of this approach is still developing, although quite a number of papers have already been published.

In order to study the method and the associated difficulties (approximation of the differential equation, stability and convergency, initial conditions, computational procedures), consider the following simple model problem, which represents, to some extend, the basic features of unsteady boundary layers

$$\frac{\partial u}{\partial t} + U(y) \frac{\partial u}{\partial x} = \sigma \frac{\partial^2 u}{\partial y^2} \tag{58}$$

where $u(x,y,t)$, say the velocity, is the single unknown, t,x,y are the independent variables, $U(y)$ is a prescribed function and σ is a positive constant. The domain of integration may be given as

$$B = \{x,y,t \mid x \geq 0, \qquad t \geq 0, \qquad 0 \leq y < 1\} \tag{59}$$

together with boundary and initial conditions, which we do not specify at this point.

Although general solutions of the linear differential equation are not known, we easily derive a special solution for u, if U is assumed to be constant, say U_0. This solution writes

$$u(x,y,t) = \sum_{\rho} A_\rho(\xi) \, e^{-\sigma k_\rho^2 t} \, e^{\sqrt{-1} \, k_\rho y} \tag{60}$$

where the lines $\xi = x - U_0 t$, $\xi = $ const, form the characteristics along which the solution is convected and where the $A_\rho(\xi)$ are functions defined by the initial conditions, the k_ρ being constants. The important physical aspects of this simple solution are that the flow quantity u is convected along the characteristic line $\xi = x - U_0 t$ and attenuated with increasing time, due to the (double) parabolic nature of the equation. In the more general case, the characteristics are given by

$$\frac{dx}{dt} = U(y) \tag{61}$$

so that in each plane y = const. the slope of the characteristic is different. The maximum turning angle between these characteristic direction is given by the formula

$$\phi = \arc \tan \frac{U_{max} - U_{min}}{1 + U_{max} U_{min}} \quad .$$

The limiting characteristic directions at a point P of the x-t plane form the domain of dependence and the domain of influence, which must be taken into consideration in order to specify correctly the initial conditions. In the simpler case of $U = U_0$, the turning angle is zero, or physically speaking, the dispersion is zero.

Another important feature of our simple equation with $U = U_0$ may be seen as follows. If each term of the equation 59 is multiplied by u and integrated with respect to y between the boundaries $y = 0$ and $y = 1$, the following equation is obtained :

$$\frac{\partial}{\partial t} \int_0^1 u^2 dy + U_0 \frac{\partial}{\partial x} \int_0^1 u^2 dy = 2\sigma u \left. \frac{\partial u}{\partial y} \right|_0^1 - 2\sigma \int_0^1 \left(\frac{\partial u}{\partial y} \right)^2 dy \qquad (63)$$

where no boundary conditions have been inserted. This equation represents in a sense the balance of the "energy" of the system. The first term on the right hand side represents the work of the "shear stress" at the boundaries and the latter one the "dissipation". In order to have a bounded solution of our problem, the boundary conditions cannot be chosen arbitrarily; the right hand side of equation 62 must always be less than zero. Fortunately, in boundary layer problems, this condition seems to be always fullfilled.

Many of the features mentioned so far are quite similar to those associated with the problem of integrating three dimensional steady boundary layers. For a discussion of these problems, the reader may refer to Krause (Ref. 33).

In the following chapter we will discuss a few finite difference schemes, which solve our basic equation 58 numerically. A grid system of equal mesh size in the domain B is introduced in writing

$$
\begin{aligned}
x_i &= i\Delta x , & i &= 0, 1, 2, \ldots \\
y_j &= j\Delta y , & j &= 0, 1, 2, \ldots, N \qquad (64) \\
t_k &= k\Delta t , & k &= 0, 1, 2, \ldots
\end{aligned}
$$

where Δx, Δy, Δt are the space and time increments, respectively. The approximate solution is denoted by

$$u_{i,k}^j \approx u(x_i, t_k, y_j)$$

and in addition we introduce two step parameters :

$$\lambda_j = \frac{\Delta t}{\Delta x} U(y_j), \qquad s = \sigma \frac{\Delta t}{(\Delta y)^2}, \qquad s \text{ being always positive.} \qquad (65)$$

Although the methods to be presented now solve only our model problem, the extension to the more complicated unsteady boundary layers is straightforward.

In figure 2 we have depicted five different schemes, two explicit and three implicit ones, which we will further investigate.

By taking first order differences for the x and t derivatives and second order centered differences for the diffusion term, the explicit schemes may be written as

$$u_{i+1,k+1}^j = E_j \, u_{i,k+1}^j + F_j \, u_{i+1;k}^j + G_j \, u_{i,k}^j \qquad (66)$$

where the operators E_j, F_j and G_j, which depend on the y-coordinate, are given through

Scheme a) $E_j = 0$

$\qquad\qquad F_j = \left((1-\lambda_j)\delta + s\delta_y^2 \right)$ $\qquad\qquad\qquad\qquad\qquad (67)$

$\qquad\qquad G_j = \lambda_j$

Scheme b) $E_j = \frac{1}{\lambda_j} \left((\lambda_j - 1)\delta + s\delta_y^2 \right)$

$\qquad\qquad F_j = 0$ $\qquad\qquad\qquad\qquad\qquad\qquad\qquad\qquad (68)$

$\qquad\qquad G_j = \frac{1}{\lambda_j} .$

In these equations, the unity operator δ and the usual central difference operator with respect to the y-coordinate, δ_y^2, have been employed.

The truncation error of these consistent schemes can be shown to be of the order $O(\Delta x + \Delta t + \Delta y^2)$. In order to guarantee the convergence of these step by step procedures, the stability must be investigated. If we omit from the analysis the influence of the boundary conditions, a rather simple way of obtaining sufficient stability limits of the above schemes is the requirement that all operators E_j, F_j, G_j must be positive. This condition leads to the following equations

Scheme a) $\lambda_j > 0$, $s \leq \frac{1}{2} (1-\lambda_{max})$,

and additionally   (69)

$\lambda \leq 1$

Scheme b) $\lambda_j > 0$; $s \leq \frac{1}{2} (\lambda_{min}-1)$

and   (70)

$\lambda > 1$.

The second condition concerning λ expresses the fact that the domain of
dependence of the numerical scheme must include the domain of dependence
of the partial differential equations, i.e., the fullfillment of the
Courant-Friedrichs-Lewy (CFL) condition.

    Because of the above limitations, explicit schemes are very
seldom employed. In order to avoid these difficulties, only implicit
schemes are practically used for the integration of the unsteady boundary layers.

    The implicit consistent schemes for our model problem may gene-
rally be written as (see Fig. 2, schemes c),d),e)) :

$$T_j \ u^j_{i+1,k+1} = E_j \ u^j_{i;k+1} + F_j \ u^j_{i+1,k} + G_j \ u^j_{i,k} \ . \qquad (71)$$

The various operators are collected in Table 1. The first one (scheme c)
is a Laassonnen type scheme with a truncation error of the order
$O(\Delta x+\Delta t+\Delta y^2)$. The second one is a Crank-Nicholson type scheme (d) with
a truncation error of the order $O(\Delta x^2+\Delta t^2+\Delta y^2)$, while the third one is
a Mehrstellen-scheme (e) with the highest order of truncation error,
being of order $O(\Delta x^2+\Delta t^2+\Delta y^4)$. In contrast to the explicit schemes, the
finite difference equation 71, written down for all values of $1 \leq j \leq N-1$,
forms a system of linear equations for the unknown values $u^j_{i+1,k+1}$ with
a tridiagonal matrix, which must be inverted at each point in the x-t
plane.

    The investigation of the stability of these schemes is now
carried out with the von Neumann analysis, which is strictly valid only,
inside the domain B, for difference equations with constant coefficients.
Thus any influence of the boundary conditions on the stability is excluded from this analysis. The amplification factors, multiplied with
their complex conjugate, are the following ones :

Laasonnen - scheme :

$$\rho\bar{\rho} = \frac{1}{(1+4s\ \sin^2\ \frac{k_2}{2}\ \Delta y)^2 + 4\lambda(1+\lambda+4s\ \sin^2\ \frac{k_2}{2}\ \Delta y^2)\sin^2\ \frac{k_1}{2}\Delta x} \tag{72}$$

Crank-Nicholson - scheme :

$$\rho\bar{\rho} = \frac{\cos^2\ \frac{k_1}{2}\ \Delta x\ (1-2s\ \sin^2\ \frac{k_2}{2}\ \Delta y) + \lambda^2\sin^2\ \frac{k_1}{2}\ \Delta x}{\cos^2\ \frac{k_1}{2}\ \Delta x\ (1+2s\ \sin^2\ \frac{k_2}{2}\ \Delta y) + \lambda^2\sin^2\ \frac{k_1}{2}\ \Delta x} \tag{73}$$

Mehrstellen - scheme :

$$\rho\bar{\rho} = \frac{\cos^2\ \frac{k_1}{2}\ \Delta x\ (1-2s\ \sin^2\ \frac{k_2}{2}\ \Delta y)^2 + \lambda^2\sin^2\ \frac{k_1}{2}\ \Delta x(1-\frac{1}{3}\ \sin^2\ \frac{k_2}{2}\ \Delta x)^2}{\cos^2\ \frac{k_1}{2}\ \Delta x(1+2s\ \sin^2\ \frac{k_2}{2}\ \Delta y)^2 + \lambda^2\sin^2\ \frac{k_1}{2}\ \Delta x\ (1-\frac{1}{3}\ \sin^2\ \frac{k_2}{2}\ \Delta y)^2} \tag{74}$$

where k_1, k_2 are arbitrary wave numbers. By inspecting these amplification factors, we see that all schemes are unconditionally stable, λ
being constant. Note that the stability analysis of the explicit schemes
does not require the assumption of constant values of λ.

If we consider now again the equations at the beginning of this
chapter (eqs. 47 to 50) subjected to any of the transformations mentioned
in chapter three, the following type of partial differential equation
(or equations) is obtained

$$A_1 F_{\eta\eta} + A_2 F_\eta + A_3 F + A_4 + A_5 F_\xi + A_6 F_\tau = 0 \tag{75}$$

where the coefficients A_1 to A_6 may depend on the independent variables
ξ, η, τ, various given parameters and the solution F which introduces
a nonlinearity. In compressible flows, a deak dependency of the coefficients A_1 to A_6 on the temperature is introduced if F represents the
normalized velocity for example.

This equation (or equations) may now be discretized with finite
difference methods following the ideas outlined before. The resulting
system of (linear) equations with a tridiagonal matrix is readily solved
by the well known Thomas algorithm. Because of the nonlinearities in
equation 75, iterations at each point in the $\xi-\tau$ plane are usually
required.

5. DISCUSSION OF RESULTS

In order to demonstrate the applicability of the computational methods to predict the development of unsteady boundary layers, we will discuss in this last chapter a selected number of solved problems.

As a first example consider the problem of a boundary layer flow over an impulsive started flat plate. The motion of the plate is parallel to itself, impulsively started from rest with uniform velocity. Because of the breakdown of the boundary layer assumptions near the leading edge, attention is confined to a region sufficiently far downstream from the leading edge.

The development of the flow in time has two simple features. Initially, the convective terms are negligible, so that the flow is described by the Rayleigh solution (eq. 11). At large times the flow settles down to a steady state, described by the Blasius solution. The problem is to explore the development of the flow from the initial to the steady state.

This problem has been studied by Stewartson (Ref. 34), Schuh (Ref. 29), Oudart (Ref. 35), Cheng (Ref. 36), Cheng and Elliot (Ref. 37), Lam and Crocco (Ref. 38), Hall (Ref. 39), Tani and Neng-Jong Yu (Ref. 40) and Piquet (Ref. 52).

The governing equations, written in a dimensionless form, are

$$\frac{\partial u}{\partial x} + \frac{\partial v}{\partial y} = 0 \tag{76}$$

$$\frac{\partial u}{\partial t} + u \frac{\partial u}{\partial x} + v \frac{\partial u}{\partial y} = \frac{\partial^2 u}{\partial y^2} \tag{77}$$

subjected to the boundary conditions

$$y = 0 \qquad u = v = 0 \tag{78}$$

$$y \to \infty \qquad u = 1 \quad .$$

Schuh, Oudart and Tani obtained their results with an integral method, assuming a one parameter family of curves, while Hall developed a Crank-Nicholson type difference scheme applied to the above equations. Finally, Piquet solved the equations, subjected first to the Crocco transformation, employing also a Crank-Nicholson type difference scheme.

The initial data in Hall's procedure is obtained by making use of the known similarity solutions, i.e., the Rayleigh and the Blasius solutions, in an iterative manner.

In figure 3 the local wall shear stress is plotted as a function of the dimensionless time. The collected results differ only slightly from each other, so that an assessment of the various methods with respect to this problem seems to be useless.

Next, we consider the problem of a flat plate in a free stream with small harmonic velocity oscillations about a steady mean in an incompressible fluid. This problem has attracted also many investigators, see for instance Schlichting and Gersten (Ref. 2). It has been treated sufficiently general, most with integral methods, by Lighthill (Ref. 41), Teipel (Ref. 42) and Miller (Ref. 31), while Farn and Arpaci (Ref. 43) developed an explicit finite difference scheme.

The free stream velocity U varies with time in the following complex form

$$U_\infty = U_0 \ (1+\varepsilon e^{i\omega t})$$

where ε is some small quantity, ω being the frequency. In order to facilitate the solution of the problem, the following expressions for the dimensionless velocity components u and v are usually assumed

$$u = u_0(x,y) + \varepsilon u_1(x,y)e^{i\omega t} + O(\varepsilon^2)$$

$$v = v_0(x,y) + \varepsilon v_1(x,y)e^{i\omega t} + O(\varepsilon^2)$$

(79)

Inserting these equations into the integral equations, a system of ordinary differential equations in the streamwise direction x is obtained, which is integrated with the Runge-Kutta method (Miller, Ref. 31).

In figures 4 and 5 the results of this calculation are depicted together with experimental data, obtained by Hill and Stenning (Ref. 44). The accordance between computed results and experimental data is remarkable. The dimensionless coordinates η, ζ, and y are given through

$$\eta = \frac{y}{\zeta} \qquad \text{with} \qquad \zeta = \sqrt{\frac{\omega X}{U_0}} \ , \qquad \text{and} \qquad y = \sqrt{\frac{\omega}{\nu}} \ Y \ ,$$

where X,Y are the streamwise and normal coordinates, ν being the kinmatic viscosity.

In two dimensional laminar boundary layers, the vanishing of the skin friction is usually the criterion for separation. However, in unsteady boundary layers, this criterion is not any longer meaningful, as Moore (Ref. 45), Rotta (Ref. 46) and Sears (Ref. 47) seem to have pointed out first. In order to study this rather striking feature of unsteady boundary layers, Phillips and Ackerberg (Ref. 48) and Telionis et al. (Ref. 49) carried out some numerical investigations with finite difference schemes of the Crank-Nicholson type, solving the differential equations for two dimensional, incompressible unsteady boundary layers, where regions of back flow occur. The flow situation near the region of back flow is illustrated schematically in Fig. 6, while in Fig. 7 the skin friction versus the streamwise coordinate s for a steady flow with a moving wall is plotted. It is clearly observed that the skin friction goes through a point of zero without any sign of singularity. This feature has been observed also elsewhere. The criterion of separation being proposed by Moore, Rott and Sears is defined by the two conditions $\frac{\partial u}{\partial N} = 0$ at a point inside the flow, where u is, in addition, zero.

In order to continue the calculation in reversed flow regions physically meaningfull, an "upwind difference" scheme for the streamwise coordinate has been established by Telionis et al., trying to take into account the appropriate data in the region of the field where the velocity profile is reversed. This concept, however, needs to be explored further.

A heat transfer problem in connection with unsteady boundary layers has recently been investigated in order to study some unsteady effects in combined forced and free convection. A finite difference scheme of the Laasonnen type to solve numerically the following dimensionless set of equations, describing the unsteady boundary layer flow along vertical walls has been developed by Wirz and Elsholz (Ref. 50) :

$$\frac{\partial u}{\partial x} + \frac{\partial v}{\partial y} = 0$$

$$\frac{\partial u}{\partial t} + u \frac{\partial u}{\partial x} + v \frac{\partial u}{\partial y} = -\frac{\partial p}{\partial y} + \frac{\partial^2 u}{\partial y^2} + \varepsilon T \qquad (80)$$

$$\frac{\partial T}{\partial t} + u \frac{\partial T}{\partial x} + v \frac{\partial T}{\partial y} = \frac{1}{Pr} \frac{\partial^2 T}{\partial y^2} \quad ,$$

subjected to the following boundary conditions

$$y = 0 \qquad u = v = 0 \qquad T = T_w(x,t)$$

$$y \to \infty \qquad u = u(x,t) \qquad T = 0 \ . \tag{81}$$

In these equations, u,v are the velocities, T is the temperature, ε is a parameter, either +1 or -1 according to the flow situation (aiding or opposing flows) and Pr is the Prandtl number.

Based on a set of similarity solutions of the above set of equations, obtained by Wirz (Ref. 51), a check of the finite difference scheme employed was easy to perform, giving sufficient agreement between the exact similarity solution and the first numerical results, plotted as points in Fig. 8.

To close this chapter, an approach to solve a rather complicated three dimensional and time dependent problem will be considered. A finite difference method of the Laasonnen type has been developed by Dwyer (Ref. 18) and successfully applied to a rotating flat plate in forward flight, which is of direct interest to helicopter rotors. A sketch of the flow geometry is given in Fig. 9. The appropriate laminar, incompressible boundary layer equations for this problem in terms of the x, y, z coordinates are :

Continuity :

$$\frac{\partial u}{\partial x} + \frac{\partial v}{\partial y} + \frac{\partial w}{\partial y} = 0$$

x-Momentum :

$$\frac{\partial u}{\partial t} + u \frac{\partial u}{\partial x} + v \frac{\partial u}{\partial y} + w \frac{\partial u}{\partial z} - 2\Omega w = -\frac{1}{\rho} \frac{\partial p}{\partial x} + \nu \frac{\partial^2 u}{\partial y^2} + \Omega^2 X \tag{82}$$

z-Momentum :

$$\frac{\partial w}{\partial t} + u \frac{\partial w}{\partial x} + v \frac{\partial w}{\partial y} + w \frac{\partial w}{\partial z} + 2\Omega u = -\frac{1}{\rho} \frac{\partial p}{\partial z} + \nu \frac{\partial^2 w}{\partial y^2} + \Omega^2 Z$$

where u, v, and w are the boundary velocities in the x, y, z directions respectively, t is the time and ν the kinematic viscosity. These equations are subjected to the following transformations :

$$\xi = x \; ; \qquad \eta = y \sqrt{\frac{U_e}{2x}} \; ; \qquad \zeta = z \; ; \qquad T = t \; ; \qquad f' = \frac{u}{U_e} \qquad (83)$$

yielding a system of equations where the formulation of initial conditions is facilitated, although it is still a formidable problem. Some of the results obtained by Dwyer are collected in Fig. 9. Significant changes of the flow quantities appear, near the retreating blade portion of the cycle, where retreating blade stall has been a recurring problem. These results promise a detailed understanding of the complicated flow structure, although their validity has not been proven yet.

REFERENCES

1. EICHELBRENNER, E. (Ed.): Recent research on unsteady boundary layers.
       Proc. IUTAM Symposium 1971, Quebec 1972.
2. SCHLICHTING, H.: Grenzschichttheorie, 5. Aufl., 1965.
       G. Braun, Germany.
3. ROTT, N.: Theory of time dependent flows, in
       Theory of laminar flows, F.K. Moore ed., Princeton U.Press, 1964
4. STUART, J.T.: Unsteady boundary layers, in
       Laminar boundary lauers, L. Rosenhead, ed., Oxford U.Press, 1963
5. STEWARTSON, K.: The theory of unsteady laminar boundary layers, in
       Advanced in Applied Mechanics, Vol. VI, H.L Dryden and Th. von
       Karman, eds., Academic Press, N.Y., 1960.
6. STUART, J.T.: Inaugural lectures.
       Imperial College, 109, 1967.
7. STUART, J.T.: Unsteady boundary layers, in
       Recent Research on Unsteady Boundary Layers, E. Eichelbrenner,
       ed., Quebec, 1972.
8. CEBECI, T. and KELLER, H.B.: On the computation of unsteady turbulent
       boundary layers, in
       Recent Research on Unsteady Boundary layers, E. Eichelbrenner,
       ed., Quebec, 1972.
9. PATEL, V.C. and NASH, J.F.: Some solutions of the unsteady two dimen-
       sional turbulent boundary layer equations, in
       Recent Research on Unsteady Boundary Layers, E. Eichelbrenner,
       ed., Quebec, 1972
10. TOWNSEND, A.A.: Equilibrium layers and wall turbulence.
       J. Fluid Mechanics, Vol. 11, 1961.
11. BRADSHAW, P., FERRIS, D.H., ATWELL, N.P.: Calculation of boundary
       layer development using the turbulent energy equation.
       J. Fluid Mechanics, Vol. 28, 1967.
12. KARLSSON, S.K.F.: An unsteady turbulent boundary layer.
       J. Fluid Mechanics, Vol. 5, 1959.
13. MILLER, J.A.: Heat transfer in the oscillating turbulent boundary
       layer.
       ASME Paper 69-GT-34
14. ROSENHEAD, L., ed.: Laminar boundary layers.
       Oxford U. Press, 1963.
15. PIQUET, J., ZEYTOUNIAN, R.Kh.: Recherches récentes dans le domaine
       des couches limites instationnaires, in
       Recent Research on Unsteady Boundary Layers, e. Eichelbrenner,
       ed., Quebec, 1972.
16. McCROSKEY, W.J., YAGGY, P.F.: Laminar boundary layers on helicopter
       rotors in forward flight.
       AIAA J., Vol. 6, No 10, 1968.

17. DORODNITZYN, A.A.: Prikl. Math. Meh. 6, 1942.
18. DWYER, H.A.: Calculations of unsteady and three dimensional boundary
    layer flows.
    AIAA.P. 72-109, 1972.
19. CROCCO, L.: A characteristic transformation of the equations of the
    boundary layer in gases.
    ARC, London, 4582, 1939.
20. SCHUH, H.: Uber die ähnlichen Lösungen der instationären laminaren
    Grenzschichtgleichungen in inkompressibler Strömung, in
    50-Jahre Grenzschichtforschung, H. Görtler und W. Tollmien, eds.,
    Vieweg, Braunschweig, Germany, 1955.
21. GEIS, T.: Bemerkungen zu den ähnlichen instationären laminaren
    Grenzschichtströmungen.
    ZAMM, Bd 36, Nr 9/10, 1956.
22. YANG, K.T.: Unsteady laminar boundary layers in an incompressible
    stagnation flow.
    J. Appl. Mechanics, Vol. 25, 1958.
23. YANG, W.J. and HUAG, H.S.: Unsteady compressible laminar boundary
    layer flow over a flat plate.
    AIAA J., Vol. 7, 1969.
24. GABBERT, C.H.: Similarity of unsteady compressible boundary layers.
    AIAA J., Vol. 5, 1967.
25. WIRZ, H.J.: Similarity solutions of unsteady, compressible plane and
    axisymmetrical laminar boundary layers, in
    Recent Research on Unsteady Boundary Layers, E. Eichelbrenner,
    ed., Quebec, 1972.
26. TANI, I.: An example of unsteady laminar boundary layer flow.
    Grenzschichtforschung - IUTAM Symposium, Freiburg, 1957.
27. HASSAN, H.A.: On unsteady laminar boundary layers.
    J. Fluid Mechanics, Vol. 9, 1960.
28. HAYASI, N.: On semi-similar solutions of the unsteady quasi-two-
    dimensional incompressible laminar boundary layer equations.
    J. Phys. Soc. Japan, Vol. 16, 1961.
29. SCHUH, H.: Calculation of unsteady boundary layers in two dimensional
    laminar flow.
    ZFW, Vol. 5, 1953.
30. YANG, K.T.: Unsteady laminar boundary layers over an arbitrary
    cylinder with heat transfer in an incompressible flow.
    J. Appl. Mech., June, 1959.
31. MILLER, R.W.: Analysis of unsteady thermal boundary layers.
    NASA TN D-7054, 1972.
32. KOOB, S.J. and ABOTT, D.E.: Investigation of a method for the general
    analysis of time dependent two dimensional laminar boundary
    layers.
    ASME Transact., Series D, J. Basic Engrg, Vol. 90, 1968.
33. KRAUSE,E.: Numerical treatment of boundary layer problems.
    AGARD, LS 64, 1973.
34. STEWARTSON, K.: On the impulsive motion of a flat plate in a
    viscous fluid.
    Qua. J. Mech. & Appl. Math., Vol. 4, 1951.
35. OUDART, A.: Mise en régime de la couche limite de la plaque plane
    dans l'impulsion brusque à partir du repos.
    Recherche Aérospatiale, No 31, 1953.
36. CHENG, S.I.: Some aspects of unsteady laminar boundary layer flows.
    Quart. Appl. Math., Vol. 16, 1957.
37. CHENG, S.I. and ELLIOT, D.: The unsteady laminar boundary layer on
    a flat plate.
    ASME Transact., Vol. 79, 1957.
38. LAM, S.H. and CROCCO, L.: Shock-induced unsteady laminar compres-
    sible boundary layers on a semi-infinite flat plate.
    Princeton U., Report 428, 1958.

39. HALL, M.G.: The boundary layer over an impulsively started flat
    plate.
        Proc. Roy. Soc., Vol. A 310, 1969.
40. TANI, I. and YU, Neng-Jong: Unsteady boundary layers over a flat
    plate started from rest, in
        Recent Research on Unsteady Boundary Layers, E. Eichelbrenner,
        ed., Quebec, 1972.
41. LIGHTHILL, M.J.: The response of laminar skin friction and heat
    transfer to fluctuations in the stream velocity.
        Proc. Roy. Soc., Vol. A 224, 1954.
42. TEIPEL, I.: Ein Integralverfahren zur Berechnung von inkompressiblen
    oszillierenden Grenzschichtströmungen.
        DLR FB 69-09, Germany, 1969.
43. FARN, C.L.E. and ARPACI, V.S.: On the numerical solution of unsteady
    laminar boundary layers.
        AIAA J., Vol. 4, 1966.
44. HILL, P.G. and STENNING, A.H.: Laminar boundary layers in oscilla-
    tory flows.
        ASME Transact., Series D., J. Basic Engrg, Vol. 82, 1963.
45. MOORE, F.K.: Research on rotating stall in axial flow compressors.
        WADC TR 59-75, Part II, 1959.
46. ROTT, N., see Ref. 3.
47. SEARS, W.R., unpublished notes.
48. PHILLIPS, J.H. and ACKERBERG, R.C.: A numerical method for inter-
    preting the unsteady boundary layer equations when there are
    regions of back flow.
        J. Fluid Mechanics, Vol. 58, No 3, 1973.
49. TELIONIS, D.P., TSAHALIS, D.Th., WERLE, M.J.: Numerical investiga-
    tion of unsteady boundary layer separation.
        Physics of Fluids, Vol. 16, No 7, 1973.
50. WIRZ, H.J. and ELSHOLZ, E.: The numerical solution of unsteady
    boundary layers with combined forced and free convection.
        Unpublished report.
51. WIRZ, H.J.: Unsteady combined forced and free convection in boun-
    dary layers.
        Proceedings of the 5th International Heat Transfer Conference,
        September 1974, Tokyo.
52. PIQUET, J.: Calcul numérique de couches limites laminaires compres-
    sibles en régime instationnaire.
        J. de Mécanique, Vol. 11, No 1, 1972.

TABLE 1 - IMPLICIT FINITE DIFFERENCE SCHEMES FOR UNSTEADY BOUNDARY LAYERS

$$T_j^{\ j} u_{i+1,k+1} = E_j^{\ j} u_{i,k+1} + F_j^{\ j} u_{i+1,k} + G_j^{\ j} u_{i,k}$$

Scheme	T_j	E_j	F_j	G_j
c) Laasonnen	$(1+\lambda_j)\delta - s\delta_y^2$	λ_j	1	0
d) Crank-Nicholson	$(1+\lambda_j)\delta - \dfrac{s}{2}\delta_y^2$	$(\lambda_j-1)\delta + \dfrac{s}{2}\delta_y^2$	$(1-\lambda_j)\delta + \dfrac{s}{2}\delta_y^2$	$(1+\lambda_j)\delta + \dfrac{s}{2}\delta_y^2$
e) Mehrstellen	$S_y + S_y^* - \dfrac{s}{2}\delta_y^2$	$S_y^* - S_y + \dfrac{s}{2}\delta_y^2$	$S_y - S_y^* + \dfrac{s}{2}\delta_y^2$	$S_y + S_y^* + \dfrac{s}{2}\delta_y^2$

Operators :

$$\delta\phi_j = \phi_j$$

$$\delta_y^2\phi_j = \phi_{j-1} - 2\phi_j + \phi_{j+1}$$

$$S_y\phi_j = \frac{1}{12}(\phi_{j-1} + 10\phi_j + \phi_{j+1})$$

$$S_y^*\phi_j = \frac{1}{12}(\lambda_{j-1}\phi_{j-1} + 10\lambda_j\phi_j + \lambda_{j+1}\phi_{j+1})$$

$$s = \sigma \frac{\Delta t}{\Delta y^2}$$

$$\lambda_j = \frac{\Delta t}{\Delta x} U(y_j)$$

Addendum
Lecture Notes in Physics, Vol. 41
By mistake the following 9 figures have not been bound at the end of the contribution:
H.J.Wirz,Computation of Unsteady Boundary Layers.   Pages 442 - 471.
We therefore attach the remaining sheets 473 - 476 with the figures 1 - 9.

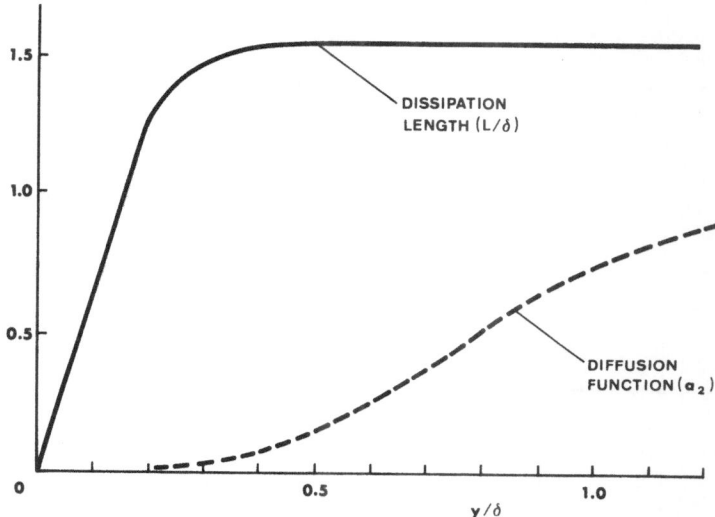

FIGURE I EMPIRICAL FUNCTIONS

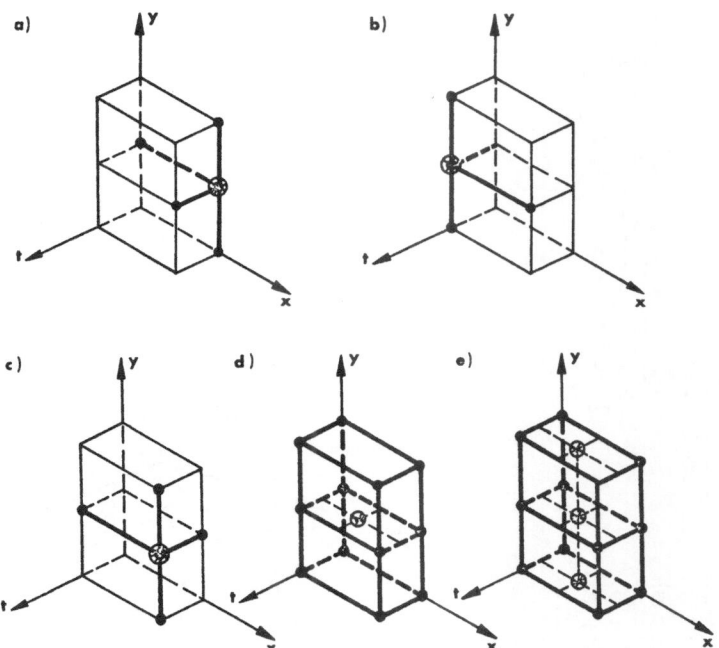

**FIG.II FINITE DIFFERENCE SCHEMES FOR UNSTEADY
        BOUNDARY LAYERS**

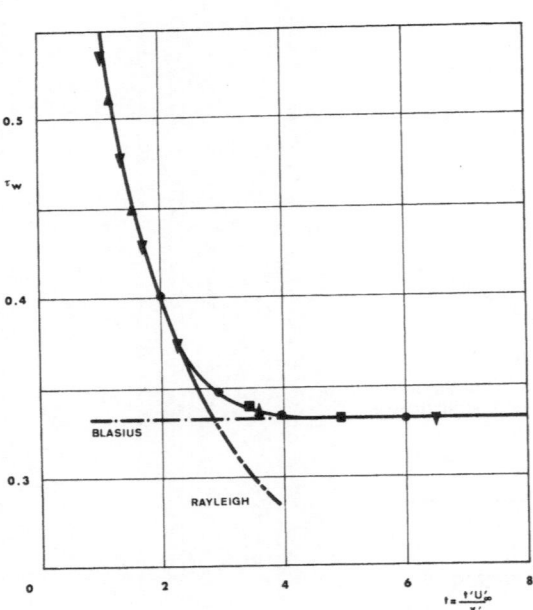

FIG.III WALL SHEAR STRESS

- ● HALL
- ■ LAM and CROCCO
- ▼ PIQUET
- ▲ TANI

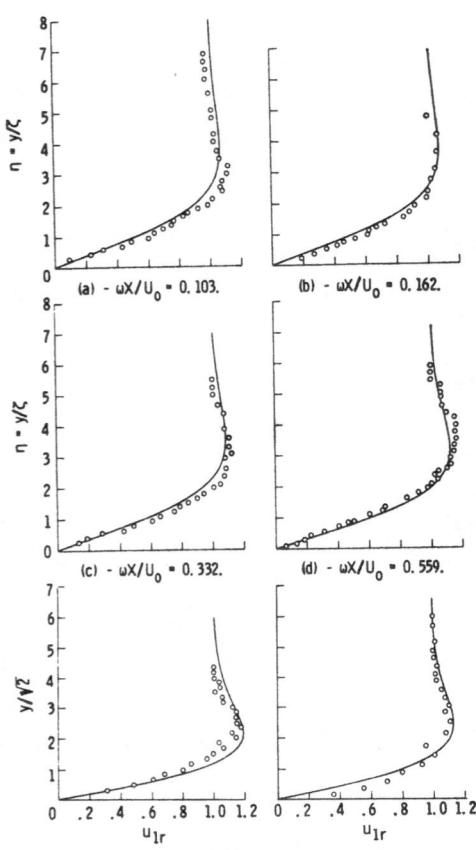

Fig. 4 Development of velocity component magnitude along flat plate with oscillating free stream. Circles are experimental data of Hill and Stenning [44].

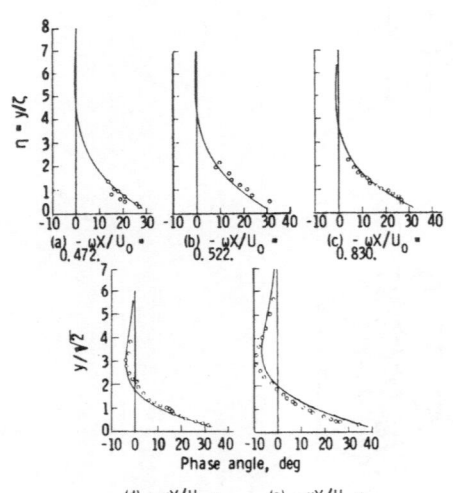

Fig. 5 Development of phase angle between in-plane and out-of-phase velocity components along flat plate with oscillating free stream. Circles are experimental data of Hill and Stenning [44].

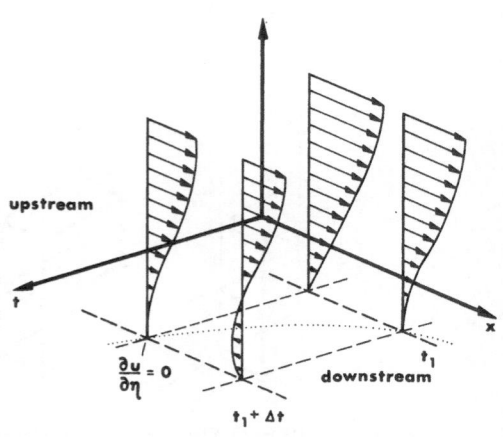

Fig.: 6 Schematic flow field

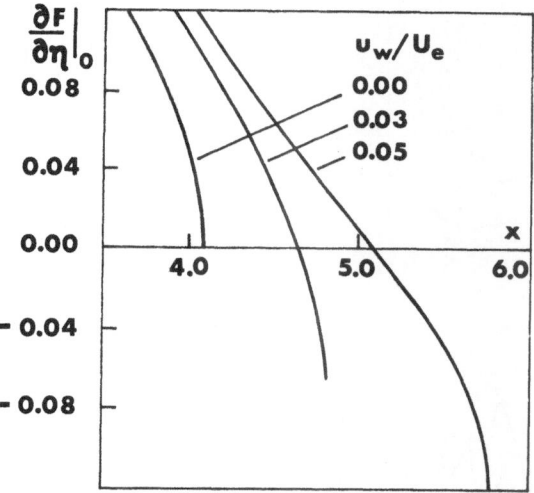

Fig.:7 Skin friction function versus the distance x for steady flow over fixed or moving walls.

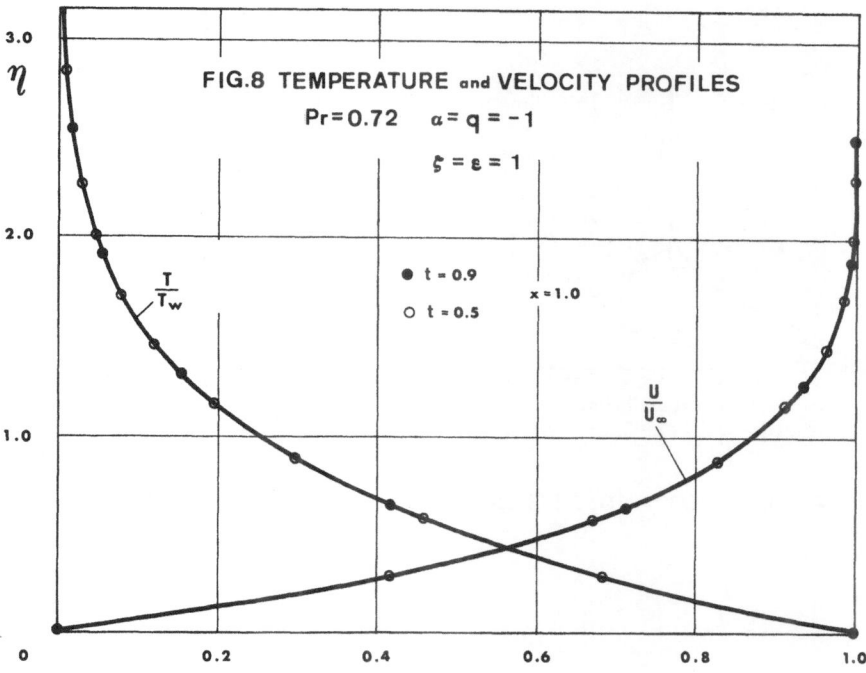

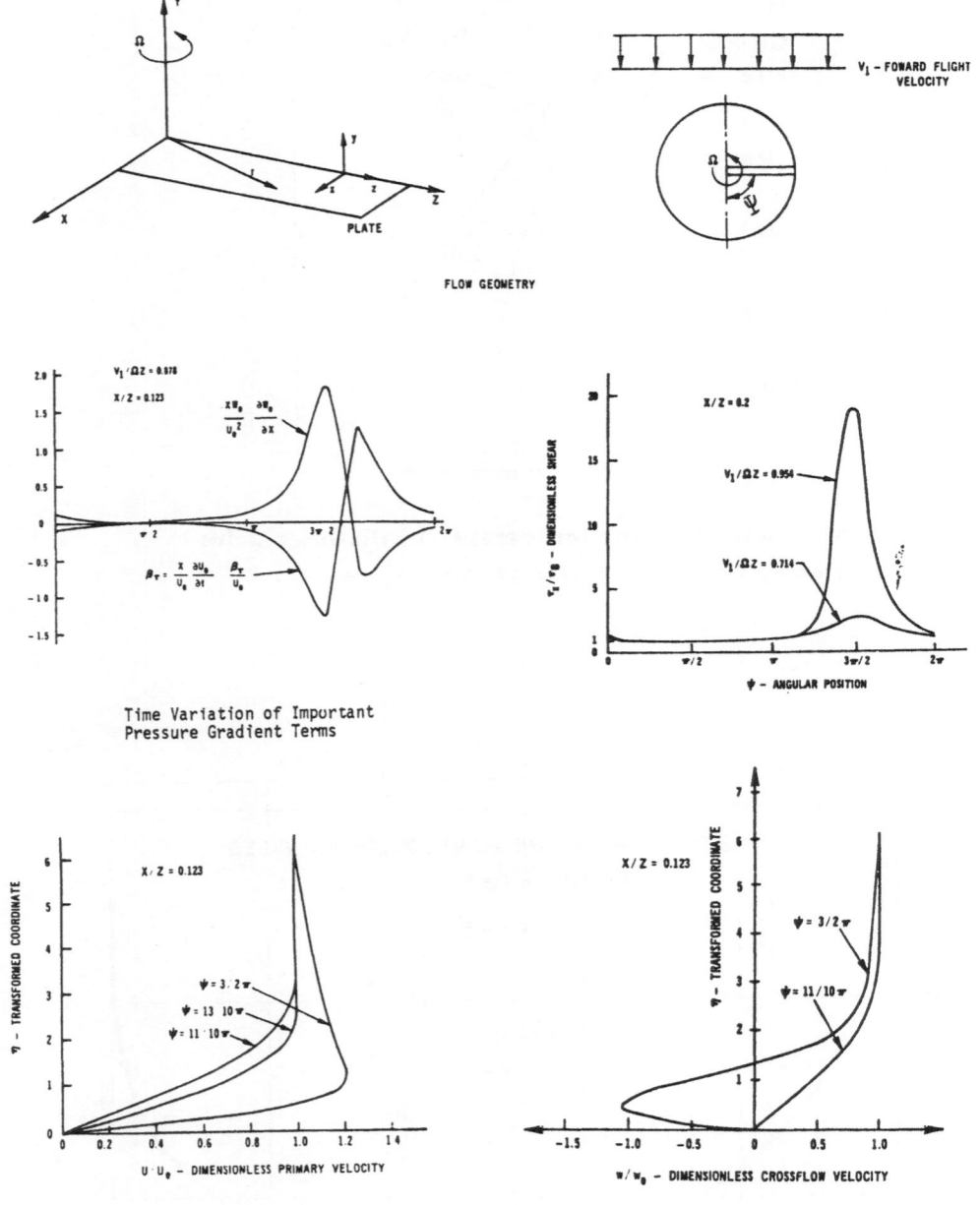

FIG.9 UNSTEADY AND THREE-DIMENSIONAL BOUNDARY LAYER FLOWS

Lecture Notes in Physics

SPRINGER TRACTS IN MODERN PHYSICS

Ergebnisse der exakten Naturwissenschaften

Editor: G. Höhler

Associate Editor:
E.A.Niekisch

Editorial Board:
S. Flügge, J. Hamilton,
F. Hund, H. Lehmann,
G. Leibfried, W. Paul

Springer-Verlag
Berlin
Heidelberg
New York

Selected Issues from

Lecture Notes in Mathematics